EATCS
Monographs on Theoretical Computer Science
Volume 10

EATCS Monographs on Theoretical Computer Science

Vol. 1: K. Mehlhorn: Data Structures and Algorithms 1: Sorting and Searching. XIV, 336 pages, 87 figs. 1984.

Vol. 2: K. Mehlhorn: Data Structures and Algorithms 2: Graph Algorithms and NP-Completeness. XII, 260 pages, 54 figs. 1984.

Vol. 3: K. Mehlhorn: Data Structures and Algorithms 3: Multi-dimensional Searching and Computational Geometry. XII, 284 pages, 134 figs. 1984.

Vol. 4: W. Reisig: Petri Nets. An Introduction. X, 161 pages, 111 figs. 1985.

Vol. 5: W. Kuich, A. Salomaa: Semirings, Automata, Languages. IX, 374 pages, 23 figs. 1986.

Vol. 6: H. Ehrig, B. Mahr: Fundamentals of Algebraic Specification 1. Equations and Initial Semantics. XI, 321 pages. 1985.

Vol. 7: F. Gécseg: Products of Automata. VIII, 107 pages, 18 figs. 1986.

Vol. 8: Kröger: Temporal Logic of Programs. VIII, 148 pages. 1987.

Vol. 9: K. Weihrauch: Computability. X, 517 pages. 1987.

Vol. 10: H. Edelsbrunner: Algorithms in Combinatorial Geometry. XV, 423 pages, 93 figs. 1987.

Herbert Edelsbrunner

Algorithms in Combinatorial Geometry

Springer-Verlag
Berlin Heidelberg New York
London Paris Tokyo

Author

Prof. Dr. Herbert Edelsbrunner
University of Illinois at Urbana-Champaign
Dept. of Computer Science
1304 West Springfield Avenue, Urbana, Illinois 61801, USA

Editors

Prof. Dr. Wilfried Brauer
Institut für Informatik, Technische Universität München
Arcisstrasse 21, 8000 München 2, Germany

Prof. Dr. Grzegorz Rozenberg
Institute of Applied Mathematics and Computer Science
University of Leiden, Niels-Bahr-Weg 1, P.O. Box 9512
2300 RA Leiden, The Netherlands

Prof. Dr. Arto Salomaa
Department of Mathematics, University of Turku
20500 Turku 50, Finland

ISBN 3-540-13722-X Springer-Verlag Berlin Heidelberg New York
ISBN 0-387-13722-X Springer-Verlag New York Berlin Heidelberg

Library of Congress Cataloging-in-Publication Data. Edelsbrunner, Herbert. Algorithms in combinatorial geometry. (EATCS monographs on theoretical computer science ; v. 10). Bibliography: p. 1. Combinatorial geometry. 2. Geometry–Data processing. I. Title. II. Series. QA167.E34 1987 516'.13 87-16655
ISBN 0-387-13722-X (U.S.)

Offsetprinting: Druckhaus Beltz, Hemsbach
Binding: J. Schäffer GmbH & Co. KG, Grünstadt
2145/3140-543210

to my wife Eva

PREFACE

Computational geometry as an area of research in its own right emerged in the early seventies of this century. Right from the beginning, it was obvious that strong connections of various kinds exist to questions studied in the considerably older field of combinatorial geometry. For example, the combinatorial structure of a geometric problem usually decides which algorithmic method solves the problem most efficiently. Furthermore, the analysis of an algorithm often requires a great deal of combinatorial knowledge. As it turns out, however, the connection between the two research areas commonly referred to as computational geometry and combinatorial geometry is not as lop-sided as it appears. Indeed, the interest in computational issues in geometry gives a new and constructive direction to the combinatorial study of geometry.

It is the intention of this book to demonstrate that computational and combinatorial investigations in geometry are doomed to profit from each other. To reach this goal, I designed this book to consist of three parts, a combinatorial part, a computational part, and one that presents applications of the results of the first two parts. The choice of the topics covered in this book was guided by my attempt to describe the most fundamental algorithms in computational geometry that have an interesting combinatorial structure. In this early stage geometric transforms played an important role as they reveal connections between seemingly unrelated problems and thus help to structure the field. These transforms led me to believe that arrangements of hyperplanes are at the very heart of computational geometry – and this is my belief now more than ever.

As mentioned above, this book consists of three parts: **I. Combinatorial Geometry, II. Fundamental Geometric Algorithms**, and **III. Geometric and Algorithmic Applications.** Each part consists of four to six chapters. The non-trivial connection pattern between the various chapters of the three parts can be somewhat untangled if we group the chapters according to four major computational problems. The construction of an arrangement of hyperplanes is tackled in Chapter 7 after Chapters 1, 2, and 5 provide preparatory investigations. Chapter 12 is a collection of applications of an algorithm that constructs an arrangement. The construction of the convex hull of a set of points which is discussed in Chapter 8 builds on combinatorial results presented in Chapter 6. Levels and other structures in an arrangement can be computed by methods described in Chapter 9 which bears a close relationship to the combinatorial studies undertaken in Chapter 3. Finally, space cutting algorithms are presented in Chapter 14 which is based on the combinatorial investigations of Chapter 4 and the computational results of Chapter 10. The above listing of relations between the various chapters is by no means exhaustive. For example, the connections between Chapter 13 and the other chapters of this book come in too many shapes to be described here. Finally, Chapter 15 reviews the techniques used in the other chapters of this book to provide some kind of paradigmatic approach to solving computational geometry problems.

Prerequisites together with notational conventions followed in this book are collected at the end in Appendices A and B. Each chapter includes a set of exercises of various degrees of difficulty. I have tried to estimate the difficulty of each problem and expressed my opinion in terms of numbers 1 through 5 with the meaning defined as follows:

1 if the problem is trivial or very easy,
2 if the problem is easy but it may be tedious to solve it,
3 if the problem is of moderate difficulty,
4 if the problem is very difficult, and
5 if the problem is still unsolved; this does not necessarily mean that it is very difficult.

Of course, the assignment of these numbers is purely subjective except for the number 5 which is used to mark research problems. One of the purposes of these collections of exercises and open problems is to give results that extend the material presented in the corresponding chapters; another purpose is to point out related open problems. Each chapter also contains a collection of bibliographic notes which occasionally give pointers to places in the literature where solutions to some of the more difficult problems in the exercise section can be found.

I would like to acknowledge the help of many colleagues and friends without whom this book would never have been written. I thank Raimund Seidel for providing the original notes for Section 8.4 and for thoroughly reading earlier versions of parts of the book. I also thank Carlos Bhola, Jeffrey Salowe, Emo Welzl, and an anonymous referee who suffered through earlier versions of all chapters and provided many valuable suggestions, comments, and corrections. Thanks also to Bernard Chazelle, Friedrich Huber, Ernst Mücke, Harald Rosenberger, and Steven Skiena for carefully reading earlier versions of various chapters. Many of the algorithms presented in this book have been implemented during projects at the Technical University of Graz and the University of Illinois at Urbana-Champaign. For these implementations I thank Barbara Geymayer, Michael Hirschböck, Friedrich Huber, Hartwig Huemer, Tom Madej, Ernst Mücke, Harald Rosenberger, Gerd Stöckl, and Roman Waupotitsch who sacrificed many of their valuable hours to do the job. For discussions on topics found in this book, I thank Franz Aurenhammer, Bernard Chazelle, David Dobkin, Jacob Goodman, Branko Grünbaum, Leonidas Guibas, David Haussler, David Kirkpatrick, Hermann Maurer, Kurt Mehlhorn, Ernst Mücke, Joseph O'Rourke, Janos Pach, Richard Pollack, Franco Preparata, Harald Rosenberger, Jean-Pierre Roudneff, Franz Josef Schnitzer, Raimund Seidel, Micha Sharir, William Steiger, Gerd Stöckl, Jan van Leeuwen, Roman Waupotitsch, Emo Welzl, Douglas West, Derick Wood, and Frances Yao. I am also grateful to Hans Wössner and Gillian Hayes from Springer-Verlag, Heidelberg, for the pleasure it was to work with them. Last, but not least, I thank Heidrun Kaiser at Graz and Janet Shonkwiler and June Wingler at Urbana for valuable assistance in typing this book.

My most particular thanks go to my wife Eva whose patience and encouragement as well as help in preparing this book made it possible for me to finish the task.

Urbana, May 1987 Herbert Edelsbrunner

TABLE OF CONTENTS

PART I

COMBINATORIAL GEOMETRY

The art of counting and estimating is at the heart of combinatorics – and it is a necessary prerequisite for analyzing algorithms and for deciding which algorithms are the most efficient ones. Part I of this book presents several combinatorial geometry problems and solutions using a variety of techniques.

CHAPTER 1

FUNDAMENTAL CONCEPTS IN COMBINATORIAL GEOMETRY

Two of the main subjects studied in combinatorial geometry and therefore in this book are finite sets of points and finite sets of hyperplanes. Not all questions about finite sets of points or hyperplanes are combinatorial, though, and one has to keep in mind that a strict classification into combinatorial and non-combinatorial problems is neither reasonable nor desirable. Nevertheless, there are a few characteristics that identify a problem as combinatorial. For example, a typical combinatorial question that can be asked about a set P of n points in d-dimensional Euclidean space E^d is the following:

"How many partitions of P into two subsets can be defined by hyperplanes?"

If H is a set of n hyperplanes in the same space, then it is a combinatorial question if one asks

"What is the number of cells the space is cut into by the hyperplanes in H?"

We will investigate both problems and many related ones in this chapter and, more generally, in this book.

A useful means in this study are geometric transforms that map geometric objects to different geometric objects. We will see such a transform which reveals that the two questions raised above are essentially identical. Other geometric transforms can be used to show that finite sets of points or hyperplanes are closely related to geometric concepts such as convex polytopes, zonotopes, and Voronoi diagrams. Convex polytopes as well as Voronoi diagrams are well studied geometric objects and there is a rich literature available on both types of structures. Zonotopes are special convex polytopes that realize a particularly close relation to finite sets of hyperplanes, and much of what is known about sets of hyperplanes has interesting consequences for zonotopes.

Below, we give an outline of this chapter. Sections 1.1 through 1.3 discuss cell complexes defined by finite sets of hyperplanes - they are called arrangements. Section 1.4 shows how finite point sets correspond to finite sets of hyperplanes. In Section 1.5, we consider a classic problem for finite point sets in the plane known as Sylvester's problem. In the light of the correspondence between points and lines in the plane, this problem has an interesting interpretation for arrangements defined by lines. Sections 1.6 through 1.8 discuss convex polytopes, zonotopes, and Voronoi diagrams and their relationship to arrangements of hyperplanes and to finite point sets. In each case, we introduce a new geometric transform to establish the connection which is of independent interest.

1.1. Arrangements of Hyperplanes

A finite set H of hyperplanes in E^d defines a dissection of E^d into connected pieces of various dimensions. We call this dissection the *arrangement* $\mathcal{A}(H)$ *of* H. For example, a finite set of lines in two dimensions dissects the plane into connected pieces of dimensions two, one, and zero. On the surface, it seems that an arrangement of hyperplanes has more structure to it than a finite set of points has. We will see later that this is only a superficial impression and that the structure of both, arrangements of hyperplanes and finite point sets, is equally rich. Nevertheless, it is true that the structure inherent in arrangements is to a larger extent explicit than in point sets. This latter distinction is the major reason for the tendency of this book to explain results in terms of arrangements rather than in terms of point sets.

In this section, we introduce the basic definitions and terminology needed in our study of arrangements. For $0 \le k \le d$, a *k-flat* in E^d is defined as the affine hull of $k+1$ affinely independent points. Thus, a 0-flat is a point, a 1-flat is commonly called a *line*, a 2-flat is called a *plane*, and a $(d-1)$-flat is called a *hyperplane*. There is only one d-flat in E^d, namely E^d itself, and, for convenience, we define the empty set as the only (-1)-flat. Equivalently, a k-flat can be defined as the intersection of $d-k$ hyperplanes whose normal vectors are linearly independent. Using standard Cartesian coordinates $x_1, x_2, \ldots, x_d$, we say that a k-flat is *vertical* if it contains a line parallel to the x_d-axis.

For every non-vertical hyperplane h, there are unique real numbers $\eta_1, \eta_2, \ldots, \eta_d$ such that h consists of all points $x = (x_1, x_2, \ldots, x_d)$ that satisfy

$$x_d = \eta_1 x_1 + \eta_2 x_2 + \ldots + \eta_{d-1} x_{d-1} + \eta_d .$$

We say that a point $p = (\pi_1, \pi_2, \ldots, \pi_d)$ is *above*, *on*, or *below* h if π_d is greater than, equal to, or less than $\eta_1 \pi_1 + \eta_2 \pi_2 + \ldots + \eta_{d-1} \pi_{d-1} + \eta_d$, respectively, and we write h^+ for the set of points above h and h^- for the set of points below h. Both, h^+ and h^- are open half-spaces.

The distinction between the half space above a non vertical hyperplane and the half-space below it allows us to specify the location of a point relative to the hyperplanes of a set H. To this end, let $h_1, h_2, \ldots, h_n$ be the hyperplanes in H, all non-vertical, and for a point p define

$$v_i(p) = \begin{cases} +1 & \text{if } p \in h_i^+, \\ 0 & \text{if } p \in h_i, \text{ and} \\ -1 & \text{if } p \in h_i^-, \end{cases}$$

for $1 \leq i \leq n$. The vector $u(p) = (v_1(p), v_2(p), \ldots, v_n(p))$ is called the *position vector of* p. Where convenient, we will write position vectors as strings in $\{+,0,-\}^n$ (see Figure 1.1). If $u(p) = u(q)$, then we say that points p and q are *equivalent*, and the equivalence classes thus defined are called the *faces of* arrangement $\mathcal{A}(H)$. Consider the one-dimensional case as an example. A hyperplane in E^1 is a point. To reflect the above/below relation, we imagine the line that represents E^1 drawn vertically, and we label the n points that define an arrangement consecutively from the bottom upwards. The position vector of a point $p \in E^1$ that lies above ℓ of the n points and below $n-\ell$ of them begins with ℓ +'es and ends with $n-\ell$ −'es. Thus, two points are equivalent if they lie in a common interval defined by the n points. The faces of the arrangement are therefore the n points and the $n+1$ intervals defined by them.

If f is a face and p is a point of f, then we set $v_i(f) = v_i(p)$, for $1 \leq i \leq n$, and we call $u(f) = (v_1(f), v_2(f), \ldots, v_n(f))$ the *position vector of* f. A face f is called a *k-face* if its dimension is k, that is, if f belongs to a k-flat but not to a $(k-1)$-flat. Special names are used to denote k-faces for special values of k: a 0-face is called a *vertex*, a 1-face is an *edge*, a $(d-1)$-face is a *facet*, and a d-face is called a *cell*. A face f is said to be a *subface of* another face g if the dimension of f is one less than the dimension of g and f is contained in the boundary of g; it follows that $v_i(f) = 0$ unless $v_i(f) = v_i(g)$, for each $1 \leq i \leq n$. If f is a subface of g, then we also say that f and g are *incident* (*upon* each other) or that they define an *incidence*.

An arrangement $\mathcal{A}(H)$ of $n \geq d$ hyperplanes in E^d is called *simple* if any d hyperplanes of H have a unique point in common and if any $d+1$ hyperplanes have no point in common. If the number n of hyperplanes in H does not exceed $d-1$, then we call $\mathcal{A}(H)$ *simple* if the common intersection of the n hyperplanes is a $(d-n)$-flat. It follows from this definition that $\mathcal{A}(H)$ is simple only if any $d-k \leq d$ hyperplanes of H intersect in a common k-flat. Figure 1.1 illustrates some of the concepts introduced above: it shows a simple arrangement of five lines in the plane. The position vectors of three cells, one edge, and one vertex are shown. The reader is asked to convince himself or herself that there are strings in $\{+,0,-\}^5$ that do not appear as position vector of any face in the arrangement displayed.

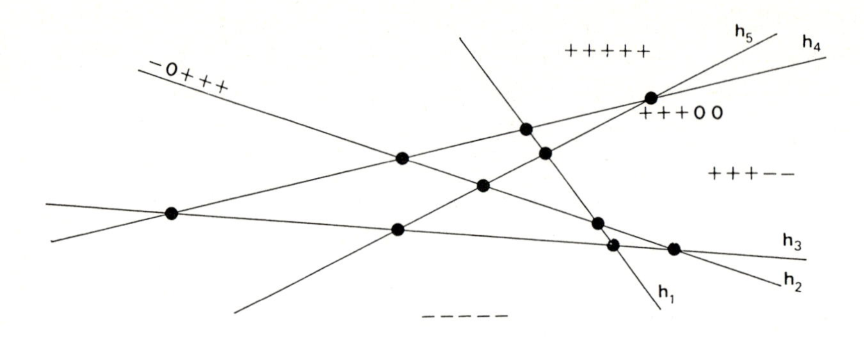

Figure 1.1. A simple arrangement of five lines.

Simple arrangements play an important role in this book for several reasons. For one thing, their structure is simpler than the structure of non-simple arrangements. For example, if $\mathcal{A}(H)$ is simple and f is a k-face of $\mathcal{A}(H)$, then the position vector $u(f)$ contains exactly $d-k$ components equal to 0. Furthermore, f is a subface of another face g only if $v_i(f) \neq v_i(g)$, for exactly one value of index i, and $v_i(f)=0$, for this value of i. Another reason for the importance of simple arrangements is that they maximize the number of faces and incident pairs of faces. This statement will be proved in the next section where we count the number of faces and incidences in simple and non-simple arrangements. Thus, simple arrangements constitute worst-case examples for an algorithm whose complexity increases with the number of faces and incidences.

Notice that our definition of faces and incidences applies in a strict sense only to arrangements of non-vertical hyperplanes. We will not bother to generalize these concepts to sets that also contain vertical hyperplanes. Part of the reason for neglecting vertical hyperplanes is convenience, another part is that in the majority of all cases where we apply arrangements to solve some problem only non-vertical hyperplanes are encountered. In any case, a redefinition of the coordinate system can be used to make all hyperplanes of a given arrangement non-vertical.

1.2. Counting Faces and Incidences

The number of faces in an arrangement $\mathcal{A}(H)$ is a measure of its complexity which is important for combinatorial as well as for computational reasons. Another such measure is the number of incidences, that is, the number of incident pairs of faces in $\mathcal{A}(H)$. To stress the importance of both measures we mention that Chapter 7 will present a data structure that needs storage of a size proportional to the number of faces and incidences to represent an arrangement.

In this section, we demonstrate tight upper bounds on the number of faces and the number of incidences in arrangements that depend only on the number of hyperplanes, the dimension of the space, and the dimension of the faces considered. Let H be a set of n hyperplanes in E^d. We denote by $f_k(H)$ the number of k-faces of $\mathcal{A}(H)$, for $0 \leq k \leq d$, and by $i_k(H)$ the number of incidences between k-faces and $(k+1)$-faces, for $0 \leq k \leq d-1$. Furthermore, we define

$$f_k^{(d)}(n) = \max\{f_k(G)\} \text{ and } i_k^{(d)}(n) = \max\{i_k(G)\},$$

where the maxima are taken over all sets G of n hyperplanes in E^d. First, we count the number of faces and incidences of simple arrangements. For this purpose, we need the following result.

Lemma 1.1: Let H be a set of d non-vertical hyperplanes in E^d that intersect in a unique point. Then $\mathcal{A}(H)$ contains

$$\binom{d}{d-k} \cdot 2^k$$

k-faces, for $0 \leq k \leq d$, and each cell of $\mathcal{A}(H)$ contains $\binom{d}{d-k}$ k-faces in its boundary, for $0 \leq k \leq d-1$.

Proof: Notice that the ith hyperplane cuts each face of the arrangement determined by the first $i-1$ hyperplanes into three pieces, one piece on each side of it and the piece contained in it. Thus, for each string $u \in \{+,0,-\}^d$ there is a face f in $\mathcal{A}(H)$ whose position vector is equal to u. If u contains $d-k$ 0's, then f is a k-face. The first part of Lemma 1.1 follows since there are

$$\binom{d}{d-k} \cdot 2^k$$

strings in $\{+,0,-\}^d$ that contain $d-k$ 0's. The second part is a consequence of the fact that a k-face f belongs to the boundary of a cell g if and only if

$$v_i(f) = 0 \text{ if } v_i(f) \neq v_i(g) \quad .$$

and that $v_i(f)$ disagrees with $v_i(g)$ for $d-k$ values of index i. □

For example, two non-parallel lines in the plane dissect the plane into one vertex, four edges, and four regions. The boundary of each region contains two edges and one vertex. Next, we count the number of faces and incident pairs of faces in a simple arrangement.

Lemma 1.2: Let H be a set of n hyperplanes in E^d such that $\mathcal{A}(H)$ is simple. Then

$$f_k(H) = \sum_{i=0}^{k} \binom{d-i}{k-i}\binom{n}{d-i}, \text{ for } 0 \leq k \leq d, \text{ and}$$

$$i_k(H) = 2(d-k) f_k(H), \text{ for } 0 \le k \le d-1.$$

Proof: Notice that a k-face in a simple arrangement is incident upon $2(d-k)$ $(k+1)$-faces, for $0 \le k \le d-1$. This implies the second part of Lemma 1.2. The first part of the assertion, the formula for the number of k-faces, is obvious for $n \le d$ since in this case $f_k(H) = \binom{n}{n-k} \cdot 2^k$ and there is a one-to-one correspondence between the faces of $\mathcal{A}(H)$ and all possible position vectors of n components.

We prove the case $n > d$ by induction on the number of dimensions. The assertion is trivial in one dimension where n points (that is, hyperplanes) dissect E^1 into $n+1$ intervals (that is, edges). Thus, assume that the assertion holds for dimensions less than d and let H be a set of n hyperplanes in E^d such that $\mathcal{A}(H)$ is simple. Any d hyperplanes define a vertex of $\mathcal{A}(H)$, and therefore

$$f_0(H) = \binom{n}{d},$$

which agrees with the assertion for $k = 0$. The rest of the argument makes use of a new hyperplane $h(t)$: $x_1 = t$ that sweeps through $\mathcal{A}(H)$ as the parameter t runs from $-\infty$ to $+\infty$. Without loss of generality, assume that no hyperplane in H is vertical and that no two vertices of $\mathcal{A}(H)$ share the same x_1-coordinate. Let

$$t_1 < t_2 < \ldots < t_m$$

be the x_1-coordinates of the $m = f_0(H)$ vertices of $\mathcal{A}(H)$. Vertices with x_1-coordinates less than t are said to *lie behind* $h(t)$, and vertices with x_1-coordinates greater than t *lie in front of* $h(t)$. Let $\mathcal{A}_t(H)$ denote the intersection of $\mathcal{A}(H)$ with $h(t)$. Unless $t = t_i$, for some $1 \le i \le m$, $\mathcal{A}_t(H)$ is a simple $(d-1)$-dimensional arrangement of n $(d-2)$-flats. By induction hypothesis, $\mathcal{A}_t(H)$ contains

$$\sum_{i=0}^{k-1} \binom{d-1-i}{k-1-i} \binom{n}{d-1-i}$$

$(k-1)$-faces, for $1 \le k \le d$. This is in particular true for $t < t_1$ and for $t > t_m$. Furthermore, each $(k-1)$-face in $\mathcal{A}_t(H)$ is contained in a unique k-face of $\mathcal{A}(H)$, and a k-face f of $\mathcal{A}(H)$ meets $h(t)$ during an open interval (t_a, t_b), for some $0 \le a < b \le m+1$ where $t_0 = -\infty$ and $t_{m+1} = +\infty$. The vertex with x_1-coordinate t_a (t_b) is called the *leftmost* (*rightmost*) vertex of f, if it exists, and f *lies behind* (*in front of*) $h(t)$ if $t \ge t_b$ ($t \le t_a$).

Below, we use Lemma 1.1 to keep track of the number of k-faces of $\mathcal{A}(H)$ that intersect $h(t)$ or lie behind $h(t)$, for $1 \le k \le d$. Initially, no k-face lies behind $h(t)$. As $h(t)$ passes a vertex of $\mathcal{A}(H)$ one more cell comes to lie behind $h(t)$ and, by the second part of Lemma 1.1, the number of k-faces that lie behind $h(t)$ increases by $\binom{d}{d-k}$ while the number of intersecting k-faces remains unchanged. During the entire sweep, $h(t)$ passes $\binom{n}{d}$ vertices which implies that ultimately $\binom{d}{d-k}\binom{n}{d}$ k-faces lie behind $h(t)$. The remaining k-faces intersect $h(t)$, $t > t_m$. Consequently,

$$f_k(H) = \binom{d}{d-k}\binom{n}{d} + \sum_{i=0}^{k-1}\binom{d-1-i}{k-1-i}\binom{n}{d-1-i} =$$

$$\binom{d}{k}\binom{n}{d} + \sum_{i=1}^{k}\binom{d-i}{k-i}\binom{n}{d-i} = \sum_{i=0}^{k}\binom{d-i}{k-i}\binom{n}{d-i}$$

for $1 \le k \le d$. □

The above proof technique that uses a distinguished hyperplane sweeping the arrangement is not the only one that can be used to prove Lemma 1.2. Another elegant argument can be obtained by using two levels of induction, the inner level on n and the outer level on d. We illustrate this idea when we prove that the numbers $f_k(H)$ and $i_k(H)$ are smaller for non-simple arrangements than for simple arrangements.

The induction basis is established by the cases $d=1$, n arbitrary, and $n=2$, d arbitrary. So assume that the statement holds for all n up to $d-1$ dimensions and for up to $n-1$ hyperplanes in d dimensions. Next, we add a new hyperplane h_n to the d-dimensional arrangement defined by $G=\{h_1,h_2,\ldots,h_{n-1}\}$. Hyperplane h_n intersects $\mathcal{A}(G)$ in a $(d-1)$-dimensional arrangement $\mathcal{A}(G')$ defined by at most $n-1$ $(d-2)$-flats. Each k-face f of $\mathcal{A}(G')$ is also a k-face of $\mathcal{A}(H)$, $H = G \cup \{h_n\}$, and if f is not a face in $\mathcal{A}(G)$, then f cuts a $(k+1)$-face of $\mathcal{A}(G)$ into two $(k+1)$-faces. Thus, we have

$$f_k(H) \le f_k(G) + f_k(G') + f_{k-1}(G'),$$

for $0 \le k \le d$, if we set $f_{-1}(G') = f_d(G') = 0$. If $\mathcal{A}(G)$ and $\mathcal{A}(G')$ are simple and if h_n does not contain any face of $\mathcal{A}(G)$, then the above is an equality that leads to Lemma 1.2. In our case, we have $\mathcal{A}(H)$ non-simple which implies that at least one of $\mathcal{A}(G)$ and $\mathcal{A}(G')$ is non-simple or that h_n contains at least one face of $\mathcal{A}(G)$. In any case, we have strict inequality and thus $f_k(H)$ smaller than

$$\sum_{i=0}^{k}\binom{d-i}{k-i}\binom{n-1}{d-i} + \sum_{i=0}^{k}\binom{d-1-i}{k-i}\binom{n-1}{d-1-i} + \sum_{i=0}^{k-1}\binom{d-1-i}{k-1-i}\binom{n-1}{d-1-i} = \sum_{i=0}^{k}\binom{d-i}{k-i}\binom{n}{d-i}.$$

A k-face f is a subface of a $(k+1)$-face g in $\mathcal{A}(H)$ only if one of the following disjoint cases occurs:

(i) f and g are faces of $\mathcal{A}(G)$,
(ii) f and g are faces of $\mathcal{A}(G')$,
(iii) f is a face of $\mathcal{A}(G')$ but not of $\mathcal{A}(G)$ and $g \ne g'$ is a connected component of $g' - h_n$, g' a $(k+1)$-face of $\mathcal{A}(G)$, or

(iv) there are faces f' and g' in $\mathcal{A}(G)$ that are cut by h_n and f and g are connected components of $f'-h_n$ and $g'-h_n$ on the same side of h_n.

Thus, we have

$$i_k(H) \le i_k(G) + i_k(G') + 2f_k(G') + i_{k-1}(G')$$

which can be used to derive the second part of Lemma 1.2 if we assume that $\mathcal{A}(H)$ is simple. Since $\mathcal{A}(H)$ is non-simple, we get

$$i_k(H) < 2(d-k)\sum_{i=0}^{k}\binom{d-i}{k-i}\binom{n-1}{d-i} + 2(d-1-k)\sum_{i=0}^{k}\binom{d-1-i}{k-i}\binom{n-1}{d-1-i} +$$

$$2\sum_{i=0}^{k}\binom{d-1-i}{k-i}\binom{n-1}{d-1-i} + 2(d-k)\sum_{i=0}^{k-1}\binom{d-1-i}{k-1-i}\binom{n-1}{d-1-i} =$$

$$2(d-k)\sum_{i=0}^{k}\binom{d-i}{k-i}\binom{n}{d-i}.$$

This implies the following theorem which summarizes the results of this section.

Theorem 1.3: For $d \ge 1$ and $n \ge 1$, we have

$$f_k^{(d)}(n) = \sum_{i=0}^{k}\binom{d-i}{k-i}\binom{n}{d-i}, \text{ for } 0 \le k \le d, \text{ and}$$

$$i_k^{(d)}(n) = 2(d-k)f_k^{(d)}(n), \text{ for } 0 \le k \le d-1.$$

Furthermore, $f_k(H) = f_k^{(d)}(n)$ ($i_k(H) = i_k^{(d)}(n)$), for H a set of n hyperplanes in E^d, if and only if $\mathcal{A}(H)$ is simple.

In the Euclidean plane, the face-count formulas of Theorem 1.3 become

$$f_0(n) = \binom{n}{2}, \quad f_1(n) = 2\binom{n}{2} + n, \quad \text{and} \quad f_2(n) = \binom{n}{2} + n + 1.$$

These formulas agree with the number of faces that can be found in the arrangement shown in Figure 1.1: there are $\binom{5}{2} = 10$ vertices, $2\binom{5}{2} + 5 = 25$ edges, and $\binom{5}{2} + 5 + 1 = 16$ regions. If d, the number of dimensions, is considered to be a constant, then we have

$$f_k^{(d)}(n) = \Theta(n^d) \text{ and } i_k^{(d)}(n) = \Theta(n^d)$$

which is sometimes more handy a statement than Theorem 1.3.

1.3. Combinatorial Equivalence

We have seen in Section 1.2 that the number of faces and incident face pairs in a simple arrangement only depends on the number of hyperplanes and the number of dimensions. This is somewhat surprising if one recognizes that two arrangements defined by the same number of hyperplanes in the same space can look rather different even if both are simple. This section makes some effort to concretize how two arrangements can be different in a combinatorial sense.

Let H_1 and H_2 be two sets of n non–vertical hyperplanes in E^d each. The two arrangements $\mathcal{A}(H_1)$ and $\mathcal{A}(H_2)$ are said to be (*combinatorially*) *equivalent* if there is a one–to–one correspondence between the hyperplanes of H_1 and of H_2 and between the k–faces of $\mathcal{A}(H_1)$ and the k–faces of $\mathcal{A}(H_2)$, for $0 \leq k \leq d$, that preserves position vectors. If there is no danger of confusion, we will call two arrangements equal if they are equivalent in this sense, and different, otherwise. Figure 1.2 illustrates the concept of combinatorial equivalence: (a) shows all different arrangements of three lines in the plane, two of them simple, and (b) gives the exhaustive list of different simple arrangements defined by four lines in the plane. We remark that all simple arrangements of four lines would be equivalent if we allowed the rotation of an arrangement. This is not true for all simple arrangements of five lines.

Now let $C^{(d)}(n)$ designate the number of different arrangements of n hyperplanes in E^d, and let $C_S^{(d)}(n)$ count those that are simple. Trivially, $C_S^{(d)}(n) \leq C^{(d)}(n)$, and we have strict inequality if $n \geq 2$ and $d \geq 2$ since there is at least one non–simple arrangement, namely, the one defined by n parallel hyperplanes. The determination of $C^{(d)}(n)$ and of $C_S^{(d)}(n)$ turns out to be an extremely difficult problem. For small values of n and d, the following straightforward results can be given:

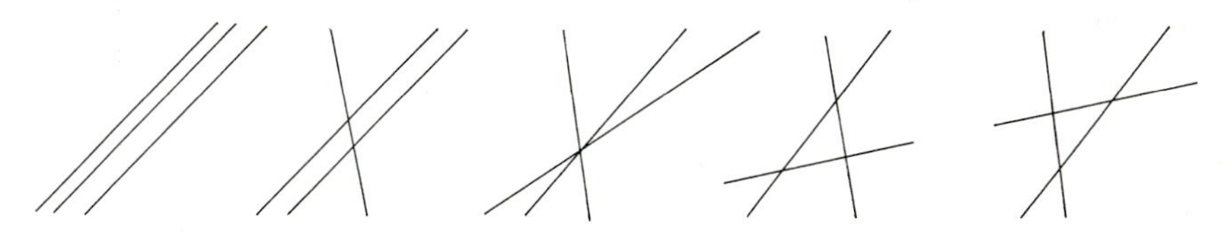

(a) Non–equivalent arrangements of three lines.

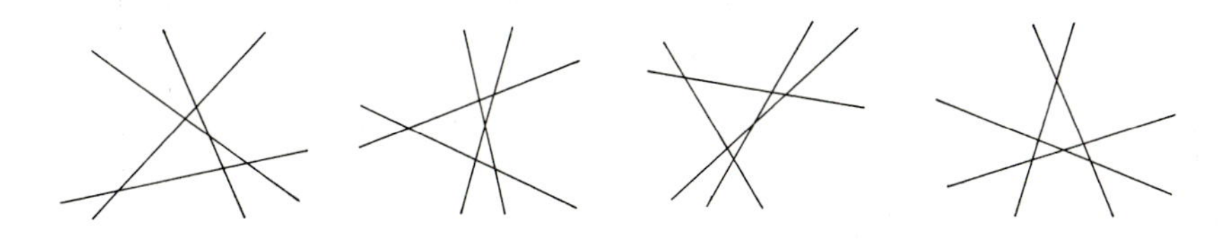

(b) Non–equivalent simple arrangements of four lines.

Figure 1.2. Exhaustive lists of non–equivalent arrangements.

$$C^{(1)}(n) = C_S^{(1)}(n) = 1,$$

$$C_S^{(d)}(n) = 1, \text{ for } 1 \leq n \leq d,$$

$$C^{(d)}(n) = C^{(d-1)}(n), \text{ whenever } n < d,$$

$$C^{(2)}(3) = 5, \text{ by Figure 1.2(a), and}$$

$$C_S^{(2)}(3) = 2 \text{ and } C_S^{(2)}(4) = 4, \text{ by Figures 1.2(a) and (b).}$$

For general values of n and d, the following theorem is the best upper bound we will be able to prove in this book.

Theorem 1.4: For every $d \geq 2$, there is a positive constant c such that $C^{(d)}(n) < 2^{cn^d}$.

Clearly, this upper bound holds also for the number of different simple arrangements. At this stage, we are not able to prove Theorem 1.4. We therefore postpone its proof until the end of Chapter 7 which presents an algorithm for constructing arrangements. Theorem 1.4 will follow from the analysis of this algorithm.

It seems worthwhile to note that the notion of combinatorial equivalence as defined above differs from the one commonly used which requires only the preservation of incidences rather than of position vectors. Our decision to base the definition of combinatorial equivalence on the notion of position vectors does conform with our choice of treating the vertical direction as a distinguished one.

1.4. Configurations of Points

As mentioned above, there is a close relationship between finite sets of points and finite sets of hyperplanes. More specifically, for each finite set P of points in d-dimensional Euclidean space E^d there is a set H of hyperplanes in E^d and a one-to-one correspondence γ between P and H such that certain statements about P are true if and only if their corresponding dual statements about H are true. For example, if E^d is the plane, that is, if $d = 2$, then the statement

> "Points p, q, and r in P are collinear, that is, they lie on a common line."

corresponds to

> "Lines $\gamma(p)$, $\gamma(q)$, and $\gamma(r)$ in H are concurrent, that is, they intersect in a common point or they are mutually parallel."

As a matter of fact, for a given set P there are many sets H for which such a one-to-one correspondence exists. This section introduces one specific geometric transform that can be used to map a given point set to a set of hyperplanes, or vice versa, such that the dual relationship holds. This geometric transform will allow us to be specific about the dual relationship between sets of points and sets of hyperplanes rather than to discuss it on an abstract level. We will adopt the convention of calling a finite set P of points a *configuration* if we are concerned with properties of P that have dual interpretations for finite sets of hyperplanes.

We denote by $\mathcal{D}$ the specific geometric transform to be defined. It maps a point in E^d to a non-vertical hyperplane in E^d, and vice versa.

> Let $p=(\pi_1,\pi_2,...,\pi_d)$ be a point in E^d. Transform $\mathcal{D}$ maps p to the hyperplane $\mathcal{D}(p)$ given by the equation
>
> $$x_d=2\pi_1x_1+2\pi_2x_2+...+2\pi_{d-1}x_{d-1}-\pi_d,$$
>
> and vice versa, that is, it maps a non-vertical hyperplane h to the point $\mathcal{D}(h)$ such that $\mathcal{D}(\mathcal{D}(h))=h$.

In the natural way, $\mathcal{D}$ is extended to sets of points and sets of hyperplanes, that is, we define $\mathcal{D}(P)=\{\mathcal{D}(p)|p\in P\}$ and $\mathcal{D}(\mathcal{D}(P))=\{\mathcal{D}(h)|h\in\mathcal{D}(P)\}=P$.

Notice that $\mathcal{D}(h)$ is not defined if h is a vertical hyperplane. This is fine if the set of hyperplanes $\mathcal{D}$ is applied to is derived from a set of points using $\mathcal{D}$ in the first place. On some rare occasions, the restriction of $\mathcal{D}$ to non-vertical hyperplanes is a serious nuisance, in which case we will use another transform $\mathcal{D}_o$ to be defined in Section 1.6. Note also that $\mathcal{D}(p)$ and $\mathcal{D}(q)$ are two parallel hyperplanes if and only if the first $d-1$ coordinates of points p and q match, which is equivalent to p and q lying on a common vertical line.

Next, we examine the relationship between the geometric transform $\mathcal{D}$ and the unit paraboloid

$$U\colon x_d=x_1^2+x_2^2+...+x_{d-1}^2.$$

If a point $p=(\pi_1,\pi_2,...,\pi_d)$ does not lie above U, that is, if $\pi_d\leq\pi_1^2+\pi_2^2+...+\pi_{d-1}^2$, then hyperplane $\mathcal{D}(p)$ intersects U exactly in those points x that allow supporting hyperplanes h such that

$$h\cap U=x \text{ and } p\in h.$$

Figure 1.3 illustrates this property for point p_1 below U and for point p_2 on U. To construct line $\mathcal{D}(p_3)$ for point p_3 above U, we use the fact that $\mathcal{D}(p_3)$ is parallel to any line obtained for a point on the same vertical line as p_3. It is not hard to prove that the vertical projection of $\mathcal{D}(p)\cap U$ onto the hyperplane h_0: $x_d=0$ is a sphere whose points $x=(x_1,x_2,...,x_{d-1},0)$ satisfy

$$(x_1-\pi_1)^2+(x_2-\pi_2)^2+...+(x_{d-1}-\pi_{d-1})^2=\pi_1^2+\pi_2^2+...+\pi_{d-1}^2-\pi_d.$$

Notice that the vertical projection of p onto h_0 is the center of the sphere and that the radius is equal to the distance from p to its vertical projection onto U. The sphere that corresponds to point p in the way described above will come into

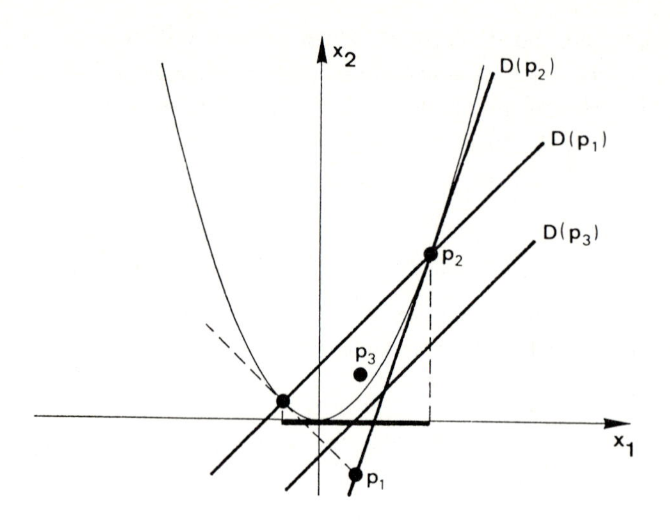

Figure 1.3. The dual transform $\mathcal{D}$ in the plane.

play in Section 1.6 and to a larger extent in Chapter 13 when we talk about power diagrams. Figure 1.3 shows the one-dimensional sphere (that is, the interval) that corresponds to point p_1.

Next, we discuss two fairly straightforward but fundamental properties of $\mathcal{D}$ which turn out to be of the utmost importance for the investigations undertaken in this book. We present the properties without proof since it requires only trivial algebra to verify them.

Observation 1.5: Let p be a point in E^d and let h be a non-vertical hyperplane in E^d.

(i) *Incidence preservation:* point p belongs to hyperplane h if and only if point $\mathcal{D}(h)$ belongs to hyperplane $\mathcal{D}(p)$.

(ii) *Order preservation:* point p lies above (below) hyperplane h if and only if point $\mathcal{D}(h)$ lies above (below) hyperplane $\mathcal{D}(p)$.

Consider Figure 1.3 as an example. Point p_1 lies below lines $\mathcal{D}(p_1)$ and $\mathcal{D}(p_3)$ and it lies on line $\mathcal{D}(p_2)$. It follows that points p_1 and p_3 must lie below line $\mathcal{D}(p_1)$ and that point p_2 must lie on $\mathcal{D}(p_1)$ which is in fact the case. Notice that the incidence preserving property of $\mathcal{D}$ implies that three collinear points in E^2 correspond to three concurrent lines as mentioned at the beginning of this section: three collinear points lie on a common line l and this line l corresponds to a point that lies on all three lines corresponding to the three points, unless l is vertical, in which case the three lines are parallel.

In addition to the above fact, Observation 1.5 implies a number of interesting correspondences between a configuration P and the set $\mathcal{D}(P)$ of hyperplanes. Below, we give a short list of such correspondences without proving them. In each case, a straightforward proof can be found if one uses Observation 1.5.

Corollary 1.6: Let P be a set of n points in E^d and set $H=\mathcal{D}(P)$, the dual set of hyperplanes.

(i) A non-vertical hyperplane h contains d affinely independent points of P if and only if $\mathcal{D}(h)$ is a vertex of $\mathcal{A}(H)$.

(ii) The affine hull of the points of P that lie on a non-vertical hyperplane h is a k-flat if and only if $\mathcal{D}(h)$ is contained in a $(d-k-1)$-face of $\mathcal{A}(H)$.

(iii) Point p is contained in a cell of $\mathcal{A}(H)$ and $H_a=\{h \in H \mid p \in h^-\}$ if and only if no point of P lies on hyperplane $\mathcal{D}(p)$ which partitions P into $\mathcal{D}(H_a)$ and $\mathcal{D}(H-H_a)$. We say that the cell c that contains p *defines* the partition of P into $\mathcal{D}(H_a)$ and $\mathcal{D}(H-H_a)$.

(iv) Two cells $c_1 \neq c_2$ in $\mathcal{A}(H)$ define different partitions of P unless c_1 is in h^+ if and only if c_2 is in h^-, for each hyperplane h in H. In this exceptional case, c_1 and c_2 are unbounded.

(v) Arrangement $\mathcal{A}(H)$ is simple if and only if every $2 \leq k \leq d$ points of P lie on a unique non-vertical $(k-1)$-flat and no $d+1$ points of P lie on a common hyperplane.

The reader is encouraged to extend the above list of correspondences between P and H according to his or her interest.

1.5. Sylvester's Problem

Historically one of the first combinatorial problems that had to do with the existence of certain constellations was posed in 1893 by J.J. Sylvester. His problem deals with configurations of points in the plane and with lines joining the points. For a configuration P of points in the plane, we call a line *ordinary* if it contains exactly two points of P. Sylvester's problem can now be formulated as the following amusing question.

> "Is it true that every configuration P in the plane admits an ordinary line, unless all points of P are collinear?"

Before providing an answer to this question, we discuss its interpretation in the dual world which hosts the set of lines $H=\mathcal{D}(P)$. Notice that there is no loss of generality if we assume that no two points of P lie on a common vertical line; in this case no two lines in H are parallel. The condition that not all points of P are collinear now implies that not all lines of H go through a common point. If

P admits an ordinary line, then $\mathcal{A}(H)$ has a vertex that belongs to exactly two lines.

Below, we answer Sylvester's question in the affirmative by proving the following theorem.

Theorem 1.7: Every non-collinear configuration of points in the plane admits an ordinary line.

Proof: We show that the negation of the assertion leads to a contradiction. Thus, assume that P is a configuration of $n \geq 3$ points, not all collinear, such that each line through two points of P also contains a third point. For any line h through three points of P, we let $p(h)$ denote a point of P that does not lie on h but that is closest to h among all such points. Call $p(h)$ the *closest point of* h, and let h_0 be a line that minimizes the distance to its closest point. Furthermore, let p, q, and r be three points of P that lie on h_0. Without loss of generality, we assume that h_0 is horizontal, that p is to the left of q and q is to the left of r, that $p(h_0)$ is above h_0, and that the x_1-coordinate of $p(h_0)$ is less than or equal to the x_1-coordinate of q (see Figure 1.4). But then q is closer to line h' through r and $p(h_0)$ than point $p(h_0)$ is to h_0, a contradiction. □

By duality, Theorem 1.7 implies that every finite set H of lines, not all concurrent, contains a pair of lines that intersect in a point avoiding all other lines or that are parallel to each other but not to any other line in H.

1.6. Convex Polytopes and Convex Polyhedra

Convex polytopes and convex polyhedra are well studied objects in combinatorial geometry. Both bear a close relationship to arrangements of hyperplanes and to configurations of points. Formally, a (*convex*) *polyhedron* is defined as the intersection of a finite number of closed half-spaces. A polyhedron $\mathcal{P}$ in E^d is called a

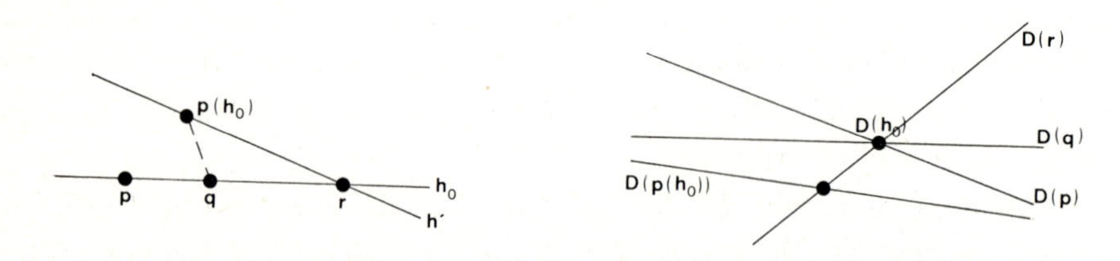

Figure 1.4. Ordinary lines in primal and in dual space.

k-polyhedron if its dimension is k, that is, if k is the smallest integer number such that $\mathcal{P}$ is contained in a k-flat. Notice that a polyhedron with non-empty interior in E^d is necessarily a d-polyhedron. If a polyhedron is bounded, then it is also referred to as a (*convex*) *polytope* and, more specifically, as a *k-polytope* if its dimension is k.

The most obvious connection between polyhedra and arrangements of hyperplanes is that the closure of any k-face in an arrangement is a k-polyhedron. In fact, for every polyhedron $\mathcal{P}$ there is an arrangement $\mathcal{A}(H)$ such that $\mathcal{P}$ is the closure of some face f of $\mathcal{A}(H)$. If f is a k-face, then all i-faces in $\mathcal{A}(H)$, $0 \leq i \leq k$, contained in the closure of f are called *faces of* $\mathcal{P}$. As usual, the 0-faces among these faces are called the *vertices of* $\mathcal{P}$, the 1-faces are the *edges of* $\mathcal{P}$, and the $(k-1)$-faces are the *subfaces of* $\mathcal{P}$. If $\mathcal{P}$ is a d-polytope, then there is a unique smallest set H of hyperplanes such that the interior of $\mathcal{P}$ is a cell of $\mathcal{A}(H)$ – of course, H is the set of affine hulls of the facets of $\mathcal{P}$.

If $\mathcal{P}$ is a convex polytope, then it coincides with the convex hull of its vertices. This suggests an alternate definition of convex polytopes as convex hulls of finite sets of points. Now let P be a finite set of points in E^d and let $\mathcal{P}$ be its convex hull. The vertices of $\mathcal{P}$ are called the *extreme points of* P, and we write ext$P$ for the set of extreme points of P. Obviously, ext$P$ is a subset of every point set Q with $\mathcal{P}=\text{conv}Q$. This implies that ext$P$ is the smallest point set with the property that its convex hull is equal to $\mathcal{P}$.

The above discussion suggests that the dual correspondence between configurations of points and arrangements of hyperplanes implies some kind of duality between certain pairs of polytopes. The remainder of this section demonstrates that such pairs of dual polytopes in fact exist. For reasons that will become apparent later, we introduce a new geometric transform $\mathcal{D}_o$ to construct specific examples of dual pairs of polytopes. Transform $\mathcal{D}_o$ resembles $\mathcal{D}$ introduced in Section 1.4, only its definition is based on the unit sphere in E^d rather than on the unit paraboloid.

To specify $\mathcal{D}_o$, we need to know that a hyperplane h can be written as the set of points $x=(\xi_1,\xi_2,...,\xi_d)$ such that the scalar product

$$\langle u,x\rangle = v_1\xi_1+v_2\xi_2+...+v_d\xi_d$$

is equal to some real number α, where $u=(v_1,v_2,...,v_d)$ is a non-zero vector. As a matter of fact, u is a normal vector of h and α is the distance between the origin and h if u is chosen such that its length $(v_1^2+v_2^2+...+v_d^2)^{1/2}$ is equal to 1. With this notation, we can specify $\mathcal{D}_o$ as follows.

Let $p=(\pi_1,\pi_2,...,\pi_d)$ be a point in E^d different from the origin. Transform $\mathcal{D}_o$ maps point p to the hyperplane

$$\mathcal{D}_o(p)\colon \langle p, x\rangle = 1,$$

and vice versa, that is, we define $\mathcal{D}_o(\mathcal{D}_o(p)) = p$.

Thus, $\mathcal{D}_o$ is not defined for p equal to the origin and for all hyperplanes that contain the origin.

The relationship between $\mathcal{D}_o$ and the unit sphere

$$S\colon x_1^2 + x_2^2 + \ldots + x_d^2 = 1$$

is the same as between transform $\mathcal{D}$ and the unit paraboloid U: if a point $p = (\pi_1, \pi_2, \ldots, \pi_d)$ does not lie inside S, that is, if $\pi_1^2 + \pi_2^2 + \ldots + \pi_d^2 \geq 1$, then hyperplane $\mathcal{D}_o(p)$ intersects S exactly in those points x that admit supporting hyperplanes h such that

$$h \cap S = x \quad \text{and} \quad p \in h$$

(see Figure 1.5). Furthermore, $\mathcal{D}_o$ is incidence and order preserving. To express in what way $\mathcal{D}_o$ preserves order, we need to assign an orientation to the hyperplanes that avoid the origin. For each such hyperplane h let h^{pos} be the open half-space that contains the origin and let h^{neg} be the other open half-space. Then we have the following fundamental property of $\mathcal{D}_o$.

Observation 1.8: Let $p \neq o$ be a point in E^d and let h be a hyperplane that avoids o.

(i) *Incidence preservation:* point p belongs to hyperplane h if and only if point $\mathcal{D}_o(h)$ belongs to hyperplane $\mathcal{D}_o(p)$.

(ii) *Order preservation:* point p lies in half-space h^{pos} (h^{neg}) if and only if point $\mathcal{D}_o(h)$ lies in half-space $\mathcal{D}_o(p)^{pos}$ ($\mathcal{D}_o(p)^{neg}$).

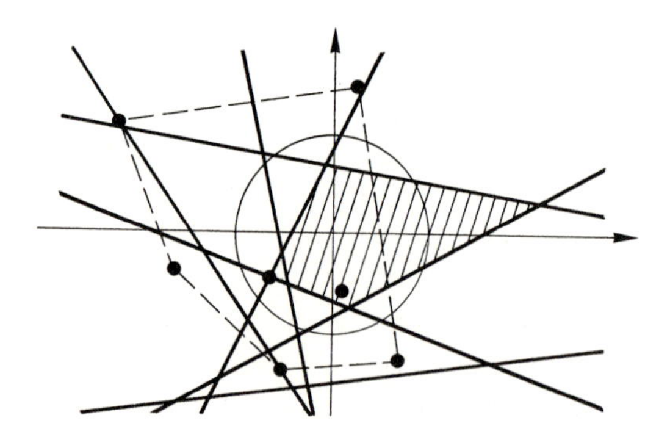

Figure 1.5. The dual transform $\mathcal{D}_o$ in the plane.

Consider Figure 1.5 as an example. It shows seven points and their corresponding lines.

We extend D_o from points and hyperplanes to sets of points and sets of hyperplanes in the natural way, that is, we define $D_o(P)=\{D_o(p)|p\in P\}$ and $D_o(H)=\{D_o(h)|h\in H\}$, for sets of points P and sets of hyperplanes H. However, primarily we are interested in extending D_o to convex polytopes. Let P be a convex polytope with non-empty interior int$P$ and assume that the origin o is contained in int$P$. Then, $D_o(P)$ is an infinite set of hyperplanes that avoid some convex set around o. We define the closure of this set as the *dual polytope* $\overline{P}$ *of* P, that is, we define

$$\overline{P}=\mathrm{cl}\,\mathrm{compl}(\bigcup_{h\in D_o(P)} h).$$

Notice that we also have $\overline{P}=\mathrm{cl}(\bigcap_{h\in D_o(P)} h^{pos})$. Since D_o is self-inverse, P is the dual polytope of $\overline{P}$. Figure 1.5 shows a dual pair of convex polytopes (that is, polygons) in the plane. One polygon is constructed as the convex hull of seven points, the other one is the intersection of seven half-planes bounded by the lines that correspond to the points.

Below, we present a short list of correspondences between convex polytopes and their duals. We include no proofs since all statements are straightforward implications of Observation 1.8.

Corollary 1.9: Let P be a convex polytope whose interior contains o, and let $p\neq o$ be a point in E^d.

(i) Point p belongs to int$P$, bd$P$, or compl$P$ if and only if hyperplane $D_o(p)$ avoids $\overline{P}$, avoids int$\overline{P}$ but not $\overline{P}$, or intersects int$\overline{P}$, respectively.

(ii) Point p belongs to a k-face of P if and only if the relative interior of $D_o(p)\cap\overline{P}$ is a $(d-k-1)$-face of $\overline{P}$.

(iii) There is a one-to-one correspondence γ between the k-faces of P and the $(d-k-1)$-faces of $\overline{P}$, for $0\leq k\leq d-1$, such that f is a subface of g in P if and only if $\gamma(g)$ is a subface of $\gamma(f)$ in $\overline{P}$.

We encourage the reader to spend some time on visualizing Corollary 1.9(ii). For example, imagine a convex polytope P in three dimensions and its dual $\overline{P}$. If we trace the boundary of P with a point p, then the plane $D_o(p)$ moves continuously such that it always remains in contact with $\overline{P}$ but never intersects its interior.

1.7. Zonotopes

Zonotopes are a particular kind of convex polytopes that are interesting to us due to their special relationship with arrangements of hyperplanes in one dimension lower. In this section, we specify when a convex polytope is called a zonotope, we show several properties of zonotopes, and we discuss the correspondence between zonotopes and arrangements.

For two points $p=(\pi_1,\pi_2,\ldots,\pi_d)$ and $q=(\psi_1,\psi_2,\ldots,\psi_d)$ in d dimensions, we define $-p=(-\pi_1,-\pi_2,\ldots,-\pi_d)$ and $p+q=(\pi_1+\psi_1,\pi_2+\psi_2,\ldots,\pi_d+\psi_d)$. For example, $p+(-p)=p-p=o$. Now let P be a set of points $p_1,p_2,\ldots,p_n$ in E^d. We define

$$P^*=\{\alpha_1p_1+\alpha_2p_2+\ldots+\alpha_np_n \mid \alpha_i\in\{-1,+1\} \text{ for } 1\leq i\leq n\},$$

and we set $Z(P)=\text{conv}P^*$. Obviously, $Z(P)$ is a centrally symmetric convex polytope whose center coincides with the origin. It follows that for each face f of $Z(P)$, its centrally symmetric image $-f$ is also a face of $Z(P)$. We call $\{f,-f\}$ an *antipodal face pair of* $Z(P)$. A polytope Z is called a *zonotope* if it is a translate of a polytope $Z(Q)$, for some finite set of points Q in E^d. If Z is a k-polytope, then it is also referred to as a *k-zonotope*. For example, a 0-zonotope is a point, a 1-zonotope is a closed line segment, and a 2-zonotope is a centrally symmetric polygon.

There is an alternate definition of zonotopes by means of the *Minkovsky sum of* sets defined as follows: if A and B are sets in E^d, then

$$A+B=\{p+q \mid p\in A \text{ and } q\in B\}.$$

A polytope Z is a zonotope if it is the Minkovsky sum of a finite number of closed line segments. If each line segment of this set is translated such that the origin is its midpoint, then $Z=Z(P)$, where P contains one endpoint of each translated line segment. Figure 1.6 illustrates both definitions of zonotopes: the hexagon shown is the Minkovsky sum of three line segments as well as the convex hull of $2^3=8$ points.

It is very instructive to visualize the inductive construction of a zonotope $Z_n=s_1+s_2+\ldots+s_n$. For convenience, we assume that no d of the line segments can be translated such that they lie in a common hyperplane. Let Z_{n-1} be the Minkovsky sum of the first $n-1$ line segments, and let p and q be the two endpoints of s_n. Then Z_n is the subset of E^d that is swept out by $Z_{n-1}+x$ if we move point x from p to q. It is fairly clear from this picture that for each face f of Z_{n-1} either $f+p$ or $f+q$ is a face of Z_n. Between the faces $f+p$ and the faces $f+q$ there is a "zone" of faces $f+s_n$, where f is a face of Z_{n-1} that belongs to a supporting hyperplane of Z_{n-1} parallel to the line through s_n. Other "zones" of Z_n can be obtained by changing the order of the segments.

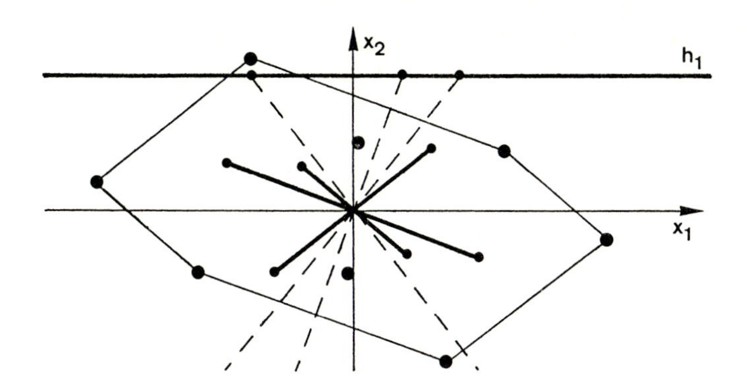

Figure 1.6. A planar zonotope with corresponding one-dimensional arrangement.

We remark that the changes that occur when we add a line segment to a zonotope have characteristics similar to those of the changes that occur when we add a hyperplane to an arrangement: both operations affect only a "zone" of the respective structure. We will see below that this similarity is not at all accidental. Indeed, there is a geometric correspondence between zonotopes in E^d and arrangements in E^{d-1} that maintains structural properties.

In order to describe a particular geometric transform that realizes this correspondence, we introduce some notation. For a point $p \neq o$ in E^d, define

$$h_p\colon \langle p,x\rangle = 0,\ \ h_p^{pos}\colon \langle p,x\rangle > 0,\ \text{ and }\ h_p^{neg}\colon \langle p,x\rangle < 0,$$

that is, h_p is the unique hyperplane through o normal to vector p, and h_p^{pos} and h_p^{neg} are the two open half-spaces bounded by h_p. Now let h_1 be the hyperplane determined by the equation $x_d = 1$, and set

$$g_p = h_1 \cap h_p,\ \ g_p^{pos} = h_1 \cap h_p^{pos},\ \text{ and }\ g_p^{neg} = h_1 \cap h_p^{neg}.$$

It is trivial to see that the transform that maps a point in E^d to a $(d-2)$-flat in h_1 has the following fundamental properties.

Observation 1.10: Let $p \neq o$ be a point in E^d.

(i) *Incidence preservation*: point q in h_1 lies on g_p if and only if point p belongs to hyperplane h_q.

(ii) *Order preservation*: point q in h_1 lies in g_p^{pos} (g_p^{neg}) if and only if point p belongs to half-space h_q^{pos} (h_q^{neg}).

With this notation, we map a zonotope $\mathcal{Z}(P)$ to the $(d-1)$-dimensional arrangement $\mathcal{A}(G_P)$ in h_1, with $G_P = \{g_p \mid p \in P\}$. Notice that there is no loss in generality if we assume that no point of P lies on the x_d-axis. In this case, each point of P corresponds to a well-defined $(d-2)$-flat in h_1. Figure 1.6 illustrates this definition: it shows a two-dimensional zonotope and the corresponding one dimensional arrangement.

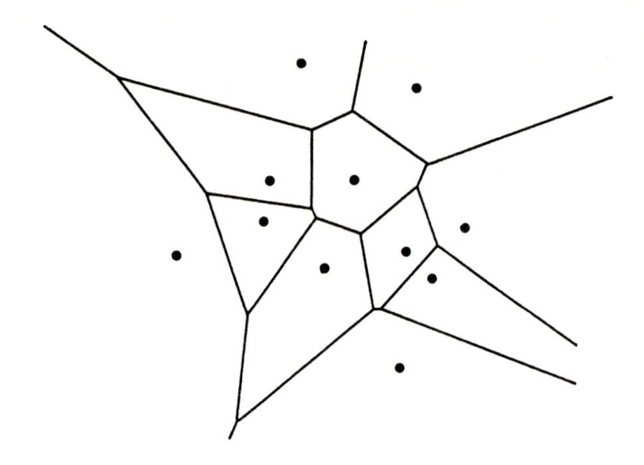

Figure 1.7. Voronoi diagram for points in the plane.

$$\mathcal{E}(p)\colon x_{d+1} = 2\pi_1 x_1 + 2\pi_2 x_2 + \ldots + 2\pi_d x_d - (\pi_1^2 + \pi_2^2 + \ldots + \pi_d^2)$$

in E^{d+1}.

It is not hard to see that $\mathcal{E}(p)$ is the unique hyperplane that touches the unit paraboloid U: $x_{d+1} = x_1^2 + x_2^2 + \ldots + x_d^2$ in the vertical projection $U(p)$ of p onto U (see Figure 1.8).

In a certain sense, $\mathcal{E}$ translates distance information in E^d into combinatorial information in E^{d+1}. To see how this works, we let $h(p)$ denote the vertical projection of a point p in h_0 onto a non-vertical hyperplane h. With this notation, we can formulate the following fundamental property of transform $\mathcal{E}$.

Observation 1.14: Let p and x be two points in E^d, and define $h = \mathcal{E}(p)$. Then $(d(x,p))^2 = d(U(x),h(x))$.

For a finite set of points P in E^d, define $H = \{\mathcal{E}(p) | p \in P\}$ and let t be the topmost cell of the arrangement $\mathcal{A}(H)$, that is,

$$t = \bigcap_{p \in P} \mathcal{E}(p)^+$$

(see Figure 1.8). By Observation 1.14, we can relate the Voronoi diagram $\mathcal{V}(P)$ with the boundary of cell t as follows. For a point x in E^d, we consider its vertical projection $U(x)$ which is a point y on U. If y drops vertically, then the first hyperplane it hits identifies the closest point of x. If y hits two or more hyperplanes at the same time, then the corresponding two or more points are equidistant from x which implies that x belongs to the boundary of two or more cells in $\mathcal{V}(P)$. This picture implies the following result.

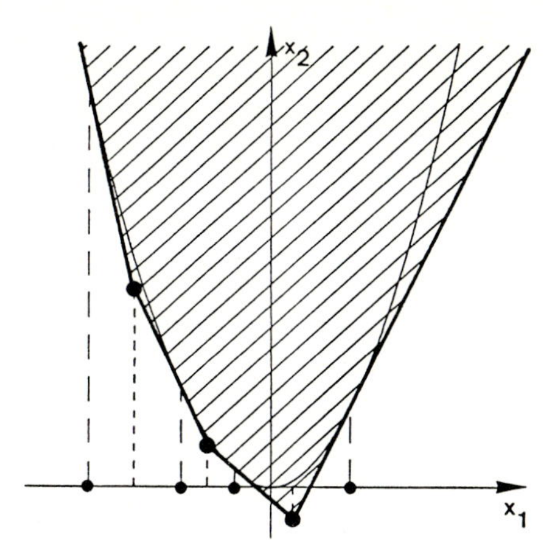

Figure 1.8. Voronoi diagram from projection.

Corollary 1.15: Let P be a finite set of points in E^d. The vertical projection onto h_0 of the faces in the boundary of the topmost cell t of arrangement $\mathcal{A}(\mathcal{E}(P))$ yields the Voronoi diagram of P.

Of course, Voronoi diagrams are also related to sets of points since arrangements of hyperplanes are. More information about this relationship and plenty of results about Voronoi diagrams can be found in Chapter 13.

1.9. Exercises and Research Problems

2 **Exercise 1.1:** (a) Prove the following combinatorial relation:

$$\binom{d-i}{k-i}\binom{n-1}{d-i}+\binom{d-1-i}{k-i}\binom{n-1}{d-1-i}+\binom{d-1-i}{k-1-i}\binom{n-1}{d-1-i}=\binom{d-i}{k-i}\binom{n}{d-i},$$

for $0 \leq i \leq k \leq d$ and $n \geq 1$. As usual assume $\binom{a}{b}=0$ if $b<0$ or $a<b$.

2 (b) Prove the following combinatorial identity:

$$\sum_{i=0}^{k}\binom{d-i}{k-i}\binom{n}{d-i}=\binom{n}{n-k}\cdot 2^k,$$

if $n \leq d$.

3 (c) Prove Lemma 1.2 by induction on n and on d, that is, prove that the number of k-faces in a simple arrangement of n hyperplanes in E^d is

$$\sum_{i=0}^{k}\binom{d-i}{k-i}\binom{n}{d-i},$$

for $0 \leq k \leq d$ (see Section 1.2.)

2 (d) Prove that the same arrangement as in (c) has

$$2\sum_{i=1}^{k}(-1)^{i+1}f_{k-i}^{(d-i)}(n)$$

unbounded k-faces, for $1 \leq k \leq d$.

2 **Exercise 1.2:** (a) Let f and g be two faces in an arrangement of n hyperplanes and let $I \neq \emptyset$ be the set of indices $1 \leq i \leq n$ for which $v_i(f) \neq v_i(g)$. Prove that f is a sub-face of g if and only if $v_i(f)=0$, for each $i \in I$, and there is no non-empty proper subset J of I such that $v_j(f')=0$, if $j \in J$, and $v_j(f')=v_j(g)$, if $j \notin J$, for any face f' of the same arrangement.

3 (b) Prove that the vertices on a line in an arrangement of n hyperplanes in E^d are incident upon at most $2\binom{n}{d-1}$ edges if $d \geq 3$, and upon at most $4n-4$ edges if $d=2$.

2 **Exercise 1.3:** (a) Compute the value of $C^{(2)}(4)$, that is, determine the number of non-equivalent arrangements of four lines in the plane.

2 (b) Draw all $C_S^{(2)}(5)$ non-equivalent simple arrangements of five lines in the plane.

4 (c) Prove $C^{(d)}(n) \leq 2^{cn\log_2 n}$, for some positive constant c depending on d.

2 **Exercise 1.4:** (a) The sequence of points (p,q,r) in E^2 is said to be a *left-turn* if r is to the left of the line through p and q that is directed from p to q. Two sets P_1 and P_2, each with n points in E^2, are said to be *equivalent* if there is a one-to-one correspondence γ such that (p,q,r) is a left-turn if and only if $(\gamma(p),\gamma(q),\gamma(r))$ is a left-turn. Show that P_1 and P_2 are equivalent if $\mathcal{A}(\mathcal{D}(P_1))$ and $\mathcal{A}(\mathcal{D}(P_2))$ are combinatorially equivalent.

1 (b) Show that the reverse of the implication in (a) is not true.

2 **Exercise 1.5:** Verify Observation 1.5 and Corollary 1.6.

Exercise 1.6: Call a hyperplane h in E^d *ordinary with respect to* a set P of n points in E^d if h contains exactly d points of P.

4 (a) Show that a set of n points in the plane defines at least $3n/7$ ordinary lines, unless all n points are collinear.

4 (b) Show that a set of n points in the plane defines at least $n/2$ ordinary lines, unless $n \in \{7,13\}$ or all n points lie on a common line.

2 (c) Show that for $d=3$ it is possible that no ordinary plane exists even if there is no common plane that contains all points of P.

1 **Exercise 1.7:** Verify the correspondences between a convex polytope and its dual polytope formulated in Corollary 1.9.

2 **Exercise 1.8:** Let P be a set of n points in E^d, no point on the x_d-axis, let $\mathcal{Z}(P)$ be the corresponding zonotope, and let $\mathcal{A}(G_P)$ be the corresponding arrangement in hyperplane h_1: $x_d=1$ (see Section 1.7).

2 (a) For each point p in P, let $H(p)$ be the set of hyperplanes that go through p and the origin and define $g(p)=\{x \mid x=\mathcal{D}(h),\ h \in H(p)\}$. Show that $G(P)=\{g(p) \mid p \in P\}$ is a set of n $(d-2)$-flats in hyperplane h_0: $x_d=0$ and prove that the $(d-1)$-dimensional arrangement defined by $G(P)$ is equal to $\mathcal{A}(G_P)$, up to a translation along the x_d-axis.

2 (b) Let $\overline{\mathcal{Z}}$ be the dual polytope of $\mathcal{Z}(P)$, as defined in Section 1.6. Prove that the central projection onto h_1 of the faces of $\overline{\mathcal{Z}}$ yields an arrangement that is equal to $\mathcal{A}(G_P)$.

3 **Exercise 1.9:** Show that the number of k-faces of a zonotope defined by n line segments in E^d is at most

$$\sum_{i=0}^{\lfloor k/2 \rfloor} \binom{d-2i}{d-k}\binom{n}{d-2i},$$

for $0 \leq k \leq d-1$. *(Hint: use the correspondence between zonotopes in E^d and arrangements in E^{d-1} described in Section 1.7 and take the "projective" view of what a face in an arrangement is, that is, identify opposite unbounded faces.)*

2 **Exercise 1.10:** Use the dual transform $\mathcal{D}_0$, defined in Section 1.6, to establish a correspondence between Voronoi diagrams of points in E^d and arrangements of hyperplanes in E^{d+1}. *(Comment: this exercise essentially asks you to demonstrate that transforms $\mathcal{D}$ and $\mathcal{D}_0$ can be substituted for each other.)*

1.10. Bibliographic Notes

There are a number of textbooks available that cover the field of combinatorial geometry to a varying extent. For example, Hadwiger, Debrunner (1959) (see also Hadwiger, Debrunner, Klee (1964)) discusses a large number of combinatorial geometry problems but considers only the two-dimensional case. Eggleston (1958), Alexandrov (1958), Yaglom, Boltyanskii (1961), and Lyusternik (1966) concentrate on questions about convex figures which include convex polygons and convex polytopes. Highly recommendable texts on the combinatorial structure of convex polytopes are Grünbaum (1967), McMullen, Shepard (1971), and Bronsted (1983).

The book by Grünbaum (1967) goes beyond the theory of convex polytopes and also discusses arrangements of hyperplanes in arbitrary dimensions. Other general sources of information about arrangements of hyperplanes are Grünbaum (1972) which presents a vast number of results for the two-dimensional case and Grünbaum (1971) which is a survey on arrangements in arbitrary dimensions. The notion of position vectors assigned to faces of an arrangement used in Section 1.1 is taken from Ringel (1956).

Formulas that count the number of faces in an arrangement are given at various places in the literature including the general sources for arrangements mentioned above. In addition, we mention Steiner (1826) for maybe the first study of this problem and Buck (1943) for formulas that cover spaces and faces of arbitrary dimensions. Zaslavsky (1975) gives a very deep treatment of the problem; he relates the number of faces with the number of degeneracies of various types defined by the hyperplanes. An analysis of the number of faces in three-dimensional arrangements that uses the space sweep approach also employed in Section 1.2 can be found in Alexanderson, Wetzel (1981).

The problem of counting non-equivalent arrangements is considered in Grünbaum (1972) and also in Klee (1938). However, both use a definition of equivalence that differs from ours which uses the vertical direction as the one that is distinguished from the other directions. The upper bound of 2^{cn^d} on the number of non-equivalent arrangements of n hyperplanes in d dimensions mentioned in Section 1.3 derives from results in Edelsbrunner, O'Rourke, Seidel (1986) which can also be found in Chapters 5 and 7 of this book. Bounds that are considerably more accurate have been found by Goodman, Pollack (1986) who use a result by Milnor (1964) in algebraic geometry. Their result is an upper bound of $2^{cn\log_2 n}$ on the number of combinatorially different arrangements in E^d, where c is a constant that depends on d. This solves Exercise 1.3(c).

Sylvester's problem originates with Sylvester (1893) and was rediscovered in the 1930s. Gallai seems to have been the first to solve the problem; a version of his proof can be found in Steinberg (1944). The proof presented in Section 1.5 is due to Leroy Kelly and reported in Coxeter (1948). Soon after settling the existence of an ordinary line for every non-collinear finite set of points in the plane, mathematicians looked at the problem of how few such lines can be defined by n points (see Exercise 1.6). An answer to this question that solves Exercise 1.6(a) can be found in Kelly, Moser (1958), and the improvement of their result indicated in Exercise 1.6(b) is given by Hansen (1981). Generalizations of the problem to dimensions higher than two are discussed in Motzkin (1951); however, his generalization to ordinary

of integer numbers. The values of these numbers will be without significance which allows us to replace any set of cardinality n by the set $\{1,2,...,n\}$. Two consecutive permutations of a circular sequence differ only slightly. This will be formalized by the notion of a so-called move which transforms one permutation to another.

Next, we describe the connection between circular sequences and configurations of n points in the plane. Suppose we look at a configuration from some direction and that what we see is the sequence of points in some order from left to right. This sequence defines a permutation of the points. As we continuously change our viewpoint, this permutation changes. Circular sequences are introduced in a formal manner below. The relationship with arrangements and configurations will be addressed in Sections 2.2 through 2.4.

For n a positive integer, we use $[n]$ to denote the set of integers $\{1,2,...,n\}$. A *permutation* Π *of* $[n]$ is a total order $\leq_\Pi$ imposed upon $[n]$. Π can also be viewed as an enumeration of $[n]$ or as a word or string $a_1a_2...a_n$, with a_i in $[n]$ and $a_i \neq a_j$ if $i \neq j$. Any string $w = a_i a_{i+1} ... a_j$, $1 \leq i \leq j \leq n$, is a *substring of* Π and $w^R = a_j a_{j-1} ... a_i$ is the *reverse of* w. A *move* reverses one or more non-overlapping substrings of Π and we write

$$\Pi \rightarrow_m \Pi'$$

if m is a move that changes Π to Π'. Move m is said to *swap* a and b of $[n]$ if it changes the relative order of a and b, that is, if we have $a \leq_\Pi b$ and $b \leq_{\Pi'} a$ or $b \leq_\Pi a$ and $a \leq_{\Pi'} b$. A sequence C of permutations Π_i of $[n]$, for i ranging over all integers, is now called a *circular sequence of* $[n]$, or an *n-sequence* for short, if

(i) there are moves m_i such that $\Pi_i \rightarrow_{m_i} \Pi_{i+1}$, and

(ii) if m_i swaps a and b and $j > i$ is minimal such that m_j again swaps a and b, then $\Pi_j = \Pi_i^R$ and m_j is the reverse of m_i, that is, we also have $\Pi_{j+1} = \Pi_{i+1}^R$.

It follows from the above definitions that there is a unique minimal positive integer number k such that $\Pi_{i+k} = \Pi_i^R$ and $\Pi_{i+2k} = \Pi_i$, for every integer number i. Consequently, any n-sequence C is determined by k consecutive permutations. We call such a sequence of k permutations a *halfperiod of* C. For convenience, we will often augment a halfperiod $\mathcal{H}$ of C with the permutation of C immediately following the last permutation of $\mathcal{H}$. As an example consider Figure 2.1 which shows halfperiods of three 4-sequences. Each move is indicated by some "X'es" between the permutations involved; one X indicates the reversal of two numbers, and three X'es in a row reverse a substring of length three. Without loss of generality, we choose $1 2 ... n$ to represent the first permutation. Notice that the last permutation of each halfperiod is therefore $n\ n-1 ... 1$.

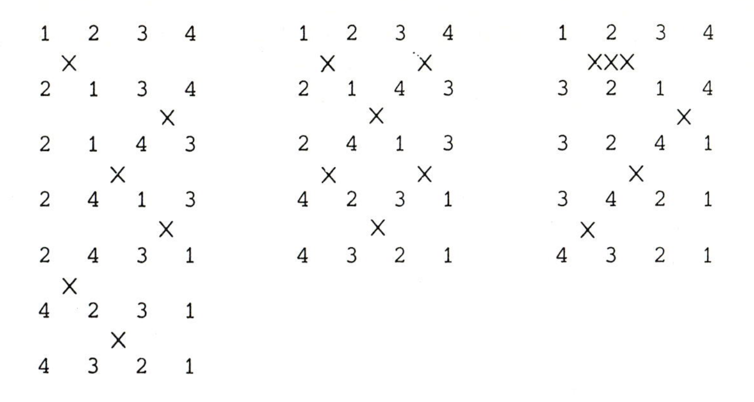

Figure 2.1. Three 4–sequences.

For some purposes, representations of a circular sequence different from the list of permutations in a halfperiod are more convenient. One such useful alternate representation is the enumeration of the moves that occur in a halfperiod. A move m, $\Pi \rightarrow_m \Pi'$, can be written as a list of substrings in Π that it reverses. In the example below, we use commas to separate substrings of a single move and we use semicolons to separate consecutive moves. With these conventions, the three 4–sequences shown in Figure 2.1 can be represented by the following three lists of moves.

12; 34; 14; 13; 24; 23.
12, 34; 14; 24, 13; 23.
123; 14; 24; 34.

Yet another possibility for describing a circular sequence is to enumerate, for each number i in the set $[n]$, the sequence of reversed substrings in a halfperiod that contain i. The first 4–sequence in Figure 2.1 can then be represented by the following four lists of three moves each.

12	12	34	34
14	24	13	14
13	23	23	24

The sequence of substrings containing some number i is called the *local sequence of* i. Note that the halfperiods of local sequences given above also represent the second 4–sequence shown in Figure 2.1. This makes it obvious that a circular sequence is not uniquely determined by its local sequences.

In the investigation of many problems concerning configurations of arrangements in the plane, we can restrict ourselves to circular sequences where every move swaps only two numbers, and these two numbers are consecutive by definition of a move. A circular sequence that satisfies this condition is called *simple*.

For instance, the first 4–sequence shown in Figure 2.1 is simple, the other two are not simple. One application of circular sequences where we are particularly interested in simple circular sequences can, for example, be found in Chapter 3. Since every number in a simple circular sequence is swapped with every other number exactly once within any halfperiod, we have the following straightforward result.

Observation 2.1: Let $\mathcal{H}$ be a halfperiod of a simple n–sequence $\mathcal{C}$.

(i) Any number $i \in [n]$ is contained in $n-1$ substrings of length two that are reversed during a halfperiod.

(ii) Halfperiod $\mathcal{H}$ consists of $\binom{n}{2}$ permutations, or of $\binom{n}{2}+1$ permutations if the reversal of the first permutation is appended to $\mathcal{H}$.

For the same reason that it is natural to call two arrangements equivalent if they share certain features, it is natural to define equivalence relations for circular sequences. Similar to the geometric counterpart, there are several possible notions of equivalence that make immediate sense. Two of the more important equivalence relations are now defined.

(1) Two n–sequences $\mathcal{C}_1$ and $\mathcal{C}_2$ are *circularly equivalent* if $\mathcal{C}_1 = \mathcal{C}_2$ up to some relabeling of the numbers in $\mathcal{C}_2$.

(2) Two n–sequences $\mathcal{C}_1$ and $\mathcal{C}_2$ are *locally equivalent* if the local sequence of some halfperiod of $\mathcal{C}_1$ agrees with the local sequence of some halfperiod of $\mathcal{C}_2$, after possibly relabeling the numbers in $\mathcal{C}_2$.

With these definitions, the 4–sequences in Figure 2.1 are pairwise not circularly equivalent. Nevertheless, the first two 4–sequences are locally equivalent. Indeed, circular equivalence is a proper refinement of local equivalence.

2.2. Encoding Arrangements and Configurations

The significance of circular sequences to the investigations of this book follows from the possibility to encode two–dimensional arrangements of lines and configurations of points. We describe how a circular sequence encodes an arrangement of lines first and then discuss the encoding of a configuration of points.

Let H be a set of n non–vertical lines in the plane denoted as $1,2,\ldots,n$, and let $t_1 < t_2 < \ldots < t_m$ be the x_1–coordinates of the vertices of the arrangement $\mathcal{A}(H)$. A vertical and upward directed line $h(t)$: $x_1 = t$ intersects line i in a point with coordinates (t, b_i). If $t \neq t_i$, for each $1 \leq i \leq m$, then we have $b_i \neq b_j$ if $i \neq j$. For such a value of t, we say that $h(t)$ *produces* the permutation $\Pi(t)$ with $i \leq_{\Pi(t)} j$ if $b_i < b_j$. Note that this permutation is the enumeration of the lines of

H in the order they intersect line $h(t)$. As t runs from $t_0 < t_1$ to $t_{m+1} > t_m$, line $h(t)$ sweeps the plane from left to right and produces a sequence of $m+1$ permutations. Each time t passes an x_1-coordinate t_i of some vertices in the arrangement, a move m exists such that

$$\Pi(t_i - \epsilon) \rightarrow_m \Pi(t_i + \epsilon),$$

for $0 < \epsilon < \min\{t_i - t_{i-1}, t_{i+1} - t_i\}$. Evidently, move m reverses j non-overlapping substrings if j vertices of $\mathcal{A}(H)$ have the same x_1-coordinate t_i. Furthermore, $a_k a_{k+1} \ldots a_\ell$ is such a substring if the lines a_k through a_ℓ are concurrent and t_i is the x_1-coordinate of their intersection. If H contains lines that are parallel to each other, then the permutations produced do not quite form a complete halfperiod. In this case, the groups of parallel lines correspond to non-overlapping substrings of $\Pi(t_{m+1})$ which can be reversed in a final move.

These concepts are illustrated in Figure 2.2 which shows arrangements that give rise to the halfperiods of Figure 2.1. Notice that the circular sequence derived from an arrangement of lines essentially determines the combinatorial type of the arrangement. The only reason that the circular sequence falls short in completely determining the combinatorial type is due to the last move which does not indicate whether it corresponds to concurrent or to parallel lines. To obtain a complete circular sequence, or at least a full period, we can repeat the sweep from left to right with $h(t)$ directed downwards.

Since circular sequences can be used to encode arrangements of lines in the plane, they accomplish the same for configurations of points in the plane. One way to specify this encoding is by means of a transformation of the point set to its dual set of lines. Although this is a sufficient specification of the mechanism, we find it worthwhile to give a direct description too.

Let P be a set of n points denotes as $1,2,\ldots,n$ in the plane and let h be a directed line rotating about the origin o. Initially, we let h coincide with the x_1-axis such that it is directed from right to left. When h is not normal to any line connecting two points of P, the orthogonal projection of P onto h defines a

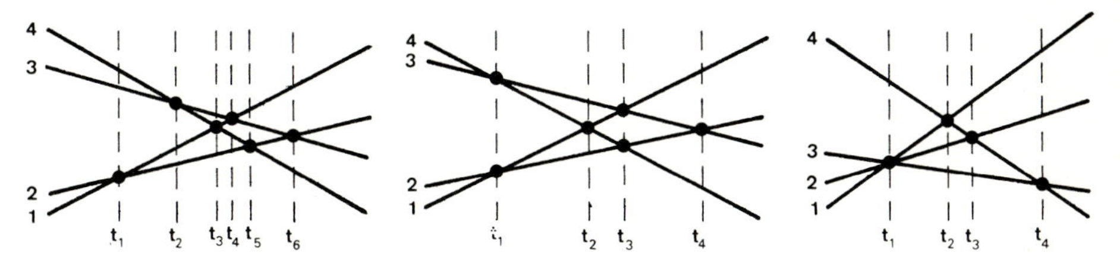

Figure 2.2. Encoding arrangements in the plane.

unique permutation of $[n]$ (see Figure 2.3). As h rotates perpendicularly through such a line, the permutation defined changes by a move. To produce a halfperiod of the circular sequence to be constructed, we rotate h counterclockwise through an angle of π. If the final position of h does not define a permutation of the points, which is the case if h is normal to some line connecting two points of P, then we add the next permutation defined by h if we continue the rotation. Figure 2.3 shows three configurations that give rise to the halfperiods of Figure 2.1. It is a straightforward but instructive exercise to show that the circular sequence obtained from the configuration P as shown above is actually the same as the one defined by its dual arrangement $\mathcal{A}(\mathcal{D}(P))$.

Circular sequences, as constructed from arrangements and configurations, reflect various convexity properties of the geometric structures that they represent. We are going to describe how circular sequences reflect such properties for the case of configurations of points. Corresponding statements about the relationship between arrangements of lines and their encoding circular sequences can be obtained by application of the dual correspondence between arrangements of lines and configurations of points.

Observation 2.2: Let P be a set of n points in the plane and let $\mathcal{C}$ be the n-sequence obtained from P.

(i) Points i,j,k in P are collinear if and only if they occur in a common substring of a move in $\mathcal{C}$.

(ii) The lines connecting points $i \neq j$ and $k \neq m$ in P, respectively, are parallel if and only if a move that swaps i and j also swaps points k and m.

(iii) Point i is extreme in P, that is, $\text{conv}P \neq \text{conv}(P-\{i\})$, if and only if there is a permutation of $\mathcal{C}$ that has i in its first position.

(iv) Points $i \neq j$ are endpoints of a common edge of $\text{conv}P$ if and only if there are consecutive permutations Π and Π' in $\mathcal{C}$ such that Π

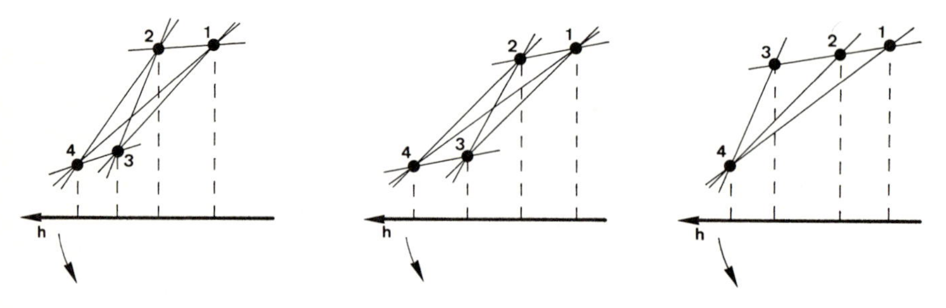

Figure 2.3. Encoding configurations in the plane.

has i in its first position and Π' has j in its first position.

Circular sequences also reflect where two lines, each connecting two points, intersect. To describe how this information can be read off a circular sequence, we need some new notation. For two points i and j, we let $r_{i,j}$ be the ray that starts at i and contains j. For a circular sequence, we write $ij < k\ell < ji$ if the move that reverses $k\ell$ properly follows the move that reverses ij so that the first move that reverses ji after ij properly follows the move that reverses $k\ell$. Given two rays $r_{i,j}$ and $r_{k,\ell}$, we can determine from the circular sequence of the points whether or not the two rays intersect. This will be particularly important in the next section where we investigate the intersections of rays defined for extreme points i, j, k and ℓ. We therefore explicitly state the condition which decides whether or not the rays of consecutive extreme points intersect.

Observation 2.3: Let i,j,k,ℓ be the counterclockwise enumeration of the vertices of a convex polygon. Ray $r_{j,i}$ intersects ray $r_{k,\ell}$ if and only if $ij < \ell k < ji$ in the corresponding 4–sequence.

The reader is encouraged to extend the list of correspondences between configurations and circular sequences according to his or her own interest in special properties of configurations or arrangements. Circular sequences turn out to be a powerful tool for obtaining and proving results on configurations and arrangements. However, not every property of circular sequences originates in a geometric counterpart, because there are circular sequences that are not derived from configurations or arrangements. This is clarified in the next two sections. First, we formally define the distinction between circular sequences that can be derived from configurations or arrangements and circular sequences that cannot be derived from these geometric entities.

An n–sequence C is called *circularly realizable* if there is a configuration P in the plane such that C is circularly equivalent to the sequence encoding P.

Of course, this definition could be equivalently formulated with P replaced by an arrangement of non–vertical lines. As an example, consider the 4–sequences shown in Figure 2.1; all three 4–sequences are circularly realizable, as is established by Figures 2.2 and 2.3.

2.3. A Circularly Non–Realizable 5–Sequence

Neither every circular sequence nor every simple circular sequence is circularly realizable. We leave it as an exercise to verify that each n–sequence is circularly realizable if $n \leq 4$. In fact, 5 is the smallest value of n that allows for a circularly non-realizable simple n-sequence C (see Figure 2.4). Below, we demonstrate the existence of a non-realizable simple 5-sequence.

Theorem 2.4: The 5-sequence C shown in Figure 2.4 is not circularly realizable.

Proof: Suppose C is circularly realizable by a configuration P in the plane. Notice that for each i, $1 \leq i \leq 5$, there is a permutation in C with i in its first position; equivalently, each halfperiod of C has a permutation with i in its first or last position. By Observation 2.2(iii), each point of P is extreme and, by Observation 2.2(iv), 1,2,3,4,5 is the counterclockwise order of the vertices of conv$P$ (see Figure 2.5). Recall that $r_{i,j}$ denotes the ray that starts at point i and

1		2		5		3		4
	X							
2		1		5		3		4
					X			
2		1		3		5		4
							X	
2		1		3		4		5
			X					
2		3		1		4		5
	X							
3		2		1		4		5
					X			
3		2		4		1		5
							X	
3		2		4		5		1
			X					
3		4		2		5		1
	X							
4		3		2		5		1
					X			
4		3		5		2		1

Figure 2.4. A non-realizable 5-sequence.

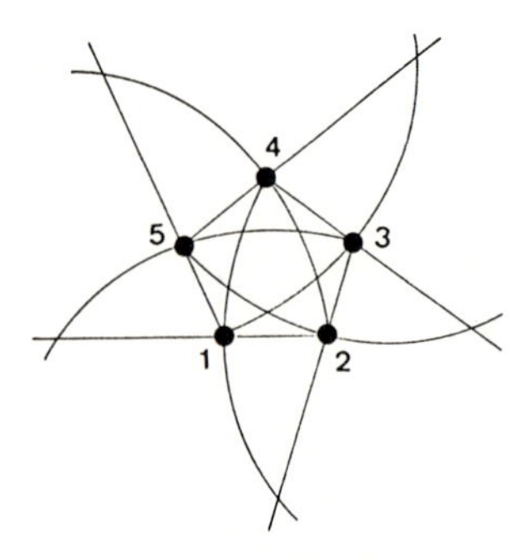

Figure 2.5. Impossible configuration.

contains j, and recall the definition of the relation "$<$" defined for simple moves in Section 2.2. Since $i(i+1)<(i-1)(i+2)<(i+1)i$, for each i and reading the numbers modulo 5, any two rays $r_{i+1,i}$ and $r_{i+2,i-1}$ intersect (see Observation 2.3 and Figure 2.5). This contradicts the possibility of positioning the five points accordingly (see Exercise 2.7). □

Obviously, C, defined in Figure 2.4, is not the only circularly non-realizable n-sequence. To obtain other non-realizable circular sequences, we say that a circular sequence C' can be *reduced to* C if C' can be made circularly equivalent to C, by repeatedly applying the following operation:

> choose some fixed number, delete it from each permutation, and omit duplicate permutations.

It is obvious that C' is not circularly realizable if it can be reduced to C.

2.4. Arrangements of Pseudo-Lines

Note that two combinatorially equivalent arrangements of lines can give rise to two circular sequences which are not circularly equivalent. This is possible since the order of the intersections as they appear from left to right is important for the circular equivalence but not for the combinatorial type of the arrangement. For many problems, the left-to-right order of the vertices can be neglected, while the order of the vertices on any one line is of central importance. In order to cast this intuitive property into a well-defined combinatorial structure, we introduce a concept that bears close relationship to the notion of local realizability which is based on the concept of local equivalence.

We begin with the presentation of some definitions. A *pseudo-line* is a connected curve that intersects every vertical line in exactly one point. A *collection S of pseudo-lines* is a set of pseudo-lines such that any two members of the set have at most one point in common and if they do that is the point at which they cross. As in the case of straight lines, the dissection of the plane induced by S is called the *arrangement $\mathcal{A}(S)$ of S*. Since we define a pseudo-line s such that it intersects any vertical line in exactly one point, an arbitrary point in the plane can be classified without ambiguity according to whether it lies above, on, or below s. Analogously, each face f in $\mathcal{A}(S)$ is uniquely specified by its position vector $u(f)$ which specifies, for each pseudo-line s in S, whether f lies above s, on s, or below s. We refer the reader to Section 1.1 for a more detailed discussion of position vectors. Let $\mathcal{A}(H)$ be an arrangement of lines in the plane. Arrangements $\mathcal{A}(S)$ and $\mathcal{A}(H)$ are called *combinatorially equivalent* if there is a one-to-one correspondence between the pseudo-lines of S and the lines of H and between the faces of $\mathcal{A}(S)$ and $\mathcal{A}(H)$ that preserves position vectors. Finally,

$\mathcal{A}(S)$ is said to be *stretchable* if there is a combinatorially equivalent arrangement of lines.

The existence of non-stretchable arrangements of pseudo-lines is a classical result in combinatorial geometry that will be henceforth described. Let S be the set of nine pseudo-lines shown in Figure 2.6: lines s_1 and s_2 are non-vertical and p_1, p_2, p_3 and q_1, q_2, q_3 are points from left to right on s_1 and s_2, respectively, such that no two points lie on a common vertical line. Point r_i is constructed by intersecting the line through p_{i-1} and q_{i+1} and the line through q_{i-1} and p_{i+1}, for $i=1,2,3$ and indices read modulo 3. The ninth pseudo-line s_9 is drawn such that r_1 and r_3 are above s_9, and r_2 is below s_9.

Theorem 2.5: The arrangement $\mathcal{A}(S)$ of pseudo-lines shown in Figure 2.6 is not stretchable.

Proof: By Pappus' theorem (see Exercise 2.9), the constructed points r_1, r_2, and r_3 are collinear. Moreover, since r_2 lies between points r_1 and r_3, there is no line that has r_1 and r_3 on one side and r_2 on the other. □

The example shown in Figure 2.6 can be modified to give a non-stretchable arrangement of nine pseudo-lines such that any two pseudo-lines intersect in a unique point. Making use of parallel lines, it is also possible to draw a non-stretchable arrangement of only eight pseudo-lines (see Exercise 2.10).

The remainder of this section briefly addresses the relationship between arrangements of pseudo-lines and circular sequences. Let $\mathcal{A}(S)$ be an arrangement of pseudo-lines. A vertical and upward directed line sweeping from left to right through $\mathcal{A}(S)$ can be used to construct a circular sequence C for $\mathcal{A}(S)$ (see

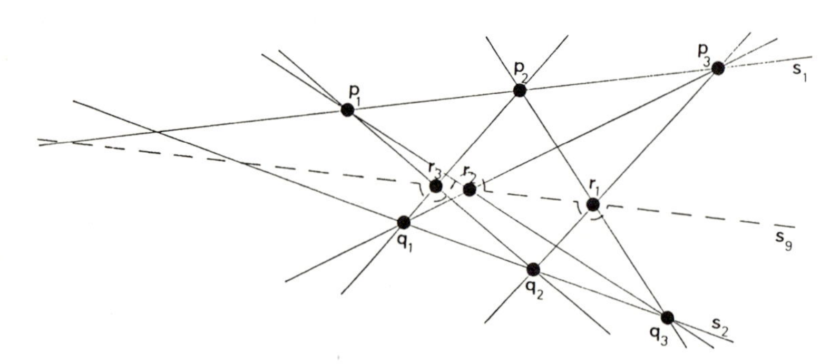

Figure 2.6. A non-stretchable pseudo-line arrangement.

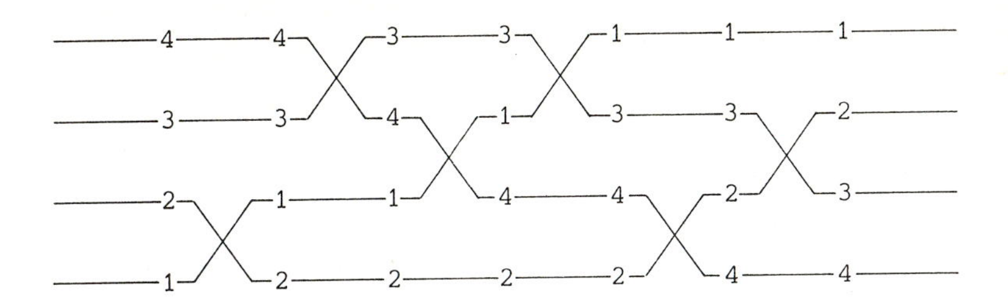

Figure 2.7. A 4–sequence and its wire diagram.

Figure 2.7). Conversely, a halfperiod of C, if displayed from left to right with permutations vertical, defines in a natural way a "wire diagram" that qualifies as an arrangement of pseudo–lines (see Figure 2.7). Notice that the stretchability of $\mathcal{A}(S)$ implies the local realizability of its corresponding circular sequence C, while C may or may not be circularly realizable.

2.5. Some Combinatorial Problems in the Plane

The goal of this section is two–fold. First, we present four combinatorial problems for finite point sets in the plane, which are interesting in their own right and have the potential to spark the reader's interest in similar questions. Second, we translate these problems into similar problems defined for circular sequences in order to demonstrate the strong relationship between finite point sets and circular sequences. We believe that such translations are useful at least for some problems, and are almost always worth trying.

Problem 1 (Convex subsets):

Let $ES(m)$ be the smallest integer such that any configuration of $ES(m)$ points in the Euclidean plane, of which no three are collinear, contains the set P of vertices of a convex m–gon, that is, $\text{card}P = m$ and $P = \text{ext}P$. It has been proven that

$$2^{m-2}+1 \leq ES(m) \leq \binom{2m-4}{m-2}+1,$$

for $m \geq 2$ (see Problem 2.11), and the conjecture is that

$$ES(m) = 2^{m-2}+1.$$

This conjecture can be equivalently stated as follows: every configuration of n points in E^2, of which no three are collinear, contains the $\lfloor \log_2(n-1) \rfloor + 2$ vertices of a convex polygon.

Next, we formulate the above problem in terms of circular sequences. To this end, we need the following straightforward result.

Observation 2.6: Let C be a circular sequence encoding the vertices of a convex m-gon. Every number of C is at the first position of at least one permutation of C.

To settle the conjecture for $ES(m)$, it is thus sufficient to prove that every $(2^{m-2}+1)$-sequence, $m \geq 2$, can be reduced to an m-sequence that satisfies the conditions of Observation 2.6. Here, we recall that an i-sequence is a reduction of some j-sequence, $i < j$, if the i-sequence can be obtained by eliminating some fixed $j-i$ numbers from each permutation of the j-sequence and finally removing all duplicate permutations. Note, however, that the combinatorial problem is slightly more general than the geometric problem since not all circular sequences are realizable.

Problem 2 (Vertices on a single line):

This combinatorial problem concerns the number of vertices contained in any one single line in an arrangement of lines in the plane. Let $DI(n)$ be the smallest integer such that every arrangement of n lines, not all concurrent or parallel, includes a line that contains at least $DI(n)$ of the vertices. Using the following straightforward result, we can formulate this problem using the language of circular sequences.

Observation 2.7: Let C be the circular sequence that encodes an arrangement $\mathcal{A}(H)$ of n lines $1,2,...,n$ in E^2. The number of vertices on line i differs by at most one from the length of a halfperiod of i's local sequence.

In terms of circular sequences, the problem thus amounts to examining the length of halfperiods of local sequences. The exclusion of arrangements with all lines concurrent or parallel corresponds to disregarding the so-called trivial n-sequence that reverses a complete permutation in one move.

Problem 3 (Directions determined by points):

To explain this problem we let a "direction" be specified by a vector v with the understanding that vector $-v$ defines the same direction. We call a direction *determined* by a configuration P in the Euclidean plane if there are two points p and q in P such that the line through p and q is orthogonal to vector v specifying the direction. Let $SC(n)$ be the smallest integer such that every n non-collinear points in the plane determine $SC(n)$ directions. For $n \geq 4$ and n even, the vertices of the regular n-gon imply $SC(n) \leq n$, and for $n \geq 5$ and n odd, the vertices of the regular $(n-1)$-gon together with its center imply $SC(n) \leq n-1$

(see Figure 2.8).

Observation 2.8: Let C be the circular sequence encoding a configuration P in the plane. There is a one–to–one correspondence between the directions determined by P and the moves in a halfperiod of C.

It follows that the problem of determining $SC(n)$ translates to calculating the minimum number of permutations that are needed for a non–trivial n–sequence.

Problem 4 (Groups of collinear points):

A *k–line of* a configuration P of points in the Euclidean plane is a line that contains k or more points of P. Let $E_k(n)$, $k \geq 2$, be the maximal number of k–lines determined by any configuration of n points in the plane which does not admit a $(k+1)$–line. Obviously, $E_2(n) = \binom{n}{2}$ and every configuration with no three points collinear achieves it.

Next, we translate the problem of counting k–lines of a configuration into a problem for circular sequences. Let C be the n–sequence that encodes a configuration of n points in the plane. By Observation 2.2(i), there is a one–to–one correspondence between the k–lines of P and the reversals of substrings of length k in a halfperiod of C. The problem is now to determine the maximal number of reversals of substrings of length k without any reversal of a longer substring.

2.6. Exercises and Research Problems

1 **Exercise 2.1:** Prove that the definition of a circular sequence given in Section 2.1 implies that every circular sequence is periodic.

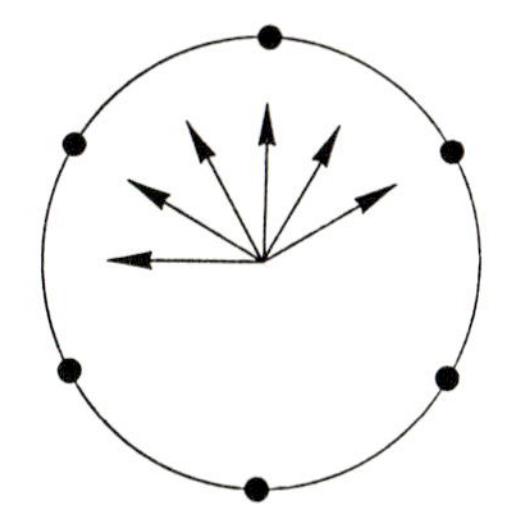

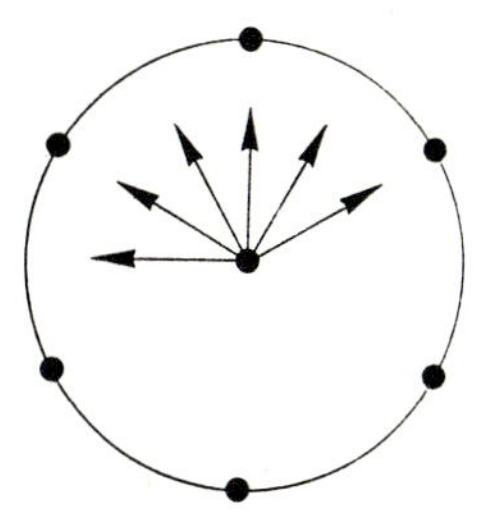

Figure 2.8. The six directions determined by six and by seven points.

1 **Exercise 2.2:** (a) Show that two n-sequences C_1 and C_2 are locally equivalent if they are circularly equivalent.

1 (b) Give an example that contradicts the inverse hypothesis.

1 **Exercise 2.3:** Verify Observation 2.1, that is, show that any number in a halfperiod of a simple n-sequence is swapped with every other number exactly once, and show that any halfperiod of a simple n-sequence consists of $\binom{n}{2}$ (or $\binom{n}{2}+1$) permutations.

2 **Exercise 2.4:** Verify that the circular sequences constructed from a configuration P and from the dual arrangement $\mathcal{A}(\mathcal{D}(P))$ are identical.

2 **Exercise 2.5:** Verify Observations 2.2 and 2.3.

2 **Exercise 2.6:** A subset P' of a configuration P in E^2 is called a *semispace of* P if there is an open half-space h with $P \cap h = P'$. How does the circular sequence of P reflect the fact that P' is a semispace of P?

3 **Exercise 2.7:** Let T be a convex polygon with vertices 1,2,3,4,5 enumerated in counterclockwise order. Let $r_{i,j}$ denote the ray that starts at vertex i and contains vertex j. Prove that it is not possible that all pairs of rays $r_{i+1,i}$ and $r_{i+2,i-1}$ intersect, where the indices are read modulo 5 (see Observation 2.3 and Figure 2.5). *(Hint: for any five points in the plane there is a containing conic, that is, a curve determined by a polynomial of degree two; it is straightforward to rule out that the points lie on a hyperbola or on a parabola; if the points lie on a circle, then this gives a contradiction on the lengths of the arcs between the points; finally, if they lie on an ellipse, then the points can be transformed to lie on a circle without changing their circular sequence.)*

2 **Exercise 2.8:** (a) Prove that all n-sequences are circularly realizable if $n \leq 4$.

3 (b) Prove that all 5-sequences, except for the one shown in Figure 2.4, are circularly realizable.

3 **Exercise 2.9:** (Pappus' theorem): Let s_1 and s_2 be two lines in the Euclidean plane, and let p_1, p_2, p_3 and q_1, q_2, q_3 be six distinct points on s_1 and s_2, respectively, such that p_2 lies between p_1 and p_3, and q_2 lies between q_1 and q_3. Show that points r_1, r_2, r_3 defined as follows are collinear: r_i is the intersection of the line through p_{i-1} and q_{i+1} and the line through p_{i+1} and q_{i-1}, where we read the indices modulo 3.

3 **Exercise 2.10:** Draw a non-stretchable arrangement of eight pseudo-lines in the Euclidean plane. *(Hint: Use Pappus' theorem with s_2 moved to infinity.)*

Problem 2.11: Call a finite set of point *convex* if all its points are extreme, and call it an *m-gon* if it is convex and its cardinality is m. Define $ES(m)$ as the smallest integer number j such that every set of j points or more, of which no three are collinear, contains an m-gon as a subset. (Compare this definition of $ES(m)$ with the definition given in Section 2.5.)

2 (a) Prove $ES(4)=5$ and $ES(5)=9$.

4 (b) Prove $ES(m) \leq \binom{2m-4}{m-2}+1$, for $m \geq 2$. *(Hint: assume without loss of generality that no two points have the same x_1-coordinate and call an m-gon an m-chain if its convex hull connects the leftmost and the rightmost point by an edge; prove now that every set of $\binom{2m-4}{m-2}+1$ points, of which no three are collinear, contains an m-chain.)*

3 (c) Prove that for every integer number $m \geq 2$, there are $\binom{2m-4}{m-2}$ points, of which no three are collinear, that do not contain an m-chain. *(Hint: call an m-chain convex if the convex hull edge that connects the leftmost and the rightmost point lies below all other points of the m-chain, and call it concave, otherwise; to construct an example that*

proves the assertion, place a set $P_{m,m-1}$ *that contains no convex m-chain and no concave* $(m-1)$*-chain in the second quadrant such that any line connecting two points of* $P_{m,m-1}$ *has positive slope, and place a set* $P_{m-1,m}$ *that contains no convex* $(m-1)$*-chain and no concave m-chain in the fourth quadrant such that any line connecting two points of* $P_{m-1,m}$ *has positive slope; repeat the argument recursively to complete the proof.)*

3 (d) Prove $ES(m) \geq 2^{m-2}+1$, for $m \geq 2$. *(Hint: construct an example of a set of* 2^{m-2} *points that contains no m-gon as follows: first, place* $m-1$ *tiny little line segments with positive slopes such that each line segment intersects the curve given by the equation* $x_2 = 2^{-x_1}$*, and second, place a point set* $P_{i+1,m+1-i}$ *arbitrarily close to the* i^{th} *line segment from the left, such that* $P_{i+1,m+1-i}$ *contains no convex* $(i+1)$*-chain and no concave* $(m+1-i)$*-chain.)*

5 (e) Prove or disprove $ES(m) = 2^{m-2}+1$, $m \geq 2$.

Problem 2.12: Define $H(m)$ as the smallest integer number j such that any set of j or more points in the Euclidean plane, of which no three are collinear, contains the vertices of a convex m-gon that contains no point of the set in the interior of its convex hull. We call such a subset an *empty convex m-gon*.

2 (a) Show $H(4)=5$.

3 (b) Prove that $H(5)$ exists. *(Hint: let the point set be large enough such that it contains a convex octagon, and let P be such an octagon that contains the smallest number of points of the set in the interior of its convex hull; next, consider the convex hull of the points in the interior of the convex hull of P and extend one of the edges of this convex hull to a line; if the line cuts off three or more points of P, then these points form an empty pentagon with the two points of the extended edge, otherwise, the two points can be used to construct a convex octagon with fewer points in the interior of the convex hull.)*

4 (c) Prove $H(5)=10$.

5 (d) Prove or disprove the existence of $H(6)$.

4 (e) Prove that $H(7)$ does not exist.

4 (f) Prove that the expected number of empty triangles is in $O(n^2)$ if n points are uniformly distributed over some convex bounded region.

3 (g) Prove that every set of n points in the plane, of which no three are collinear, contains $\Omega(n^2)$ empty convex quadrilaterals, and show that this bound is tight in the asymptotic sense.

5 (h) Prove or disprove that every set of n points in the plane, of which no three are collinear, contains $\Omega(n^2)$ empty convex pentagons.

4 (i) Give an algorithm that computes the largest empty k-gon of a point set in time proportional to the number of empty triangles.

4 **Exercise 2.13:** Define $DI(n)$ as in Section 2.5, that is, it is the smallest integer number j such that every set of n lines in the plane contains one line that intersects the other lines in at least j distinct points, unless all lines are concurrent. Prove $DI(n) \geq c.n$, for c a positive constant independent of n.

4 **Exercise 2.14:** Define $SC(n)$ as in Section 2.5, that is, it is the smallest integer number j such that every set of n points in the plane defines at least j directions. Prove $SC(n) = 2\lfloor n/2 \rfloor$.

Problem 2.15: Define $E_k(n)$ as in Section 2.5, that is, it is the smallest integer number j such that every set of n points in the plane, of which no $k+1$ points are collinear, contains at most j subsets of k collinear points.

3 (a) Show $E_3(7)=6$, $E_3(8)=7$, and $E_3(9)=10$.
3 (b) Prove $E_3(n)=\Theta(n^2)$.
4 (c) Prove $E_k(n)=\Omega(n^{(k-1)/(k-2)})$, for $k\geq 3$.
5 (d) Give non-trivial asymptotic upper bounds for $E_k(n)$, $k\geq 4$.

5 **Exercise 2.16:** Generalize the concept of a circular sequence to three dimensions and use it to derive new and non-trivial bounds on three-dimensional generalizations of the combinatorial problems discussed in this chapter.

2.7. Bibliographic Notes

The concept of a circular sequence was introduced to combinatorial geometry by Goodman, Pollack (1980a) when they corrected an error in Perrin (1881/82) who treated a similar concept. An alternate encoding of arrangements by some combinatorial structure, namely oriented matroids, can be found in Folkman, Lawrence (1978). The relationship between circular sequences and arrangements of lines or configurations of points in the plane, and the issues of circular and local realizability are treated in detail in Goodman, Pollack (1980a, 1984). Information on arrangements of pseudo-lines (although defined in a slightly different way than in Section 2.4) can be found in Ringel (1956) and in Grünbaum (1972). The theorem that every arrangement of eight pseudo-lines in the projective plane is stretchable is shown in Goodman, Pollack (1980b); this implies the same result for at most 7 pseudo-lines in the Euclidean plane. A novel presentation of Pappus' theorem and other classical results in Euclidean geometry is presented in Martin (1982). Problems 1 to 3 of Section 2.5 and their reformulations for circular sequences are taken from Goodman, Pollack (1981a). The bounds given for $ES(n)$ of Problem 1 in Section 2.5 as well as the related conjecture can be found in Erdös, Szekeres (1935 and 1960) (see Problem 2.11, (b) through (e)). A proof of $ES(5)=9$ (see Problem 2.11(a)) is given in Kalbfleisch, Kalbfleisch, Stanton (1970). Solutions to Problems 2.12(c) and 2.12(e) are given in Harborth (1978) and in Horton (1983), respectively. The elegant argument for the existence of $H(5)$ presented as a hint to Problem 2.12(b) is due to Andrzej Ehrenfeucht and can be found in Erdös (1979). Solution to Problems 2.12(f) and (g) are given in Bárány, Füredi (1986). Dobkin, Edelsbrunner, Overmars (1987) discuss the algorithmic question of enumerating all empty convex m-gons defined by a given finite point set and also give a solution to Problem 2.12(i). The first conjecture on function $DI(n)$ of Problem 2 in Section 2.5 was ventured by Dirac (1951) and modified by Erdös (1961) to $DI(n)\geq c\cdot n$, for some positive constant c. This conjecture was recently settled in the affirmative by Szemerédi, Trotter (1983) using methods different to those of circular sequences; this solves Exercise 2.13. Scott (1970) seems to be the first to have considered $SC(n)$ of Problem 3 in Section 2.5, and Ungar (1982) settled the problem, that is, he showed $SC(n)=2\lfloor n/2\rfloor$ and thus solved Exercise 2.14. According to Grünbaum (1976), the problem of calculating $E_k(n)$ of Problem 4 in Section 2.5 was first mentioned by Erdös (1962) in a paper written in Hungarian. Lower bounds on the maximum number of collinear triplets in two-dimensional point sets that contain no four collinear points are given in Burr, Grünbaum, Sloane (1974) and in Füredi, Palásti (1984). Their results imply a solution to Problem 2.15(b); the first of the two papers also presents constructions that resolve Problem 2.15(a). The currently best lower bounds on $E_k(n)$ stated in Problem 2.15(c) can be found in Grünbaum (1976).

CHAPTER 3

SEMISPACES OF CONFIGURATIONS

A non–vertical hyperplane h, disjoint from a finite set P of points, partitions P into two sets $P^+ = P \cap h^+$ and $P^- = P \cap h^-$ called *semispaces of* P, where h^+ is the open half–space above hyperplane h and h^- is the open half–space below h. Note that this definition includes the empty set and P itself as semispaces of P. Using geometric transformations and the face counting formulas for arrangements of hyperplanes presented in Chapter 1, it is not hard to find tight upper bounds on the number of semispaces of P that depend solely on the cardinality of P. Unfortunately, little is known about the maximum number of semispaces with some fixed cardinality. This chapter addresses the latter counting problem and derives non–trivial upper and lower bounds.

In our study of semispaces, we will use the dual transform $\mathcal{D}$ introduced in Section 1.4 which maps a point to a hyperplane and a hyperplane to a point. Using the order preserving property of $\mathcal{D}$, we will see that the semispaces of a configuration P correspond to certain subsets P' of the set of non–vertical hyperplanes $H = \mathcal{D}(P) = \{\mathcal{D}(p) | p \in P\}$:

> P' is a semispace of P if and only if there is a vertical ray r that intersects all hyperplanes in $H' = \mathcal{D}(P')$ but no hyperplane in $H - H'$.

This correspondence is discussed in more detail in Sections 3.1 and 3.2 where so–called levels of arrangements are introduced. In Sections 3.3, 3.4, and 3.5, we demonstrate lower bounds on the maximum number of semispaces with fixed cardinality, and in Sections 3.6 and 3.7, we prove corresponding upper bounds in two dimensions making extensive use of the concept of circular sequences introduced in Chapter 2. By duality, the bounds on the maximum number of semispaces with fixed cardinality translate to bounds on the number of faces of levels in arrangements and lead to bounds on the amount of time and storage needed by algorithms in Parts II and III of this book.

3.1. Semispaces and Arrangements

Semispaces of configurations and the hyperplanes that define semispaces correspond to certain concepts in dual arrangements of hyperplanes. This section discusses this relationship and derives tight upper bounds on the total number of semispaces of any set of n points in d dimensions.

Let P be a set of n points in E^d. A non-vertical hyperplane h disjoint from P defines two semispaces $P^+ = P \cap h^+$ and $P^- = P \cap h^-$ of P. Note that the restriction to non-vertical hyperplanes does not imply any loss of generality since any vertical hyperplane disjoint from P can be moved into a non-vertical position without changing the defined partition. Below, we make repeated use of the dual transform $\mathcal{D}$ (defined in Section 1.4) that maps a point $p = (\pi_1, \pi_2, \ldots, \pi_d)$ in E^d to the hyperplane

$$\mathcal{D}(p)\colon\ x_d = 2\pi_1 x_1 + 2\pi_2 x_2 + \ldots + 2\pi_{d-1} x_{d-1} - \pi_d$$

and vice versa, that is, the transform $\mathcal{D}$ maps a non-vertical hyperplane $h\colon x_d = \eta_1 x_1 + \eta_2 x_2 + \ldots + \eta_{d-1} x_{d-1} + \eta_d$ to the point

$$\mathcal{D}(h) = \left(\frac{\eta_1}{2}, \frac{\eta_2}{2}, \ldots, \frac{\eta_{d-1}}{2}, -\eta_d\right).$$

Now, $H = \mathcal{D}(P) = \{\mathcal{D}(p) | p \in P\}$ is a set of n non-vertical hyperplanes in E^d and $\mathcal{A}(H)$ is the arrangement defined by H. By the results of Section 1.4, $\mathcal{A}(H)$ is a dual arrangement of P and $\mathcal{D}(h)$ is a point contained in some cell c of $\mathcal{A}(H)$. By Observation 1.5(ii), a point p of P belongs to half-space h^+ or h^- if and only if point $\mathcal{D}(h)$ lies in half-space $\mathcal{D}(p)^+$ or $\mathcal{D}(p)^-$, respectively. We say that point $\mathcal{D}(h)$ *defines* the partition of H into sets $H^+ = \mathcal{D}(P^+) = \{\mathcal{D}(p) | p \in P \cap h^+\}$ and $H^- = \mathcal{D}(P^-) = \{\mathcal{D}(p) | p \in P \cap h^-\}$.

Obviously, every point in cell c defines the same partition which allows us to attribute that partition to c. If c is bounded or if it fits between two parallel hyperplanes then it is the only cell which defines the partition of H into H^+ and H^-. Otherwise, there is another unbounded cell c' in $\mathcal{A}(H)$ with position vector $u(c') = -u(c)$ that defines the same partition. Recall that the position vector of a cell has an entry for each hyperplane in H and that this entry indicates whether the cell is below, on, or above the hyperplane.

This suggests the temporary use of the projective view of $\mathcal{A}(H)$ which considers c and c' as parts of one and the same cell. Bounded cells or unbounded cells that fit between two parallel hyperplanes experience the same interpretation in projective space as in Euclidean space. Formally, we define projective faces of $\mathcal{A}(H)$ as follows. Write $H = \{h_1, h_2, \ldots, h_n\}$, and for any two points p and q define

(i) $p \sim_{pos} q$ if $p \in h_i^+$ ($p \in h_i^-$) if and only if $q \in h_i^+$ ($q \in h_i^-$) for each $1 \leq i \leq n$, and

(ii) $p \sim_{neg} q$ if $p \in h_i^+$ ($p \in h_i^-$) if and only if $q \in h_i^-$ ($q \in h_i^+$) for each $1 \leq i \leq n$.

Points p and q are *projectively equivalent* if $p \sim_{pos} q$ or $p \sim_{neg} q$, and the *projective faces of* $\mathcal{A}(H)$ are the equivalence classes of this equivalence relation.

The above discussion thus establishes a one-to-one correspondence between complementary semispaces of P and projective cells of $\mathcal{A}(H)$. Notice that the number of projective cells in a simple arrangement can be obtained by subtracting half the number of the unbounded cells from the total number of cells. The upper bounds on the number of projective cells that can now be derived from Theorem 1.3 and the solution to Exercise 1.1(d) imply an upper bound on the number of semispaces of P.

Theorem 3.1: A set of n points in E^d contains at most

$$2 \sum_{i=0}^{d} (-1)^i f_{d-i}^{(d-i)}(n)$$

semispaces, where $f_{d-i}^{(d-i)}(n) = \sum_{j=0}^{d-i} \binom{n}{j}$.

The upper bound is achieved if and only if no $d+1$ points lie in a common hyperplane or if $\dim \operatorname{aff} P = n-1$. Figure 3.1 shows a configuration of four points in E^2 together with its dual arrangement. Where space permits, the cells are labeled with the complementary pairs of semispaces they define.

3.2. k-Sets and Levels in Arrangements

In this section, we introduce combinatorial functions that count the number of semispaces of some fixed cardinality k. We also present a few straightforward results on these functions and discuss the relationship between the number of

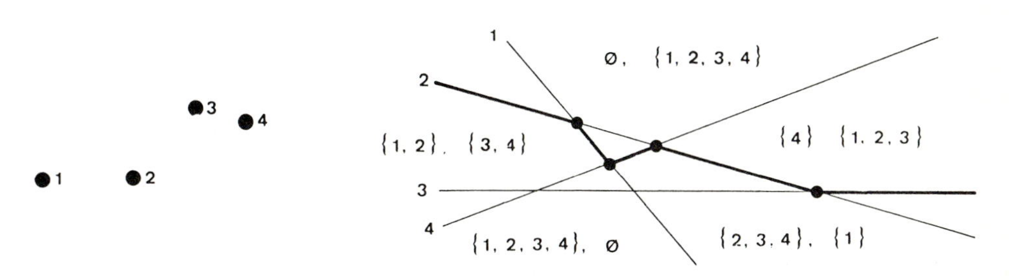

Figure 3.1. Semispaces and projective cells.

semispaces with fixed cardinality and the number of faces of so-called levels in arrangements of hyperplanes. Such levels are defined depending on an integer parameter k chosen between 1 and the cardinality of the set of hyperplanes.

We call a semispace P' of P a *k-set of P* if $k = \text{card}P'$. For instance, the empty set is the only 0-set of P and, for $n = \text{card}P$, P itself is the only n-set of P. Furthermore, a singleton set $\{p\}$ is a 1-set of P if and only if point p is extreme in P. For $0 \leq k \leq n$, we denote the number of k-sets of P by $e_k(P)$. The following results on the number of k-sets are evident:

$$e_0(P) = e_n(P) = 1,$$

$$e_1(P) = \text{card}\,\text{ext}P,$$

and, by Theorem 3.1,

$$\sum_{k=0}^{n} e_k(P) \leq 2 \sum_{i=0}^{d} (-1)^i f_{d-i}^{(d-i)}(n) = O(n^d).$$

Since each k-set has a unique complementary $(n-k)$-set, we also have

$$e_k(P) = e_{n-k}(P)$$

which allows us to restrict our attention to the range of integer numbers $0 \leq k \leq n/2$. Define $e_k^{(d)}(n) = \max\{e_k(Q) | Q$ a set of n points in $E^d\}$. By the above trivial facts we have

$$e_k^{(d)}(n) = e_{n-k}^{(d)}(n),$$

$$e_0^{(d)}(n) = 1, \text{ and}$$

$$e_k^{(d)}(n) = O(n^d) \text{ for } 0 \leq k \leq n/2.$$

The following sections of this chapter will derive non-trivial lower and upper bounds for $e_k^{(d)}(n)$. In this section, we continue the investigation into the implications of bounds on $e_k^{(d)}(n)$ to certain concepts in arrangements of hyperplanes.

Let H be a set of n non-vertical hyperplanes in E^d. For every point p we let $a(p)$ and $b(p)$ be the number of hyperplanes h in H such that p is in h^- and h^+, respectively. If c is a cell of $\mathcal{A}(H)$ and p is a point in c then the partition of H defined by p is into two sets of cardinality $a(p)$ and $b(p)$. It follows that the number of k-sets of point set $\mathcal{D}(H)$ is equal to the number of projective cells in $\mathcal{A}(H)$ that contain points p with $a(p) = k$ or $b(p) = k$. The only exception to this rule is when $2k = n$ in which case the number of k-sets is twice the number of such cells. For $1 \leq k \leq n$, define the *k-level $\mathcal{L}_k(H)$ of $\mathcal{A}(H)$* as the set of points p with

$$a(p) \leq k-1 \text{ and } b(p) \leq n-k.$$

In Figure 3.1, the 2-level of the arrangement shown is indicated by solid line segments. We will write $\mathcal{L}_k$ for $\mathcal{L}_k(H)$ if H is understood. Below, we offer a short list of basic and straightforward properties of levels in arrangements.

Observation 3.2: Let H be a set of n non–vertical hyperplanes in E^d.

(i) A point p is not contained in any k–level of $\mathcal{A}(H)$ if and only if p is not contained in any hyperplane of H.

(ii) A point p is contained in i different k–levels of $\mathcal{A}(H)$ if and only if p is contained in i hyperplanes of H.

(iii) The k–level of $\mathcal{A}(H)$, for $1 \leq k \leq n$, intersects every vertical line in exactly one point.

(iv) Let p and q be the intersections of a vertical line with the i–level and the j–level of $\mathcal{A}(H)$, respectively. If $i \leq j$ then the x_d–coordinate of p is no less than the one of q.

By Observation 3.2(iii), the k–level $\mathcal{L}_k$ of $\mathcal{A}(H)$ can be viewed as a function from E^{d-1} to E^1, that is, it maps a point $p'=(\pi_1,\pi_2,...,\pi_{d-1})$ of the hyperplane h_0: $x_d=0$ to a unique real number π_d such that point $(\pi_1,\pi_2,...,\pi_d)$ lies in $\mathcal{L}_k$. For convenience, we write $\pi_d=\mathcal{L}_k(p')$ and we say that $q=(\psi_1,\psi_2,...,\psi_d)$ is *above* , *on*, or *below* $\mathcal{L}_k$ if $\psi_d > \mathcal{L}_k(q')$, $\psi_d = \mathcal{L}_k(q')$, or $\psi_d < \mathcal{L}_k(q')$, respectively, for $q'=(\psi_1,\psi_2,...,\psi_{d-1})$. Observation 3.2(iv) implies a natural order of the levels of $\mathcal{A}(H)$: for $1 \leq i \leq j \leq n$, a point p is above $\mathcal{L}_i$ only if it is above $\mathcal{L}_j$, and p is below $\mathcal{L}_j$ only if it is below $\mathcal{L}_i$.

We define the *complexity of* level $\mathcal{L}_k$ as the number of faces of arrangement $\mathcal{A}(H)$ that are contained in $\mathcal{L}_k$. Since each cell c that defines a partition of H into k and $n-k$ hyperplanes contributes at least one facet to $\mathcal{L}_k$ or $\mathcal{L}_{n-k}$, the combined complexity of $\mathcal{L}_k$ and $\mathcal{L}_{n-k}$ is no less than $e_k(\mathcal{D}(H))$. As a consequence, the maximum complexity of a k–level of an arrangement of n hyperplanes in E^d is in $\Omega(e_k^{(d)}(n))$. Below, we prove an upper bound on the complexity of $\mathcal{L}_k$.

In order to simplify the notation, we define $e_i^{(d)}(n)=0$ if $i<0$ or $i>n$. We can now show the following result.

Theorem 3.3: Let $\mathcal{L}_k$ be the k–level of an arrangement $\mathcal{A}(H)$ of n non–vertical hyperplanes in E^d, with $1 \leq k \leq n$. The complexity of $\mathcal{L}_k$ is in $O(\max\{e_i^{(d)}(n) \mid k-d+1 \leq i \leq k+d-2\})$.

Proof: Without loss of generality, we can assume that H contains at least d hyperplanes and that $\mathcal{A}(H)$ is simple; otherwise, the hyperplanes can be perturbed appropriately without decreasing the complexity of $\mathcal{L}_k$. We can also assume that no edge of $\mathcal{A}(H)$ is contained in a hyperplane normal to the x_1–axis. By simplicity of $\mathcal{A}(H)$, each vertex v is contained in exactly d hyperplanes of H

which implies that $b(v)=n-a(v)-d$. By definition of $\mathcal{L}_k$, vertex v is contained in $\mathcal{L}_k$ if and only if

$$k-d \leq a(v) \leq k-1.$$

For the time being, we call v the *leftmost point of* a face f if v is contained in the closure of f and if it has the minimal x_1-coordinate of all such points. By the assumption above about the direction of the edges of $\mathcal{A}(H)$, the leftmost point of a face of $\mathcal{A}(H)$ either does not exist or is unique. Vertex v is the leftmost point of exactly one cell c, and at least one hyperplane through vertex v lies above (below) c. Thus, if p is any point in cell c, then

$$a(v)+1 \leq a(p) \leq a(v)+d-1.$$

Combining the above two pairs of inequalities implies

$$k-d+1 \leq a(p) \leq k+d-2.$$

The number of cells in $\mathcal{A}(H)$ that contain a point p with $a(p)$ in the above specified interval and that have a leftmost point is thus bounded from above by

$$\sum_{i=k-d+1}^{k+d-2} e_i(\mathcal{D}(H)).$$

The assertion follows since every face of $\mathcal{L}_k$ is incident upon at least one vertex of $\mathcal{L}_k$ and every vertex v is incident upon at most some constant number of faces of $\mathcal{A}(H)$ as well as of $\mathcal{L}_k$. Notice, however, that this constant number depends on the number of dimensions d. □

3.3. A Lower Bound on the Number of Bisections in the Plane

This section demonstrates a lower bound on $e_k^{(2)}(n)$ for the special case when $n=2k$. It turns out that this case is actually the most interesting one and that lower bounds on $e_k^{(2)}(n)$, for other values of k, can be derived from this bound. The lower bound for $n=2k$ is shown by constructing and analyzing a series of point sets in the plane. In the next two sections, we will generalize the lower bound to arbitrary positive integer numbers n and k and to dimensions higher than two. To help the discussion, we call a partition of a set P of $n=2k$ points into two k-sets a *bisection of* P.

In a first step, we construct a point set $P_1(n)$, where we assume that $n=2\cdot 3^m$, for some positive integer number m; thus $k=3^m$. Let r, s, and t denote three rays emanating from the origin such that r coincides with the positive x_1-axis, ray s is obtained by rotating r through $2\pi/3$ radians about the origin, and t is obtained by rotating s through the same angle (see Figure 3.2). We choose a subset $P_1^a(n)$ of $n/3=2\cdot 3^{m-1}$ points of $P_1(n)$ to lie on ray a, for $a=r,s,t$. A straightforward counting argument shows that $P_1(n)$ allows $n/2+3$

bisections which implies the following result.

Observation 3.4: $e_{n/2}(P_1(n)) = n+6$, for $n = 2 \cdot 3^m$ and $m \geq 1$.

We observe that all $n/2$–sets of $P_1(n)$ can be retained even if $P_1^r(n)$, $P_1^s(n)$, and $P_1^t(n)$ are slightly perturbed. This leads to a configuration $P_2(n)$ consisting of subsets $P_2^r(n)$, $P_2^s(n)$, $P_2^t(n)$ near r, s, t, respectively, such that

(i) no three points of $P_2^a(n)$ are collinear, for $a = r,s,t$, and
(ii) the line through any two points of $P_2^a(n)$ separates $P_2^b(n)$ from $P_2^c(n)$, for any permutation (a,b,c) of (r,s,t).

In addition to the type of bisections already occurring in $P_1(n)$, we have a new bisection for each line that bisects $P_2^a(n)$, for $a = r,s,t$. The only exception to this rule is the bisection of $P_2^a(n)$ into the 3^{m-1} points closest to the origin and the 3^{m-1} points farthest from the origin. This bisection of $P_2^a(n)$ can be extended to two different bisections of $P_2(n)$, and both extensions are already accounted for in Observation 3.4. This yields the following result on the number of $n/2$–sets of $P_2(n)$.

Observation 3.5: For $n = 2 \cdot 3^m$ and $m \geq 1$ we have $e_{n/2}(P_2(n))$ equal to
$n + e_{n/6}(P_2^r(n)) + e_{n/6}(P_2^s(n)) + e_{n/6}(P_2^t(n))$.

One obvious method of specializing point set $P_2(n)$ is to define its subsets $P_2^r(n)$, $P_2^s(n)$, and $P_2^t(n)$ recursively as follows:

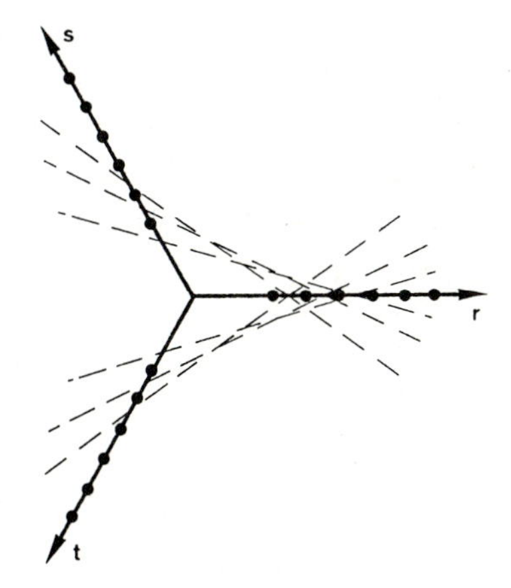

Figure 3.2. $P_1(18)$ and, if perturbed, $P_2(18)$.

> $P_3(n)$ is chosen like $P_2(n)$ and consists of subsets $P_3^r(n)$, $P_3^s(n)$, and $P_3^t(n)$ near the rays r, s, and t, respectively, such that $P_3^a(n)$ is derived by suitably translating, rotating, and stretching point set $P_3(n/3)$, for $a=r,s,t$.

It is important to observe that the transformations used in the construction of $P_3^a(n)$ do not change the combinatorial structure of the point set they are applied to. In particular, the number of k-sets of the point set remains the same, for each value of k. The formula in Observation 3.5 now specializes to the recurrence relation

$$e_{n/2}(P_3(n)) = n + 3e_{n/6}(P_3(n/3)).$$

Together with

$$e_3(P_3(6)) = e_3(P_1(6)) = 12$$

this implies the following result.

Observation 3.6: $e_{n/2}(P_3(n)) = n(\log_3(n/2)+1)$, for $n = 2 \cdot 3^m$ and $m \geq 1$.

The result stated in Observation 3.6 can be extended to arbitrary positive even integer numbers n as follows.

> For any positive even integer number n let m be the largest integer number such that $2 \cdot 3^m \leq n$. We define the configuration $P_4(n)$ to consist of three subconfigurations $P_4^a(n)$, $P_4^o(n)$, and $P_4^b(n)$ of $n/2-3^m$, $2 \cdot 3^m$, and $n/2-3^m$ points, respectively. Set $P_4^o(n)$ is chosen like $P_3(2 \cdot 3^m)$ with the additional constraint that no two points lie on a common vertical line. All points of $P_4^a(n)$ are chosen to lie above all lines connecting any two points of $P_4^o(n)$, and all points of $P_4^b(n)$ are chosen below all lines connecting any two points of $P_4^o(n)$.

For each 3^m-set P' of $P_4^o(n)$ we have either $P' \cup P_4^a(n)$ or $P' \cup P_4^b(n)$ as an $n/2$-set of $P_4(n)$. Consequently, $P_4(n)$ contains at least $2 \cdot 3^m(m+1)$ $n/2$-sets which implies the main result of this section.

Theorem 3.7: $e_{n/2}^{(2)}(n) = \Omega(n \log n)$, for even $n \geq 2$.

3.4. Lower Bounds on the Number of k–Sets in the Plane

We continue with an extension of Theorem 3.7 to arbitrary integer numbers n and to integer numbers k not necessarily equal to $n/2$. Any configuration of n points realizing a high number of bisections is shown to lead to configurations of n points with many k-sets, for $1 \leq k \leq n/2$. This stresses the central role of bisections in the analysis of the number of k–sets and thus substantiates the importance of Theorem 3.7. We are going to prove the following specific result.

Lemma 3.8: Let $b(m)$ be a positive–valued function with the property that $e_m^{(2)}(2m) \geq 2m \cdot b(m)$, for $m \geq 1$. Then $e_k^{(2)}(n) = \Omega(n \cdot b(k))$, for $1 \leq k \leq n/2$.

Proof: Let $P_5(2m)$ be a configuration of $2m$ points in E^2 with at least $2m \cdot b(m)$ m–sets. Using $P_5(2m)$ as a building block, we construct a configuration $P_6(n,k)$ which implies the assertion. With $j = \lfloor n/2k \rfloor$, point set $P_6(n,k)$ is designed as the disjoint union of subsets $P_6^0(n,k), P_6^1(n,k), \ldots, P_6^j(n,k)$, where $P_6^i(n,k)$ contains $2k$ points, for $0 \leq i \leq j-1$, and $P_6^j(n,k)$ contains the remaining $n-2kj$ points. Each subset $P_6^i(n,k)$, for $0 \leq i \leq j-1$, is obtained by suitably translating, rotating, and stretching point set $P_5(2k)$ such that the following conditions are satisfied:

(i) $P_6^i(n,k)$ is contained in rectangle R_i of length $\delta > 0$ and width $\epsilon > 0$, for $0 \leq i \leq j-1$,

(ii) rectangle R_i contains no point enclosed by the unit circle $x_1^2 + x_2^2 = 1$ such that the midpoint of one of R_i's long sides coincides with point $(\sin 2i\pi/j, \cos 2i\pi/j)$ on this circle (see Figure 3.3),

(iii) the real numbers ϵ and δ are chosen such that any line through two points of $P_6^i(n,k)$, for $0 \leq i \leq j-1$, has all rectangles R_ℓ on the same side, for $0 \leq \ell \leq j-1$ and $\ell \neq i$, and

(iv) the points of $P_6^j(n,k)$ are enclosed by the unit circle.

By construction, at least half of the k–sets of $P_6^i(n,k)$ are also k–sets of $P_6(n,k)$,

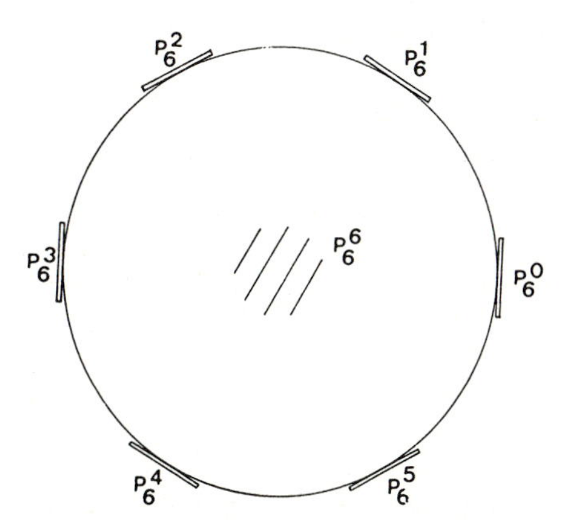

Figure 3.3. Configuration $P_6(n.k)$ with $j=6$.

for $0 \leq i \leq j-1$, which implies that $P_6(n,k)$ realizes no less than

$$j \cdot k \cdot b(k) = \Theta(n \cdot b(k))$$

k-sets. □

By Theorem 3.7, we can choose $b(m)$ in $\Omega(\log(m+1))$ which implies the following lower bound on the maximum number of k-sets of any set of n points in the plane.

Theorem 3.9: $e_k^{(2)}(n) = \Omega(n \log(k+1))$, for $1 \leq k \leq n/2$.

In Section 3.7, we will prove that $O(n\sqrt{k})$ is an upper bound on $e_k^{(2)}(n)$. It is obvious that there is a large gap between both bounds and we challenge the interested reader to close this gap.

3.5. Extensions to Three and Higher Dimensions

This section demonstrates lower bounds on the maximum number of bisections of any set of n points in three and higher dimensions. To this end, we construct point sets in E^d which have a large number of bisections. In the construction, we make use of two-dimensional point sets that realize the number of bisections indicated in Theorem 3.7.

The following lemma is similar in flavor to Lemma 3.8; it will be used to prove lower bounds on the maximum number of bisections of any set of $n = 2k$ points in E^d.

Lemma 3.10: Let $b(m)$ be a positive-valued function with the property that $e_m^{(2)}(2m) \geq 2m \cdot b(m)$, for $m \geq 1$. Then $e_{n/2}^{(d)}(n) = \Omega(n^{d-1} \cdot b(n/6))$, for n a multiple of 6 and $d \geq 3$.

Proof: Let $P_5(2m)$ be a configuration of $2m$ points in E^2 with at least $2m \cdot b(m)$ m-sets. Below, we construct a configuration $P_7(n,d)$ in E^d which consists of disjoint subsets $P_7^a(n,d)$, $P_7^o(n,d)$, and $P_7^b(n,d)$ of $n/3$ points each. Point set $P_7^a(n,d)$ is chosen in the $(d-2)$-flat which is the intersection of the two hyperplanes defined by the equations

$$x_1 = 0 \text{ and } x_2 = 1.$$

The points of $P_7^a(n,d)$ are selected such that no $d-1$ points lie in a common $(d-3)$-flat, and we define $P_7^b(n,d) = -P_7^a(n,d)$ (see Figure 3.4 for a schematic drawing of the case $d=3$). Note that any hyperplane that contains the x_1-axis and that avoids point set $P_7^a(n,d)$ also avoids point set $P_7^b(n,d)$ and that it partitions $P_7^a(n,d) \cup P_7^b(n,d)$ into two equal-sized subsets. Point set $P_7^o(n,d)$ is a suitably stretched copy of $P_5(n/3)$ placed sufficiently close to the x_1-axis in the plane spanned by the x_1-axis and the x_d-axis.

Now, let h be an arbitrary hyperplane that contains the x_1-axis and avoids $P_7^a(n,d)$ and $P_7^b(n,d)$ and let h partition $P_7^a(n,d) \cup P_7^b(n,d)$ into sets A and B. If $P_7^o(n,d)$ is sufficiently close to the x_1-axis, then any bisection of $P_7^o(n,d)$ into sets C_1 and C_2 can be defined by a line close enough to the x_1-axis. It follows that $C_1 \cup A$ and $C_2 \cup B$ or $C_1 \cup B$ and $C_2 \cup A$ define a bisection of the entire set $P_7(n,d)$. The assertion follows since any bisection of $P_7^o(n,d)$ can be extended to form a bisection of $P_7(n,d)$ in $\Theta(n^{d-2})$ ways (see Theorem 3.1). □

By Theorem 3.7, we can choose $b(m)$ in $\Omega(\log(m+1))$ which implies the following result.

Theorem 3.11: For positive integer numbers $d \geq 2$ and n a multiple of 6, we have $e_{n/2}^{(d)}(n) = \Omega(n^{d-1}\log n)$.

Using ideas similar to the ones presented in Section 3.4, it is possible to generalize Theorem 3.11 to arbitrary integer numbers n and to integer numbers k which are not necessarily equal to $n/2$. Such a generalization is indicated in Problem 3.3, (c) and (d).

3.6. Semispaces and Circular Sequences

Chapter 2 introduced circular sequences as a combinatorial representation of point sets in the plane and demonstrated that convexity properties of a point set P, including the realization of semispaces, can conveniently be read off the

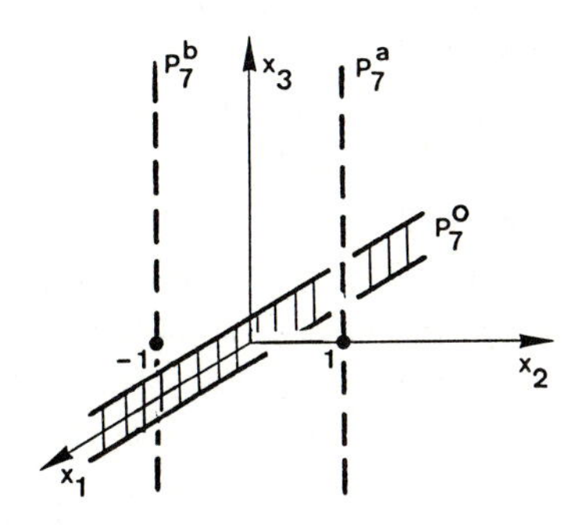

Figure 3.4. Configuration $P_7(n,3)$.

circular sequence of P. Recall that the circular sequence of P (also called the n-sequence of P, for $n = \text{card}P$) is the list of permutations obtained by projecting P onto a directed line which rotates about some point. This section discusses the notion of semispaces and of k-sets in the context of circular sequences. This serves as a preparation to further investigations of upper bounds for $e_k^{(2)}(n)$ in Section 3.7.

Let $\mathcal{C}$ be an n-sequence and let C be a subset of $\{1,2,...,n\}$. We call C a *semispace of* $\mathcal{C}$ if there is a permutation in $\mathcal{C}$ that begins with any ordering of the numbers in C. If $\text{card}C = k$ then C is also termed a *k-set of* $\mathcal{C}$. The following results extend Observation 2.2 (see also Exercise 2.6) and are obvious from the definitions.

Observation 3.12: Let $P = \{1,2,...,n\}$ be a set of points in the plane, and let $\mathcal{C}$ be the n-sequence obtained from P.

(i) A subset P' of P, with $k = \text{card}P'$, is a k-set of P if and only if P' is a k-set of $\mathcal{C}$.

(ii) The number of k-sets of $\mathcal{C}$ is equal to $e_k^{(2)}(P)$.

For $\mathcal{C}$ an n-sequence and for $0 \leq k \leq n$, we define $g_k(\mathcal{C})$ as the number of k-sets of $\mathcal{C}$, and we write $g_k(n)$ for the maximum of $g_k(\mathcal{D})$ taken over all n-sequences $\mathcal{D}$. Since every configuration in E^2 defines a circular sequence we have

$$e_k^{(2)}(n) \leq g_k(n).$$

It is, however, not known whether or not $e_k^{(2)}(n) = g_k(n)$, for there are circular sequences that are neither circularly nor locally realizable (see Sections 2.3 and 2.4).

In order to examine $g_k(n)$, it is sufficient to consider simple n-sequences $\mathcal{C}$, that is, we can assume that every permutation Π_2 of $\mathcal{C}$ is obtained by swapping two consecutive numbers of its preceding permutation Π_1. If these two numbers occupy the positions k and $k+1$ of Π_1 then we call this move a *k-swap*. It is not hard to prove the following facts about the relationship between k-sets and k-swaps in circular sequences.

Observation 3.13: Let $\mathcal{C}$ be a simple n-sequence.

(i) Let Π_2 be a permutation of $\mathcal{C}$ obtained by a k-swap from its predecessor Π_1 and let $C_{i,j}$ be the set containing the j leftmost numbers of Π_i, for $i = 1,2$ and $0 \leq j \leq n$. Then $C_{1,j} \neq C_{2,j}$ if and only if $j = k$.

(ii) For $1 \leq k \leq n-1$, the number of k-swaps in a period of $\mathcal{C}$ is equal to $g_k(\mathcal{C})$.

We will use Observation 3.13(ii) to derive upper bounds on the number of

k-sets of n-sequences. Obviously, there are exactly $2\binom{n}{2}$ swaps in any period of a simple n-sequence C which implies

$$\sum_{k=1}^{n-1} g_k(C) = 2\binom{n}{2}.$$

Be aware of the fact that the above sum does not include the one 0-set and the one n-set of C. It is instructive to compare this result with Theorem 3.1 for the special case $d=2$. Although it seems elusive to get any reasonably tight upper bounds on $g_k(n)$, unless k is very small, it is not difficult to derive tight upper bounds on the sum of $g_i(n)$, taken over all integer numbers i between 0 and some positive integer $k < n/2$. We therefore define

$$G_k(C) = \sum_{i=0}^{k} g_i(C)$$

and

$$G_k(n) = \max\{G_k(D) \mid D \text{ an } n\text{-sequence}\},$$

for $0 \le k \le n$. Bounds on $G_k(n)$ turn out to have applications in the analysis of generalized Voronoi diagrams (see Chapter 13). We prove the following tight upper and lower bounds on $G_k(n)$.

Theorem 3.14: $G_k(n) = kn+1$, for $0 \le k < n/2$.

Proof: The n-sequence defined by the n vertices of a convex polygon realizes one 0-set and n i-sets, for each $1 \le i \le k$. We therefore conclude that $kn+1$ is a lower bound on $G_k(n)$.

To prove the upper bound note that any number m in a simple n-sequence C participates in exactly $2n-2$ swaps of any period of C since m swaps with every other number exactly twice. By reasons of symmetry, we know that m participates in exactly j i-swaps if and only if it participates in exactly j $(n-i)$-swaps. Let p be the leftmost position of any permutation in C that m ever occupies and consider a period of C that begins with m at position p. If $p \le k$ then m crosses the "corridor" of columns $k+1, k+2, \ldots, n-k$ from position p to position $n-p+1$ and back. Thus, this corridor consists of $n-2k$ columns, and to cross it twice, m has to participate in at least two i-swaps, for each $k+1 \le i \le n-k-1$ (see Figure 3.5). This accounts for at least $2n-4k-2$ swaps involving m. Consequently, the number of i-swaps, with $i \le k$ or $i \ge n-k$ is smaller than or equal to

$$2n-2-(2n-4k-2) = 4k.$$

By symmetry, this implies that the total number of i-swaps, for all integer numbers between 1 and k including the limits, that involve m is at most $4k/2 = 2k$. Hence, the total number of i-swaps involve at most $2kn$ numbers,

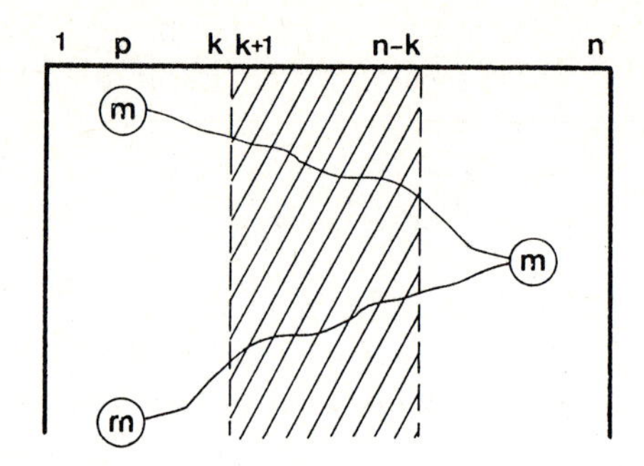

Figure 3.5. Number m crossing the corridor of columns.

and, by Observation 3.13(ii) and the fact that a swap involves two numbers, we have $G_k(n) \leq kn+1$.

Notice that the above range for number i is empty if $2k+1 \geq n$. Still, the argument above is correct unless $2k+1 > n$, which is equivalent to $k \geq n/2$ and thus excluded by the assumption in the statement of the theorem. □

The n–sequence that shows the lower bound of Theorem 3.14 is trivially realizable which implies the following result.

Corollary 3.15: Let P be a configuration of n points in E^2. Then we have

$$\sum_{i=0}^{k} e_i(P) \leq kn+1,$$

for $0 \leq k < n/2$, and the bound is tight.

It is worthwhile to note that Theorem 3.14 and Corollary 3.15 do not hold for the case $k \geq n/2$. In fact, if they did hold for $k = n/2$, then we could easily derive a tight upper bound on $g_k(n)$, for $k = n/2$. This is indicated in Exercise 3.10.

3.7. An Upper Bound on the Number of k–Sets in the Plane

Making use of Observation 3.13(ii), we will prove that $g_k(n)$ is in $O(n\sqrt{k})$ which implies the same upper bound for $e_k^{(2)}(n)$. The essential idea for the proof is to take the numbers of an n–sequence in groups of size about $\sqrt{k}$ and to analyze the interactions within each group separately from the interactions between groups.

We begin with a technical lemma. Let $\mathcal{H}$ be a halfperiod of some n–sequence and recall that we can assume that this n–sequence is simple. Let Π be the first permutation of $\mathcal{H}$, and let Y be a substring of Π, that is, there exist strings X and Z such that $\Pi = XYZ$. We let y denote the length of substring Y.

Lemma 3.16: Let $\mathcal{H}$ and Y be as defined above, and let k be an integer number between 1 and $n-1$ including the limits. Then at most $\binom{y}{2}+\min\{n-y,2k\}$ of the k–swaps in $\mathcal{H}$ involve a number of Y.

Proof: There are $\binom{y}{2}$ swaps involving two numbers of Y altogether, and at most all of them are k–swaps. This argument accounts for the k–swaps that involve two numbers of Y each.

To get an upper bound on the number of k–swaps involving exactly one number of Y, we color all numbers in X white and all numbers in Z black. We also imagine that we cannot distinguish two equally colored numbers and therefore do not recognize a swap involving two numbers in X or two numbers in Z. We now repeatedly remove one of two consecutive permutations if they are indistinguishable with our limited power of observation, and we get a new sequence of permutations $\mathcal{H}^*$. Finally, we relabel the white numbers in $\mathcal{H}^*$ such that they appear in the same order in each permutation, and we do the same for the black numbers. The net effect of this transformation is that $\mathcal{H}^*$ is a prefix of a halfperiod of some new circular sequence which has the property that each swap between two white or two black numbers succeeds all other swaps. Still, the number of k–swaps involving a number of Y is the same as in $\mathcal{H}$.

Since no number in X can move to the left and no number in Z can move to the right within $\mathcal{H}^*$, each number of X or Z can participate at most once in a k–swap. In fact, at most $2k$ do so since a number of X participates in a k–swap only if it starts at one of the k leftmost positions, and a number of Z participates in a k–swap if and only if it ends at one of the k leftmost positions. □

Using Lemma 3.16, it is easy to derive the main result of this section which is an upper bound on $g_k(n)$.

Theorem 3.17: $g_k(n) = O(n\sqrt{k})$, for $1 \leq k \leq n/2$.

Proof: Let $\mathcal{H}$ be a halfperiod of a simple n–sequence C, and let Π be the first permutation of $\mathcal{H}$. We partition Π into $j = \lceil n/\lceil\sqrt{k}\rceil\rceil$ consecutive substrings, each but one of length $\ell = \lceil\sqrt{k}\rceil$. By Lemma 3.16, the members of any one substring participate in at most

$$\binom{\ell}{2}+2k < 3k$$

k–swaps. The overall participation is thus limited to at most

$$3k \cdot j \leq 3k(n/\sqrt{k}+1) \leq 3n\sqrt{k}+3k = O(n\sqrt{k}).$$

The assertion follows since the number of k-swaps in a period of C is at most twice the number of k-swaps in a halfperiod that maximizes this number. □

We have seen in Section 3.6 that $e_k^{(2)}(n)$ is bounded from above by $g_k(n)$. Thus, Theorem 3.17 implies an upper bound on the number of k-sets of a set of n points in the plane.

Corollary 3.18: $e_k^{(2)}(n) = O(n\sqrt{k})$, for $1 \leq k \leq n/2$.

It seems obvious from the proof of Theorem 3.17 that $O(n\sqrt{k})$ is not an asymptotically tight bound for $g_k(n)$ and $e_k^{(2)}(n)$. We therefore challenge the ambitious reader to improve upon Theorem 3.17.

3.8. Exercises and Research Problems

3 **Exercise 3.1:** Prove $e_k^{(d)}(n) = O(e_{k+1}^{(d)}(n))$, for $1 \leq k \leq n/2$ and $d \geq 2$. *(Comment: this result implies that Theorem 3.3 can be simplified to "the complexity of level L_k is in $O(e_{k+d-2}^{(d)}(n))$", if $k+d-2 \leq n/2$.)*

2 **Exercise 3.2:** (a) Let P_1 be a configuration of n points in E^2 and let R be some rectangle with positive area. Prove the existence of a configuration P_2 in R which allows a one-to-one correspondence between the point of P_1 and P_2 such that a subset of P_1 is a k-set of P_1 if and only if the corresponding subset of P_2 is a k-set of P_2, for $0 \leq k \leq n$.

2 (b) Prove (a) with the additional restriction that the slopes of lines that connect any two points of P_2 must lie in some non-degenerate interval.

5 **Problem 3.3:** (a) Construct configurations of points in the Euclidean plane that show $e_{n/2}^{(2)}(n) = \omega(n \log n)$.

5 (b) Construct configurations of points in E^d that prove $e_{n/2}^{(d)}(n) = \omega(n^{d-1} \log n)$, for $d > 2$.

2 (c) Prove $e_k^{(d)}(n) = \binom{n}{k}$, for $k \leq d/2$. *(Hint: read Section 6.2.1.)*

3 (d) Construct configurations of points in d-dimensional Euclidean space that prove $e_k^{(d)}(n) = \Omega(n \cdot k^{d-2} \log(k+1) + n^{\min\{\lfloor d/2 \rfloor, k\}}/k)$, for $1 \leq k \leq n/2$. *(Hint: choose $2k$ points close to each vertex of a cyclic polytope with $\lfloor n/2k \rfloor$ vertices.)*

4 **Problem 3.4:** (a) Let P be a set of n points in the plane. Prove that

$$\sum_{k \in K} e_k(P) < n(8 \sum_{k \in K} k)^{1/2},$$

for K any non-empty subset of $\{1,2,\ldots,\lfloor n/2 \rfloor\}$.

3 (b) Prove $e_k^{(3)}(n) = O(n^2 k)$.

4 (c) Prove $e_k^{(3)}(n) = O(nk^5)$. *(Hint: prove first that the number of k-sets of a set of n extreme points in three dimensions is in $O(nk)$; see also Section 13.4.)*

5 (d) Prove or disprove $e_{n/2}^{(d)}(n) = o(n^d)$, for $d \geq 3$.

4 **Problem 3.5:** (a) For even positive integer numbers n, construct n-sequences that realize $\omega(n \log n)$ $n/2$-sets.

4 (b) Prove the existence of a constant positive real number c such that $g_k(n)=\Omega(n\cdot 2^{c\sqrt{\log k}})$, for every $1\leq k\leq n/2$.

5 (c) Decide whether or not $e_k^{(2)}(n)=g_k(n)$.

2 **Exercise 3.6:** Verify Observation 3.13(ii).

5 **Problem 3.7:** (b) Prove or disprove $g_k(n)=o(n^{1+\epsilon})$, for every $\epsilon>0$.

2 **Problem 3.8:** (a) For every $n\geq 4$, construct a set P of n points in the plane with $e_2^{(2)}(P)=\lfloor 3n/2\rfloor$.

3 (b) Use n-sequences to prove $e_2^{(2)}(n)\leq\lfloor 3n/2\rfloor$.

4 (c) For every $n\geq 20$, construct a set P of n points in E^2 with $e_3^{(2)}(P)=\lfloor 11n/6\rfloor$.

5 (d) Prove or disprove $g_3(n)\leq\lfloor 11n/6\rfloor$ up to a finite number of exceptional integer numbers n. *(Comment: recall that $g_3(n)$ is an upper bound on $e_3^{(2)}(n)$.)*

3 **Exercise 3.9:** Prove $e_2^{(3)}(n)=3n-6$, for $n\geq 3$ and $n\neq 5$, and show $e_2^{(3)}(5)=10$.

2 **Exercise 3.10:** Show $g_k(n)=2G_k(n)-n(n-1)$, if $n=2k$ (see also Theorem 3.14).

2 **Exercise 3.11:** (a) Let $\bar{x}(n)$ denote the maximal number of edges in any arrangement of n lines in E^2 such that no two of them intersect a common vertical line and the union of their closures is connected. Prove that $\bar{x}(n)=\Omega(n^2)$.

3 (b) Let $\hat{x}(n)$ denote the maximal number of turns in the union of the closures of a collection of edges in an arrangement of n non-vertical lines in the plane such that every vertical line intersects this union in exactly one point. Prove $\hat{x}(n)=\Omega(n^{3/2})$.

3.9. Bibliographic Notes

The combinatorial geometry problem discussed in this chapter has several applications to other combinatorial questions, such as counting the number of faces of levels in arrangements of hyperplanes. Some of these applications are mentioned in Edelsbrunner, Welzl (1985). Levels in arrangements and some of their applications to computational geometry are discussed for E^2 in Edelsbrunner, Welzl (1986a), and for higher dimensions in Edelsbrunner, O'Rourke, Seidel (1986). The concept of a so-called k-hull discussed in Cole, Sharir, Yap (1987) bears a close relationship to levels in arrangements. An extensive description of the relationship between arrangements and configurations in the plane on one hand and circular sequences on the other, can be found in Goodman, Pollack (1984).

The reported asymptotic upper and lower bounds on $e_k^{(2)}(n)$ were first derived by Lovász (1971) and Erdős, Lovász, Simmons, Straus (1973). The constructions in Sections 3.3, 3.4, and 3.7, however, are taken from the independent development reported in Edelsbrunner, Welzl (1985). The extension of the lower bound on the number of bisectors of a finite set of points in the plane to higher dimensions, described in Section 3.5, is an elaboration of an idea communicated by Raimund Seidel.

The proof of the upper bound on $G_k(n)$ presented in Section 3.6 is due to Alon, Győri (1986); the same result with a different proof can be found in Peck (1985). Welzl (1986) offers extensions of these results to upper bounds on the total number of k-sets, for k in some collection of not necessarily consecutive integer numbers between 1 and $n/2$; this solves Problem 3.4(a). A summary of results on $e_k^{(2)}(n)$ and $g_k(n)$ for small k, including solutions to Problems 3.8(a) and 3.8(c), is given in Stöckl (1984). A solution to Exercise 3.9 can be found in Edelsbrunner, Stöckl (1986). A solution to Problem 3.5(b) has been shown but not published by Maria Klawe, Mike Paterson, and Nicolas Pippinger. Currently, the only non-trivial upper bounds on the number of k-sets of a set of n points in three dimensions can be found in Cole, Sharir, Yap (1987) who solve Problem 3.4(b), in Chazelle, Preparata (1986) who solve Problem 3.4(c), and in Clarkson (1987). The function $\hat{x}(n)$ defined in Exercise 3.11(b) and a solution to the same exercise were communicated by Micha Sharir.

CHAPTER 4

DISSECTIONS OF POINT SETS

A hyperplane that contains no point of a finite set P partitions P into two subsets contained in the two half–spaces defined by the hyperplane. Equivalently, a point contained in a cell of an arrangement $\mathcal{A}(H)$ classifies each hyperplane h in H according to whether it is above or below h. Examples of two related combinatorial problems that are interesting to ask are

> "Is there a point p which is such that the partition of P defined by any hyperplane that contains p is reasonably balanced?"

and

> "If P is a set of points in E^d, are there d hyperplanes that partition P into 2^d subsets of approximately the same size?"

The subsequent sections of this chapter answer these and other related questions. All issues discussed are motivated by algorithmic problems, some of which are described in Chapter 14; they are concerned with efficient algorithms and data structures for dissecting configurations.

The organization of this chapter is as follows. Section 4.1 gives a proof of the classic Helly theorem for convex sets and exploits it to resolve the first of the two questions above. In Section 4.2, we prove a discrete version of the so–called ham–sandwich theorem using a topological result attributed to Borsuk. Several methods for dissecting finite two– and three–dimensional point sets in a balanced way are developed in Sections 4.3 through 4.5. Sections 4.3 and 4.5 also address recursive applications of these techniques which yield cell complexes with various interesting properties. Finally, Section 4.6 addresses the generalizability of the two– and three–dimensional results to four and higher dimensions.

4.1. Centerpoints

Let P be a configuration of n points in E^d. A point x, not necessarily in P, is called a *centerpoint of* P if no open half-space that avoids x contains more than $\frac{dn}{d+1}$ points of P. This section offers a proof of the existence of a centerpoint for every configuration using results from combinatorial geometry. Proofs of these results will be included. First, we show that the result is best possible. That is, we construct a set S of $n \geq d+1$ points in E^d such that for every point x there is an open half-space that avoids x but contains $\lfloor \frac{dn}{d+1} \rfloor$ points of S.

Let $v_0, v_1, \ldots, v_d$ be $d+1$ affinely independent points in E^d, and let c be a point in the interior of the convex hull of $v_0, v_1, \ldots, v_d$, which is a simplex in E^d. Let r_i be the ray that emanates from v_i, that belongs to the line through c and v_i, and whose direction is such that it does not contain point c (see Figure 4.1). Define $j = n-(d+1)\lfloor \frac{n}{d+1} \rfloor$ and choose $\lceil \frac{n}{d+1} \rceil$ points including v_i on ray r_i if $i < j$, and choose $\lfloor \frac{n}{d+1} \rfloor$ points including v_i if $i \geq j$. For every point x there is an index i, $0 \leq i \leq d$, such that x belongs to a closed half-space which contains r_i but avoids the other d rays. The complement of this half-space avoids x and contains at least $n - \lceil \frac{n}{d+1} \rceil = \lfloor \frac{dn}{d+1} \rfloor$ points.

We now present two classic results from geometry which will be used to prove the existence of centerpoints. The first of the two is known as Radon's theorem.

Theorem 4.1: Let P be a set of $n \geq d+2$ points in E^d. There exists a partition of P into sets P_1 and P_2 such that $\text{conv}P_1$ and $\text{conv}P_2$ intersect.

Proof: Since $n \geq d+2$ the set $P = \{p_1, p_2, \ldots, p_n\}$ is affinely dependent. This

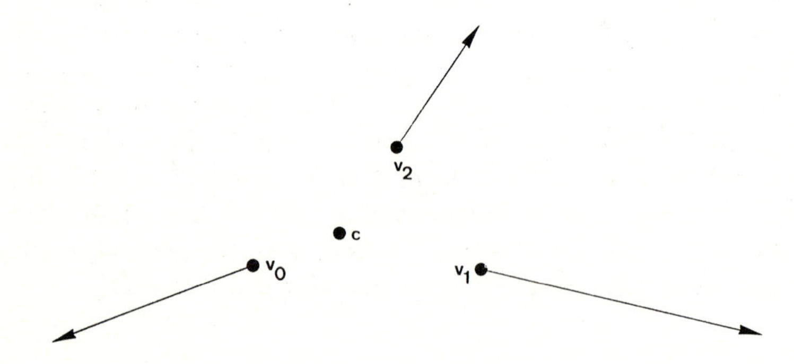

Figure 4.1. Extreme configuration in the plane.

implies that there are real numbers $\lambda_1,\lambda_2,\ldots,\lambda_n$, not all vanishing, such that

$$\sum_{i=1}^{n}\lambda_i=0 \text{ and } \sum_{i=1}^{n}\lambda_i p_i=o,$$

where o denotes the origin. Define $I_1=\{i|\lambda_i>0\}$ and $I_2=\{i|\lambda_i<0\}$, and note that both sets of indices are non-empty. Using

$$\lambda=\sum_{i\in I_1}\lambda_i=-\sum_{i\in I_2}\lambda_i$$

we define

$$q_1=\frac{1}{\lambda}\sum_{i\in I_1}\lambda_i p_i \text{ and } q_2=-\frac{1}{\lambda}\sum_{i\in I_2}\lambda_i p_i.$$

If $P_j=\{p_i|i\in I_j\}$, then convP_j contains q_j, for $j=1,2$. But now $q_1=q_2$ implies that the intersection of convP_1 and convP_2 is non-empty. □

Radon's theorem turns out to be useful in proving the next result known as Helly's theorem. In fact, both theorems are equivalent, that is, one can be used to prove the other.

Theorem 4.2: Let $S_1,S_2,\ldots,S_n$ be $n\geq d+1$ convex sets in E^d. If any $d+1$ of the sets have a non-empty common intersection, the common intersection of all sets is non-empty.

Proof: Note that the statement is trivial for $n=d+1$. We use induction on the number of sets and assume that the assertion is true for all $d+1\leq n\leq N-1$. By induction hypothesis, the sets $S_1,S_2,\ldots,S_{i-1},S_{i+1},\ldots,S_N$ have a common point p_i, for each $1\leq i\leq N$. Define $P=\{p_i|1\leq i\leq N\}$. Since $N\geq d+2$, Radon's theorem implies the existence of a partition of P into sets P_1 and P_2 such that convP_1 and convP_2 have a common point q. By definition of P, convP_j belongs to set S_i whenever p_i is not in P_j, for $j=1,2$. Consequently, q is contained in set S_i, for all indices $1\leq i\leq N$. □

Among the many interesting implications of Helly's theorem is the existence of a centerpoint for every finite configuration P. A hyperplane h defines a partition of P into set $P_0=P\cap h$ and sets P_1 and P_2 contained in different open half-spaces bounded by h. If h contains a centerpoint of P then

$$\max\{\text{card}P_1,\text{card}P_2\}\leq\frac{dn}{d+1}$$

and, equivalently,

$$\min\{\text{card}(P_0\cup P_1),\text{card}(P_0\cup P_2)\}\geq\frac{n}{d+1}.$$

Below, we prove that for every finite set of points there is a point not necessarily in the set that is a centerpoint of the set.

Theorem 4.3: Every finite set of points in d-dimensional Euclidean space admits a centerpoint.

Proof: Let P be a set of n points in E^d and assume inductively that the assertion holds in E^k, $k<d$. We can thus assume that P contains $d+1$ affinely independent points. Call an open half-space s *non-saturated* if it contains at most $\lceil \frac{n}{d+1} \rceil - 1 = n - \lfloor \frac{dn}{d+1} \rfloor - 1$ points of P, and call s *maximal* if

(i) s is non-saturated,

(ii) there are d affinely independent points of P in the boundary of s, and

(iii) the closure cl$s$ of s contains at least $\lceil \frac{n}{d+1} \rceil$ points.

Point x in E^d is a centerpoint of P if and only if no non-saturated half-space contains x; otherwise, the complement of the closure of such a non-saturated half-space contains more points than is allowed. Since every non-saturated half-space is contained in the union of suitable maximal half-spaces, it suffices that x avoids all maximal half-spaces. Now, every $d+1$ maximal half-spaces together contain at most $(d+1)(\lceil \frac{n}{d+1} \rceil - 1) < n$ points, which implies that the complement of their union is non-empty. Since the set of maximal half-spaces is finite we can apply Theorem 4.2 which implies that E^d is not covered by all maximal half-spaces. Finally, a centerpoint of P exists since the union of all non-saturated half-spaces is the same as the union of all maximal half-spaces. □

By duality, Theorem 4.3 implies that for every arrangement $\mathcal{A}(H)$ of n non-vertical hyperplanes in E^d there is a hyperplane h, not necessarily in H, such that half-space clh^+ contains the $\lceil \frac{n}{d+1} \rceil$-level of $\mathcal{A}(H)$ and half-space clh^- contains its $\lfloor \frac{dn}{d+1} \rfloor$-level (see Chapter 3 for a discussion of levels in arrangements).

4.2. Ham-Sandwich Cuts

We first formalize the notion of a bisecting hyperplane and then prove the main result of this section which guarantees the existence of a hyperplane that simultaneously bisects d sets in E^d. Some of the results in this section are formulated in terms of multisets instead of sets which has the advantage that they can be applied to higher dimensions via the inverse operation of projection.

Let P be a multiset of n points in E^d. A hyperplane h is termed a *bisector of P* if neither of the two open half–spaces defined by h contains more than $\frac{n}{2}$ points of P; in this case, we also say that h *bisects* P. Note that every bisector h contains at least one point of P if n is odd. Consider the one–dimensional case as an example. For n points $p_1 \le p_2 \le \dots \le p_n$ in E^1, point $h = p_{(n+1)/2}$ is a bisector if n is odd. If n is even, then any point h that satisfies $p_{n/2} \le h \le p_{(n+2)/2}$ is a bisector of the multiset.

Lemma 4.4: Let P be a multiset of n points in E^d and let v be a non–zero vector. There exists a bisector of P normal to v.

Proof: Let l be a line parallel to v, let $l(q)$ be the orthogonal projection of a point q onto l, and define the multiset $Q = \{l(p) | p \in P\}$. Every hyperplane h normal to l which intersects l in a bisector of Q on l (a one–dimensional scenario) bisects P. □

Lemma 4.4 relates to the fact that every vertical line intersects the $\lfloor \frac{n}{2} \rfloor$- and the $\lceil \frac{n}{2} \rceil$-level of an arrangement of n non–vertical hyperplanes. There is also a close relation to the studies undertaken in Chapter 3, which can be seen as follows. Call two bisectors *equivalent* if they partition P into the same two sets. Then the maximal number of pairwise non–equivalent bisectors of n points in E^d relates linearly to function $e^{(d)}_{\lfloor n/2 \rfloor}(n)$ defined in Chapter 3 (see also Theorem 3.3).

The argument used to prove Lemma 4.4 resembles the well–known mean–value theorem for continuous functions, where the function considered counts the number of points to the left of some point x. However, it is not quite the same since the function is not continuous. To handle extensions of Lemma 4.4 rigorously, we define a continuous family of mass–distributions $\mu_{\epsilon,\delta}$ for each finite multiset of points. This family of mass–distributions depends on the multiset and on parameters ϵ and δ. All mass–distributions considered in this section are reasonably well behaved, which has the net effect that little familiarity with the concept of mass–distributions is required to understand the discussion below.

Let P be a finite multiset of points, let ϵ and δ be two positive real numbers, and define $b(p) = \{x | d(p,x) < \epsilon,\ p \in P\}$. The *approximating mass–distribution* $\mu_{\epsilon,\delta}$ *of* P assigns to point $x = (x_1, x_2, \dots, x_d)$ the mass

$$i + \delta e^{-|x_1| - |x_2| - \dots - |x_d|},$$

where i is the number of balls $b(p)$ that cover x, and $e \approx 2.718$ is equal to Euler's constant. The *measure* $M(\epsilon,\delta)$ *of* $\mu_{\epsilon,\delta}$ is defined as

$$\int_{-\infty}^{+\infty}\int_{-\infty}^{+\infty}\cdots\int_{-\infty}^{+\infty}\mu_{\epsilon,\delta}(x)\,dx_1 dx_2 \ldots dx_d.$$

Note that $M(\epsilon,\delta)$ equals $\delta 2^d + n\epsilon^d \mu_0$, with μ_0 the measure of the unit-ball, since

$$\int_0^\infty\int_0^\infty\cdots\int_0^\infty e^{-x_1-x_2-\ldots-x_d}\,dx_1 dx_2 \ldots dx_d = 1$$

and each ball $b(p)$ contributes $\epsilon^d \mu_0$ to $M(\epsilon,\delta)$.

The measure *restricted* to a subset of E^d is the same integral as above, only taken over all points of the subset rather than those of E^d. A hyperplane g is said to *bisect* $\mu_{\epsilon,\delta}$ if the measure restricted to one side of g amounts to exactly half the total measure. Since the approximation of P improves for decreasing ϵ and δ, bisectors of $\mu_{\epsilon,\delta}$ approach bisectors of P when the parameters go to zero. The following choice of ϵ and δ suffices for our purpose.

(i) Choose $\epsilon > 0$ small enough such that any subset of the balls $b(p)$ intersect a common hyperplane only if their centers belong to a common hyperplane.

(ii) Choose δ such that $0 < \delta 2^d < \epsilon^d \mu_0$.

Parameters ϵ and δ which satisfy (i) and (ii) are called *sufficiently small*. The significance of sufficiently small ϵ and δ can be seen from the following result.

Lemma 4.5: Let P be a multiset of n points in E^d and let $\mu_{\epsilon,\delta}$ be an approximating mass–distribution with ϵ and δ sufficiently small.

(i) For every hyperplane g with open half-spaces g^+ and g^- there is a hyperplane h with open half-spaces h^+ and h^- such that point p belongs to h if ball $b(p)$ intersects g, p belongs to h^+ only if $b(p)$ lies in g^+, and p belongs to h^- only if $b(p)$ lies in g^-.

(ii) If g bisects $\mu_{\epsilon,\delta}$ then h bisects P.

Proof: Assertion (i) follows immediately from the choice of ϵ small enough. Part (ii) can now be false only if there is a bisecting hyperplane g of $\mu_{\epsilon,\delta}$ with more than $n/2$ balls $b(p)$ in one half-space. This is impossible, since $\lceil \frac{n+1}{2} \rceil \epsilon^d \mu_0 > \frac{n\epsilon^d \mu_0 + \delta 2^d}{2}$. □

Together with a fundamental result from topology, known as the Borsuk–Ulam theorem, the above technical lemmas can be used to prove the main theorem of this section. We state the Borsuk–Ulam theorem without proof.

Theorem 4.6: Let f be a continuous mapping from the unit–sphere S^{d-1} in d dimensions to E^{d-1}. Then $f(x)=f(-x)$ for some antipodal pair x and $-x$ on S^{d-1}.

We are now prepared to proceed with the presentation of a discrete version of the so–called ham–sandwich theorem which constitutes the main result of this section.

Theorem 4.7: Let $P_1,P_2,\ldots,P_d$ be d finite sets of points in E^d. There exists a hyperplane h that simultaneously bisects $P_1,P_2,\ldots,P_d$.

Proof: Let ϵ and δ be parameters sufficiently small to approximate the multiset $P=P_1 \cup P_2 \cup \ldots \cup P_d$, and let μ_i be the mass–distribution which approximates P_i with parameters ϵ and δ, for $1 \leq i \leq d$. We demonstrate the existence of a hyperplane g which bisects $\mu_1,\mu_2,\ldots,\mu_d$. Lemma 4.5 then implies that there is a hyperplane h that bisects point sets $P_1,P_2,\ldots,P_d$.

To prove the existence of g, we construct a continuous function f, from S^{d-1} to E^{d-1}; Theorem 4.6 will prove the rest.

For every point y of S^{d-1}, let $g(y)$ be the hyperplane normal to vector y that bisects μ_d. Here we take advantage of the fact that $g(y)$ is unique since μ_d is positive everywhere (see also Lemma 4.4). Next, we define half–space

$$g(y)^{pos}: \langle z,y \rangle > \langle z_0,y \rangle,$$

for z_0 any point of hyperplane $g(y)$. Function f maps y to the vector whose components are the measures of μ_1 through μ_{d-1} restricted to $g(y)^{pos}$.

Notice that f is indeed a continuous function from S^{d-1} to E^{d-1}. Obviously, antipodal points y and $-y$ of S^{d-1} correspond to coinciding hyperplanes, that is, $g(y)=g(-y)$. However, the half–spaces corresponding to y and $-y$ lie on different sides of this hyperplane, that is, $g(y)^{pos} \cap g(-y)^{pos}$ is empty. Theorem 4.6 now implies that there is a point x of S^{d-1} such that $f(x)=f(-x)$. By construction of f, $g(x)$ bisects μ_1 through μ_d. □

Theorem 4.7 implies that every sandwich filled with ham and cheese can be cut into two parts of equally much bread, ham, and cheese by a straight knife slice – even if one forgets the cheese in the refrigerator.

It is not difficult to construct sets $P_1,P_2,\ldots,P_d$ such that the simultaneous bisector is unique. At the other extreme, when $P_1=P_2=\ldots=P_d$, every bisector of P_d bisects all sets. Nevertheless, counting the number of non–equivalent simultaneous bisectors becomes a challenging and relevant question when some conditions are imposed on the sets. For example, let $t(n)$ be the maximal number of non-equivalent simultaneous bisectors that exist for two point sets P and Q in E^3, with

$$\begin{gathered} \mathrm{card} P + \mathrm{card} Q = 2n, \\ \text{both } \mathrm{card} P \text{ and } \mathrm{card} Q \text{ are odd, and} \\ \mathrm{conv} P \cap \mathrm{conv} Q = \emptyset. \end{gathered}$$

Upper bounds for $t(n)$ have applications to algorithms that simultaneously "four-sect" two point sets in E^3 (see Section 4.4). Currently, no non-trivial bounds for $t(n)$ are known which led me to venture the following conjecture.

Conjecture 4.8: $t(n) = \Theta(e^{(2)}_{\lfloor n/2 \rfloor}(n))$, that is, $t(n)$ is proportional to the maximal number of non-equivalent bisectors of n points in E^2.

4.3. Erasing Subdivisions in the Plane

The two-dimensional case of the ham-sandwich theorem (Theorem 4.7) implies the existence of two lines that cut a set P of n points such that each open sector contains at most $\frac{n}{4}$ of the points. To see this, we bisect P and then simultaneously bisect the subsets in the two open half-planes defined. Note that the first bisector can be chosen arbitrarily and we will still be able to produce the four balanced parts from the bisection. We exploit these results to construct so-called erasing subdivisions in E^2 that have few regions intersecting any one line. We formally define the concepts for arbitrary dimensions d.

Let P be a set of n points in E^d. A cell complex C with convex cells is said to *erase* P if all points of P are contained in facets and lower-dimensional faces

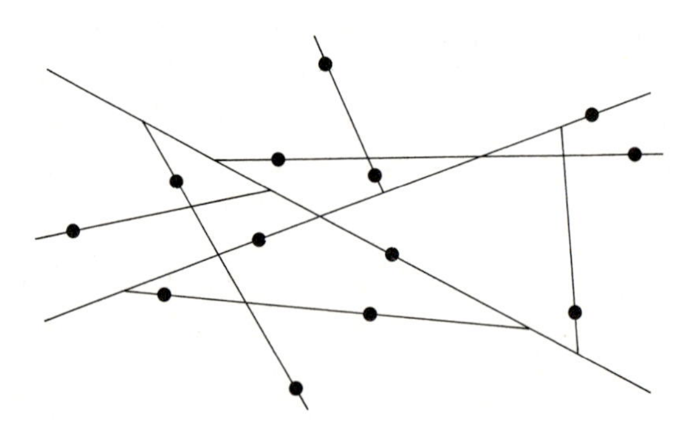

Figure 4.2. Erasing subdivision for 13 points.

of C. An erasing cell complex in E^2 is also called an *erasing subdivision*. The *stabbing number* $s(C)$ *of* C is the maximal number of cells of C that intersect any one hyperplane. Figure 4.2 shows an erasing subdivision for 13 points in E^2 with stabbing number 8. Erasing cell complexes that have a small stabbing number are relevant in the design of data structures for so-called simplicial range queries (see Chapter 14). We therefore define the *stabbing number of* P as $s(P)=\min\{s(D)|D$ an erasing cell complex of $P\}$, and $s^{(d)}(n)=\max\{s(Q)|Q$ a set of n points in $E^d\}$.

The following lemma explains why ham-sandwich cuts, as described in Theorem 4.7, are useful for constructing erasing subdivisions with low stabbing number.

Lemma 4.9: Every line intersects at most three of the open regions defined by any two lines in the plane.

Proof: If the new line coincides with one of the subdividing lines then it intersects no region. Otherwise, it intersects the subdividing lines in at most two points. Consequently, it is cut into at most three intervals which implies the assertion. □

We now give an upper bound on $s^{(2)}(n)$ using Lemma 4.9.

Theorem 4.10: $s^{(2)}(n)=O(n^\alpha)$, with $\alpha=\log_2\frac{1+\sqrt{5}}{2}<0.695$.

Proof: Let P be a set of n points in E^2. We outline an algorithm that constructs an erasing subdivision C of P with $s(C)=O(n^\alpha)$.

Algorithm 4.1 (Erasing subdivisions):

Initial step: Find a bisector l_0 of P and initialize C to be the subdivision that consists of edge l_0 and regions r_0 and r_0' on both sides of l_0. Call edge l_0 *active* and call r_0 and r_0' the *regions of* l_0.

Iterative step: Refine C as follows:

while there is an active edge e in C **do**
  Disactivate e and let r and r' be the two regions of edge e.
  unless both $P\cap r$ and $P\cap r'$ are empty **do**
    Find a line l that simultaneously bisects $P\cap r$ and $P\cap r'$. Add edges $f=l\cap r$ and $f'=l\cap r'$ to C, and declare both edges as active. In addition, let the two parts of r on both sides of f be the regions of f and let the parts of r' on both sides of f' be the regions of f'.
  endunless
endwhile.

Observe that it can be the case that we need to bisect a pair of point sets where one of the sets is empty. This case is actually simpler than if both sets are non-empty since any bisection of the non-empty set will do. Notice that Algorithm 4.1 disactivates an edge immediately after finding an intersecting bisector. This feature is necessary to obtain a reasonable stabbing number as will be shown below. The subdivision shown in Figure 4.2 can be obtained by Algorithm 4.1 which, of course, is not deterministic since bisectors, in general, are not unique.

We define $\dot{s}(n)=\max\{s(D)|Q$ a set of n points in E^2 and D an erasing subdivision of Q that can be constructed by Algorithm 4.1$\}$. Obviously, $s^{(2)}(n)\leq\dot{s}(n)$. To get an upper bound on $\dot{s}(n)$, we need to make two observations. By Lemma 4.9, any line intersects at most three of the four sectors defined by the first two lines used in any construction done by Algorithm 4.1. Second, the subdivision of each one of the sectors is an erasing subdivision for at most $n/4$ points which is of the same kind as the entire subdivision. As a matter of fact, the same is true for both half-planes defined by the first line of the subdivision, that is, the subdivision of each such half-plane is an erasing subdivision of the same kind for at most $n/2$ points. It is interesting to note that the subdivisions of the half-planes and the sectors are of the same kind as the entire subdivision only because Algorithm 4.1 chooses the directions along which it bisects in an alternating fashion. This implies

$$\dot{s}(n)\leq\dot{s}(\lfloor\frac{n}{2}\rfloor)+\dot{s}(\lfloor\frac{n}{4}\rfloor).$$

We have $\dot{s}(0)\leq 2$ since an empty region can be cut into two empty regions only if the other region of its active edge is not empty. As a consequence, we derive

$$\dot{s}(n)\leq a_{k+2},$$

with $k=\lfloor\log_2 n\rfloor$, and a_k the k-th Fibonacci number defined recursively by the rule

$$a_0=a_1=1, \text{ and } a_k=a_{k-1}+a_{k-2},$$

for $k\geq 2$. To prove the assertion we make use of the identity

$$a_k=\frac{1}{\sqrt{5}}\left(\left(\frac{1+\sqrt{5}}{2}\right)^{k+1}-\left(\frac{1-\sqrt{5}}{2}\right)^{k+1}\right).$$

Recall that we have $k\leq\log_2 n$. Since the second term of the right side of the above formula can be bounded from above by 1, we get

$$\dot{s}(n)\leq\frac{1}{\sqrt{5}}\left(\left(\frac{1+\sqrt{5}}{2}\right)^{\log_2 n+3}+1\right)=$$

$$\frac{5+2\sqrt{5}}{5}n^{\alpha}+\frac{1}{\sqrt{5}}.$$

□

It is possible to construct erasing subdivisions with stabbing numbers in $O(n^{\beta})$, $\beta<\alpha$, for every set of n points in E^2 using techniques different than those described (see Exercise 4.17). Such an improvement, however, is not possible if we try to use three lines to cut a point set into seven subsets each of approximately the same size. More specifically, it is generally not possible to cut a point set by three lines such that each region contains at most one seventh of the points. Recall that three lines in general position cut the plane into seven regions. As a counterexample, take n points on a circle, for n sufficiently large. Each line cuts the circle in at most two points, hence, the circle is cut into at most six pieces and therefore meets at most six of the seven open regions.

4.4. Simultaneous Four–Section in Three Dimensions

Using the Borsuk–Ulam theorem (Theorem 4.6), this section gives a generalization of the two–dimensional version of the ham–sandwich theorem (Theorem 4.7) which differs considerably from the ham–sandwich theorem in three dimensions. Section 4.5 will then exploit this generalization to construct erasing cell complexes in E^3. Let P and Q be sets of m and n points in E^3. Two planes g and h are said to *four–sect P and Q* if each one of the four open wedges defined by g and h contains at most $\frac{m}{4}$ points of P and at most $\frac{n}{4}$ points of Q. The proof for the existence of such a pair of planes given below considers the case when P and Q are *separable*, that is, when conv$P$ and conv$Q$ do not intersect.

First, we prove a technical lemma which asserts uniqueness of the two–dimensional ham–sandwich cut under certain conditions. To simplify the statement, define h^+: $x_2>0$ and h^-: $x_2<0$, that is, h^+ is the half–plane above the x_1–axis and h^- is the half–plane below the x_1–axis.

Lemma 4.11: Let μ^+ and μ^- be mass–distributions in E^2 with bounded total measure such that μ^+ is positive in h^+ and vanishes in h^- and the reverse is true for μ^-. Then there is a unique line which bisects both μ^+ and μ^-.

Proof: Assume that there exist two different bisectors l_1 and l_2, and assume without loss of generality that l_1 and l_2 do not intersect in h^+. Thus, l_1 and l_2 cut h^+ into three unbounded regions r_1, r_2 and r_3, with r_2 bounded by both l_1 and l_2, say. Since the measure of μ^+ restricted to r_2 is positive, l_1 and l_2 cannot both bisect μ^+. □

The main theorem of this section can now be proved with the help of this lemma.

Theorem 4.12: Let P and Q be two separable sets of m and n points in E^3. Then there exist two planes that four-sect P and Q.

Proof: Without loss of generality, we assume that P belongs to half-space h^+: $x_3>0$ and that Q belongs to h^-: $x_3<0$. We construct a mass-distribution μ that approximates $P \cup Q$ with sufficiently small parameters ϵ and δ (see Section 4.2 for a definition of the term "sufficiently small"). In fact, we choose ϵ also small enough such that no ball with radius ϵ and center in P or Q intersects the horizontal plane h: $x_3=0$. Let μ^+ be the restriction of μ to h^+ and let M^+ denote the total measure of μ^+. Similarly, let μ^- be the restriction of μ^- to h^- and let M^- denote the total measure of μ^-. By an argument similar to that used in the proof of Lemma 4.5, it suffices to demonstrate the existence of planes g_1 and g_2 which four-sect μ^+ and μ^-, that is, the measure of μ^+ restricted to any wedge defined by g_1 and g_2 amounts to $\frac{M^+}{4}$ and the measure of μ^- restricted to any such wedge is equal to $\frac{M^-}{4}$.

Clearly, g_1 and g_2 must simultaneously bisect μ^+ and μ^-. Below, we introduce the collection of all simultaneous bisectors of μ^+ and μ^- and demonstrate the existence of a four-secting pair in this collection. Let S^1 be the unit-circle in plane h, that is, S^1 is given by the equations $x_1^2+x_2^2=1$ and $x_3=0$.

Claim: For each point (or vector) u in S^1 there is a unique half-space $g(u)^{pos}$ such that its bounding plane $g(u)$ bisects μ^+ and μ^-, u is normal to the line $h \cap g(u)$, and (vector) u points into $g(u)^{pos}$.

Proof of Claim: Let $h(u)$ be the plane orthogonal to h that contains u and the origin; so $g(u)$ must be orthogonal to $h(u)$. Define $\mu^+(u)$ and $\mu^-(u)$ as the orthogonal projections of μ^+ and μ^- onto $h(u)$. That is, $\mu^+(u)$ is a mass-distribution in $h(u)$ such that its mass restricted to any subset of $h(u)$ is the mass of μ^+ restricted to the set of points in E^3 that project orthogonally onto this subset of $h(u)$, and the symmetric statement is true for μ^-. By Lemma 4.11, there is a unique line l in $h(u)$ that simultaneously bisects $\mu^+(u)$ and $\mu^-(u)$. Consequently, $g(u)$ is the unique plane orthogonal to $h(u)$ such that $l=g(u) \cap h(u)$, and $g(u)^{pos}$ is the half-space on the side of $g(u)$ determined by u. This proves the claim.

Obviously, antipodal points u and $-u$ of S^1 define complementary half-spaces and identical planes, that is, $g(u)=g(-u)$. For two points u and v of S^1, we define

$$M(u,v)=\begin{pmatrix} M^+(u,v)-M^+/4 \\ M^-(u,v)-M^-/4 \end{pmatrix},$$

with $M^+(u,v)$ the measure of μ^+ restricted to the wedge $g(u)^{pos} \cap g(v)^{pos}$, and

$M^-(u,v)$ the measure of μ^- restricted to the same wedge. Observe that $g(u)$ and $g(v)$ four–sect μ^+ and μ^- if and only if

$$M(u,v)=\begin{pmatrix}0\\0\end{pmatrix}.$$

To demonstrate the existence of such points u and v, we construct a continuous mapping w_1 from the unit–ball B^2: $x_1^2+x_2^2\leq 1$ in h to E^2 such that w_1 is *antipodal on* S^1, that is, $w_1(x)=-w_1(-x)$, for each point x of the domain S^1. From w_1, we will derive an antipodal mapping w_2 from S^2 to E^2 – the Borsuk–Ulam theorem will then complete the proof.

To construct w_1, we describe points of B^2 by means of their polar coordinates; for example, point $p=(\pi_1,\pi_2)$ in Cartesian coordinates equals $p=(\alpha,\rho)$, with $\rho=(\pi_1^2+\pi_2^2)^{1/2}$ and angle α in $[0,2\pi)$ such that $\pi_1=\cos\alpha$ and $\pi_2=\sin\alpha$. Note that the polar coordinates of point p are unique unless p coincides with the origin o. For $p\neq o$, we define points

$$u(p)=(\frac{\alpha}{2},1) \text{ and } v(p)=(\frac{\alpha}{2}+\rho\frac{\pi}{2},1)$$

(see Figure 4.3), and the mapping

$$w_1(p)=M(u(p),v(p))=\begin{pmatrix}M^+(u(p),v(p))-M^+/4\\M^-(u(p),v(p))-M^-/4\end{pmatrix}.$$

To complete the definition of w_1, we set

$$w_1(o)=\begin{pmatrix}M^+/4\\M^-/4\end{pmatrix}.$$

Mapping w_1 is antipodal on S^1 by the following argument. If point p belongs to S^1, that is, $p=(\alpha,1)$ for some angle α, then

$$u(p)=(\frac{\alpha}{2},1),\ u(-p)=(\frac{\alpha+\pi}{2},1),$$

$$v(p)=u(-p), \text{ and } v(-p)=-u(p),$$

and consequently

$$g(v(p))^{pos}=g(u(-p))^{pos},$$

$$g(u(p))=g(v(-p)), \text{ and}$$

$$g(u(p))^{pos} \text{ and } g(v(-p))^{pos} \text{ are complementary.}$$

(To see these fairly apparent observations it might be helpful to construct some examples of points p, $u(p)$, $v(p)$, $u(-p)$, and $v(-p)$ as shown in Figure 4.3.) Therefore,

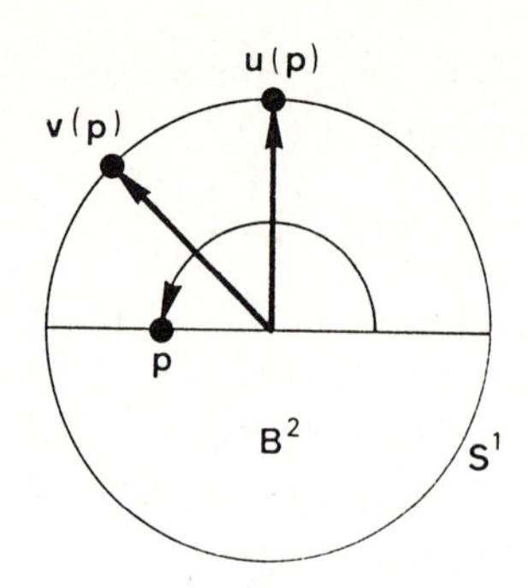

Figure 4.3. Mapping points from B^2 to $S^1 \times S^1$.

$$M^+(u(p),v(p)) = \frac{M^+}{2} - M^+(u(-p),v(-p)) \text{ and}$$

$$M^-(u(p),v(p)) = \frac{M^-}{2} - M^-(u(-p),v(-p)).$$

This implies $w_1(p) = -w_1(-p)$ which is equivalent to stating that w_1 is antipodal. Furthermore, w_1 is continuous, since we have $M(u,v) = M(-u,-v)$, for any two points u and v of S^1 and because $w_1(p)$ approaches $w_1(o)$ if p approaches o, regardless of the direction from which o is approached.

It is now easy to construct a continuous and antipodal mapping w_2 from S^2 to E^2. Let $p = (\pi_1, \pi_2, \pi_3)$ be a point of S^2. We set $p' = (\pi_1, \pi_2)$, a point in B^2, and we define

$$w_2(p) = w_1(p') \text{ if } \pi_3 \geq 0, \text{ and}$$

$$w_2(p) = -w_1(-p'), \text{ otherwise.}$$

Intuitively, w_2 is obtained by vertical projection of w_1 from B^2 onto the northern hemisphere of S^2. The values of w_2 on the southern hemisphere of S^2 are obtained by central projection through o of the northern hemisphere. Function w_2 is continuous because w_1 is continuous in B^2 and antipodal on S^1. Furthermore, w_2 is antipodal by construction. Consequently, $w_2(x) = w_2(-x)$ immediately implies that w_2 vanishes for both arguments.

Theorem 4.6 now implies that there is a point p on the northern hemisphere of S^2, with

$$w_2(p) = \begin{pmatrix} 0 \\ 0 \end{pmatrix}.$$

Hence,

$$w_1(p') = \begin{pmatrix} 0 \\ 0 \end{pmatrix},$$

and planes $g(u(p'))$ and $g(v(p'))$ exist since $p \neq (0,0,1)$, by construction of w_1. It follows that $g(u(p'))$ and $g(v(p'))$ four-sect μ^+ and μ^-. □

4.5. Erasing Cell Complexes in Three Dimensions

In pretty much the same way that Section 4.3 derived erasing subdivisions in the plane by application of the two-dimensional ham-sandwich theorem, we exploit Theorem 4.12 for the construction of three-dimensional erasing cell complexes. Note that the existence of two four-secting planes of two separated point sets in E^3 implies that one point set can be dissected by three planes such that each open cell contains at most one eighth of the points. Furthermore, the first bisector of such a construction can be chosen arbitrarily. Algorithm 4.2 below gives the details of a dissection of a finite set P of points in E^3 that is obtained by repeatedly finding pairs of four-secting planes.

Algorithm 4.2 (Repeated dissection into octants):

Initial step: Find a bisecting plane h_0 of P and initialize C to be the cell complex that consists of a single facet, h_0, and two cells c_0 and c'_0. We call h_0 *active* and we call c_0 and c'_0 the *cells of* h_0.

Iterative step: Refine cell complex C repeatedly as follows:

while there is an active facet f in C **do**
  Disactivate f and let c and c' be the cells of f.
  unless both $P \cap c$ and $P \cap c'$ are empty **do**
    Determine planes g and h that four-sect $P \cap c$ and $P \cap c'$, and let g_1, g_2, h_1, and h_2 be the four half-planes obtained from g and h by cutting them along their common intersection. Add $g_i \cap c$, $g_i \cap c'$, $h_i \cap c$, and $h_i \cap c'$, for $i=1,2$, as facets to C. Declare each new facet as active if it belongs to h, and let the two cells on both of its sides be its cells.
  endunless
endwhile.

To analyze the stabbing number of a cell complex constructed by Algorithm 4.2, we need the following result.

Lemma 4.13: Every plane intersects at most seven of the cells defined by any three planes in E^3.

Proof: The new plane intersects the three planes in at most three lines, unless it coincides with one of them in which case it avoids all cells. Three lines cut a plane into at most seven regions which implies the assertion. □

Let $\ddot{s}(n)$ denote the maximum stabbing number of any erasing cell complex C that is constructed by Algorithm 4.2 for any set of n points in E^3. Lemma 4.13 implies that any plane intersects at most seven of the eight octants defined by the first three planes used to construct C. Notice that the part of C in a single octant is not constructed in the same way as C itself. Nevertheless, the part of C restricted to a wedge defined by the first two planes is derived in the same way as the entire cell complex. Now, any plane intersects both octants of at most three such wedges and only one octant of the fourth octant. Thus, we have

$$\ddot{s} \leq 3\ddot{s}(\lfloor \frac{n}{4} \rfloor) + 2\ddot{s}(\lfloor \frac{n}{16} \rfloor).$$

When a cell is empty, then it can happen that it is further cut into four cells later but not into more than four cells. Thus,

$$\ddot{s}(0) \leq 4.$$

It follows that $\ddot{s}(n)$ is in $O(n^\beta)$, with $\beta = \log_4 \frac{3+\sqrt{17}}{2}$. By $s^{(3)}(n) \leq \ddot{s}(n)$, we derive the main result of this section.

Theorem 4.14: $s^{(3)}(n) = O(n^\beta)$, with $\beta = \log_4 \frac{3+\sqrt{17}}{2} < 0.917$.

If the three–dimensional ham–sandwich theorem is used in conjunction with the method described in Algorithm 4.2, erasing cell complexes with marginally smaller stabbing numbers can be constructed (see Exercise 4.13).

4.6. Generalizations to Higher Dimensions

Erasing cell complexes in d dimensions with sublinear stabbing number can be constructed if every set of n points can be dissected by d hyperplanes such that each open cell contains at most $\frac{n}{2^d}$ points. The existence of these d hyperplanes is unresolved for $d=4$. In E^d, $d \geq 5$, however, there are point sets for which a balanced dissection by d hyperplanes is impossible. To appreciate the following theorem recall that d hyperplanes cut E^d into 2^d cells if their normal vectors are linearly independent. For example, five five–dimensional hyperplanes dissect E^5 into up to 32 cells.

Theorem 4.15: For every positive integer n, there are n points in E^5, no six of which are contained in a common hyperplane, such that at least six of the open cells defined by any five hyperplanes contain no points.

Proof: To construct a point set that proves the assertion, we use the so called moment curve in five dimensions. The *moment curve* M_5 in E^5 contains points $(\tau,\tau^2,\tau^3,\tau^4,\tau^5)$, for all real numbers τ. Any hyperplane intersects M_5 in at most five points which will be shown in Section 6.2.1. Consequently, five hyperplanes cut M_5 into at most 26 pieces. To prove the assertion, we choose arbitrary n points on M_5. □

Notice that Theorem 4.15 does not imply $s^{(d)}(n)=\Omega(n)$, for $d\geq 5$. In fact, erasing cell complexes with a sublinear stabbing numbers can be found in every dimension (see Exercises 4.16 and 4.17).

4.7. Exercises and Research Problems

3 **Exercise 4.1:** A hyperplane h is a *transversal of* a family of convex sets if h meets every set of the family. Prove that n vertical line segments in E^d admit a transversal if every $d+1$ of them do. *(Hint: use duality and Helly's theorem (Theorem 4.2).)*

3 **Exercise 4.2:** Formulate a theorem for zonotopes which is equivalent to the existence of a centerpoint for every configuration of points (see Theorem 4.3).

2 **Exercise 4.3:** Let P be a finite multiset of points in E^d, let $\epsilon>0$ be a real number, and define $b(p)=\{x\in E^d|d(x,p)<\epsilon\}$, for each point p in P. Verify that ϵ can be chosen such that for every hyperplane g there is a hyperplane h that contains point p if g intersects $b(p)$. *(Compare with Lemma 4.5.)*

2 **Exercise 4.4:** Prove the equivalence of the following statement and the Borsuk-Ulam theorem (Theorem 4.6). Let f be a mapping from the unit-sphere S^{d-1} in d dimensions to E^{d-1} that is continuous and antipodal, that is, $f(y)=-f(-y)$, for every point y of S^{d-1}. Then there is a point x of S^{d-1} such that all components of $f(x)$ vanish.

Exercise 4.5: Let P and Q be two sets of m and n points in E^2, and let α_1, α_2, β_1, and β_2 be non-negative real numbers such that $\alpha_1+\alpha_2=\beta_1+\beta_2=1$ and $\alpha_1\neq\beta_1$. For a half-plane s, define $m_1=\text{card}(P\cap \text{int} s)$, $m_2=\text{card}(P\cap \text{int compl} s)$, $n_1=\text{card}(Q\cap \text{int} s)$, and $n_2=\text{card}(Q\cap \text{int compl} s)$.

2 (a) Produce sets P and Q such that there exists no half-plane s with $m_1\leq\alpha_1 m$, $m_2\leq\alpha_2 m$, $n_1\leq\beta_1 n$, and $n_2\leq\beta_2 n$.

3 (b) Prove that a half-plane s with the properties required in (a) exists for every proper choice of α_1, α_2, β_1, and β_2, provided P and Q are separable.

2 **Exercise 4.6:** Construct four sets of points in E^3 such that no plane bisects all four.

5 **Problem 4.7:** Prove or disprove Conjecture 4.8.

2 **Exercise 4.8:** Let the Fibonacci numbers a_i be defined as follows: $a_0=a_1=1$ and $a_i=a_{i-1}+a_{i-2}$, for $i\geq 2$. Prove the identity

$$a_i=\frac{1}{\sqrt{5}}\left(\left(\frac{1+\sqrt{5}}{2}\right)^{i+1}-\left(\frac{1-\sqrt{5}}{2}\right)^{i+1}\right).$$

Exercise 4.9: Three concurrent lines define a *six-partition of* a set of n points in E^2 if each sector contains at most $\frac{n}{6}$ points.

3 (a) Prove the existence of a six-partition for every finite set in E^2.

3 (b) Apply six-partitions iteratively to construct an erasing subdivision of any set of n points in E^2 such that the stabbing number is in $O(n^\beta)$, with $\beta=\log_6 4<0.774$.

Problem 4.10: Let P be a set of $2n$ points in E^2. A *perfect matching M of P* is a partition of P into pairs, and the *stabbing number* $\bar{s}(M)$ is the maximal number of pairs $\{p,q\}$ in M that can be separated by any one line. Define $\bar{s}(P)=\min\{\bar{s}(N)|N$ a perfect matching of $P\}$, and $\bar{s}(n)=\max\{\bar{s}(Q)|Q$ a set of $2n$ points in $E^2\}$.

2 (a) Prove $\bar{s}(n)=O(n^\alpha)$, with $\alpha=\log_2\frac{1+\sqrt{5}}{2}$. *(Hint: use ham-sandwich cuts iteratively as in Algorithm 4.1.)*

3 (b) Prove $\bar{s}(n)=\Omega(\sqrt{n})$. *(Hint: consider the points on the integer grid with coordinates between 1 and $\sqrt{2n}$.)*

5 (c) Determine the asymptotic order of $\bar{s}(n)$. *(Comment: the investigation of $\bar{s}(n)$ is interesting because of its close relationship with $s^{(2)}(n)$.)*

4 **Exercise 4.11:** Let P and Q be two finite point sets in E^3, not necessarily separable. Prove that there exist two planes that four-sect P and Q (see also Section 4.4). *(Hint: consult the paper by Hadwiger mentioned in the bibliographic notes which proves a continuous analogue of the statement.)*

Problem 4.12: Let P and Q be two finite point sets in E^3.

2 (a) Prove the existence of three planes that intersect in a common line normal to a prescribed plane such that each open wedge contains at most one sixth of the points of P. *(Hint: use Exercise 4.9(a).)*

5 (b) Decide upon the existence of six planes g_1, g_2, g_3, h_1, h_2, and h_3 that intersect in a common line such that any open wedge defined by g_1, g_2 and g_3 contains at most one sixth of the points of P, and the same is true for planes h_1, h_2, h_3 and set Q. Does it make any difference if P and Q are known to be separable?

3 (c) Prove that every plane intersects at most 10 of the 12 cells constructed as follows: take any three planes that intersect in a common line, partition the six wedges into two triplets such that no two wedges in one triplet share a common facet, add the intersections of a fourth plane with the wedges of the first triplet, and add the intersections of a fifth plane with the wedges of the second triplet.

3 **Exercise 4.13:** (a) Prove that a plane in E^3 intersects at most 11 of the 14 cells constructed as follows: take three planes that intersect in a common point, choose two disjoint triplets out of the eight octants such that no two octants of one triplet share a common facet, add the intersections of a fourth plane with the octants of the first triplet, and add the intersections of a fifth plane with the octants of the second triplet.

3 (b) Produce two separated sets P and Q of m and n points in E^3 and eight non-negative real numbers $\alpha_1+\alpha_2+\alpha_3+\alpha_4=1$ and $\beta_1+\beta_2+\beta_3+\beta_4=1$ such that there are no two planes that satisfy the following condition: if w_1, w_2, w_3, and w_4 denote the four open wedges defined by the planes, then w_i contains at most $\alpha_i m$ points of P and at most $\beta_i n$ points of Q, for $1\leq i\leq 4$.

2 **Exercise 4.14:** Describe a finite point set in E^3 such that no four planes define an arrangement with at most one fifteenth of the points in any one open cell. *(Hint: choose the points on a sphere.)*

5 **Problem 4.15:** Decide upon the existence of a finite point set in E^4 such that no four hyperplanes define 16 cells each one of which contains at most one sixteenth of the set.

2 **Exercise 4.16:** (a) Prove that a plane in E^3 intersects at most seven of the eight cells constructed as follows: take a first plane h and four additional planes that intersect in a

common line meeting h in a point called the *center*, and remove the parts of two planes on one side of h as well as the parts of the other two planes on the other side of h.

4 (b) Show that any finite set of points in E^3 allows a dissection as described in (a) such that each cell contains at most one eighth of the points.

4 (c) Extend the construction to arbitrary d dimensions, that is, choose one bisecting hyperplane h, a line l cutting h in a point p, the center, and find two such $(d-1)$-dimensional constructions in h, both with center p, such that the one gives a balanced partition if extended along l into one half-space of h, and the other one gives a balanced partition by extension along l into the other half-space of h.

4 **Exercise 4.17:** (a) Let P be a set of n points in E^d. Prove the existence of a subset Q_ϵ of size $c\frac{d}{\epsilon}\log_2\frac{d}{\epsilon}$, where c is a positive constant, such that every half-space that contains more than ϵn points of P also contains a point of Q_ϵ. *(Hint: use a counting argument.)*

3 (b) Define a *corridor in E^d* as the set of points outside the intersection of $d+1$ closed half-spaces and outside the intersection of the closures of their complements. Prove the result of (a) for corridors instead of for half-spaces. *(Hint: map the problem for corridors to a higher-dimensional instance of the problem for half-spaces.)*

3 (c) Let H be a set of n hyperplanes in E^d and let h be a hyperplane not necessarily in H. Show that there are $d+1$ half-spaces bounded by hyperplanes in H such that their corridor contains h but no hyperplane of $\mathcal{A}(H)$. *(Hint: dualize the problem.)*

3 (d) Let P be a set of n points in E^d, let Q_ϵ be a subset of P that satisfies the conditions in (b), and let H be the largest set of hyperplanes such that each hyperplane is the affine hull of some subset of Q_ϵ. Prove that the number of points that are contained in the cells of $\mathcal{A}(H)$ that intersect any one hyperplane h is at most ϵn.

3 (e) Use the result in (d) to prove $s^{(d)}(n)=O(n^{\alpha+\delta})$, for $\alpha=\frac{d(d-1)}{d(d-1)+1}$ and for every positive δ.

3 **Exercise 4.18:** Let P be a finite set of points in the plane and let c_1 be a circle such that at most half of the point of P lie inside c_1 and at most half lie outside c_1. Prove the existence of two additional circles c_2 and c_3 such that each one of the eight regions defined by c_1, c_2, and c_3 contains at most one eighth of the points of P. *(Hint: identify E^2 with the plane h_0: $x_3=0$ in E^3 and project each point in P vertically onto the paraboloid U: $x_3=x_1^2+x_2^2$; a circle in h_0 bisects P if and only if the plane through the vertical projection of the circle onto U bisects the point set on U.)*

4.8. Bibliographic Notes

Helly's and Radon's theorems presented in Section 4.1 are classic results in geometry and appear in textbooks including Yaglom, Boltyanskii (1961) and Bronsted (1983). The original publications are Helly (1923) and Radon (1921). Both theorems are equivalent and belong to the most fundamental among the non-trivial results in geometry. Goodman, Pollack (1982a) demonstrate that analogous results hold for generalizations of configurations and arrangements such as circular sequences and pseudo-line arrangements (see also Chapters 2 and 3). Applications of Helly's theorem, including the existence of centerpoints (Theorem 4.3) and solutions to the problems stated as Exercise 13.16, can be found in Yaglom, Boltyanskii (1961).

A proof of the Borsuk-Ulam theorem (Theorem 4.6) can be found in Borsuk (1933) and also in textbooks of topology such as Alexandroff, Hopf (1935) and Schurle (1979); the latter discusses related results in an informal way. Hadwiger (1959) provides a fine collection of two– and three–dimensional results similar in spirit to the Borsuk–Ulam theorem; he also includes elementary proofs. Applications of the Borsuk–Ulam theorem including the ham–sandwich theorem can be found in Schurle (1979) and in Mendelsohn (1962). A proof of the discrete version of the two–dimensional ham–sandwich theorem can be found in Willard (1982) where the generalization indicated in Exercise 4.5(b) in also included. He exploits this result in the design of erasing subdivisions with sublinear stabbing number. From a computational point of view, such subdivisions are data structures that represent finite point sets and solve simplicial range queries efficiently, that is, in linear storage and sublinear query time (see Chapter 14). Section 4.3 describes the improvement of Willard's construction given by Edelsbrunner, Welzl (1986c). A continuous version of Exercise 4.9(a) is proved in Buck, Buck (1948/49). Its discrete version is established in Edelsbrunner, Huber (1984) where the result is used in the construction of erasing cell complexes in E^3 (see also Exercise 4.12). A survey of so–called measures of symmetry including six–partitions is offered in Grünbaum (1961). Yaglom, Boltyanskii (1961) is another extensive collection of geometric results that can be proved by means of continuous functions.

The proof of Theorem 4.12 (the existence of four–secting planes) is taken from an unpublished manuscript of David Dobkin, Frances Yao, and the author of this book which includes a solution to Exercise 4.13: this result forms the basis of an erasing cell complex in E^3 with stabbing number in $O(n^{0.899})$. Hadwiger (1966) proves the existence of two four–secting planes for any two connected but not necessarily separated bodies in E^3; his techniques can be extended to solve Exercise 4.11.

Theorem 4.15 (the non–existing balanced dissection by five hyperplanes in E^5) is taken from Avis (1985). Dissections as indicated in Exercise 4.16 are given in Yao, Yao (1985); their construction implies $s^{(d)}(n)=O(n^\gamma)$, with $\gamma=\frac{\log_2(2^d-1)}{d}<1$, for every $d\geq 2$. A different construction in E^4 using parallel hyperplanes can be found in Cole (1985). Generalizations of this approach are described in Cole (1987). The currently best erasing cell complexes can be found in Haussler, Welzl (1987); their construction is indicated in Exercise 4.17.

CHAPTER 5

ZONES IN ARRANGEMENTS

In this chapter, we investigate the complexity of the boundary of a certain collection of cells in an arrangement of hyperplanes. We formalize the problem by introducing the notion of a so-called zone which is defined relative to some chosen hyperplane in the arrangement. Intuitively, the zone of a hyperplane h contains all faces in the boundaries of those cells which are supported by h. The introduction of this concept is motivated by an algorithm that constructs an arrangement incrementally, that is, the hyperplanes are inserted one after another (see Chapter 7). A more formal definition of the zone of a hyperplane in terms of visibility is as follows.

> Let h_0 be a hyperplane in an arrangement $\mathcal{A}(H)$. A k-face f for $0 \le k \le d-1$, is said to be *visible from* h_0 if there is a (possibly degenerate) line segment s that connects f and h_0 such that the relative interior ris of s is contained in h_0 or in a cell of $\mathcal{A}(H)$. It follows that all faces contained in h_0 are visible from h_0. The *zone of* h_0 *in* $\mathcal{A}(H)$ is the set of k-faces in $\mathcal{A}(H)$, $0 \le k \le d-1$, that are visible from h_0.

It is intuitively plausible (and will be established in Chapter 7) that h_0 can be inserted into $\mathcal{A}(H-\{h_0\})$ in a time proportional to the cardinality of the zone of h_0 in $\mathcal{A}(H)$. Thus, the time needed to construct an arrangement incrementally depends on the cardinality of the various zones that occur. The bounds on the size of a zone, to be established in this chapter, imply that the above-mentioned strategy leads to an optimal algorithm for constructing arrangements. Further implications of the bounds on the cardinality of zones include an upper bound on the number of different arrangements (see Section 7.6) and combinatorial results on arrangements (see Theorem 5.5 of this chapter and Theorem 6.20 in Chapter 6).

The organization of this chapter is as follows: Section 5.1 presents an alternate definition of the zone of a hyperplane and discusses its meaning for dual configurations. Bounds on the cardinality of zones are then established in

Sections 5.2 through 5.4. Finally, Section 5.5 focuses on an interesting combinatorial implication of those bounds.

5.1. Supported Cells, Zones, and Duality

Let H be a set of $n+1$ hyperplanes in E^d; for convenience we assume that H contains the hyperplane h_0: $x_d=0$. A cell c in $\mathcal{A}(H)$ is said to be *supported by* h_0 if h_0 intersects the closure $\mathrm{cl}c$ of c. Recall that two faces are said to be incident upon each other if their dimensions differ by exactly one, and if one face is contained in the boundary of the other face. It will turn out to be convenient to define a generalization of the incidence relation to faces whose dimensions differ by at least one. We say that a k-face f *bounds* a cell c if f is contained in $\mathrm{cl}c$. We call f_c a *k-border of* c and think of it as f in its character of bounding c; f_c is also said to *face* c. Subsequently, we say that f contains f_c and that any set that contains f also contains f_c. The number of k-borders of c is called the *k-degree* $deg_k(c)$ *of* c. The set of k-faces which bound cells that are supported by h_0, for $0 \leq k \leq d-1$, is called the *zone of* h_0.

This chapter is primarily concerned with the analysis of the cardinality of zones in arrangements. For convenience, we will examine a slightly different concept.

> We call a cell c *active (with respect to* h_0) if it lies above h_0 and if it is supported by h_0; a k-border f_c is termed *active* if cell c is active. Now, we define $s_k(H,h_0)$ as the number of active k-borders in $\mathcal{A}(H)$, for $0 \leq k \leq d-1$.

Figure 5.1 illustrates these definitions; the active regions are highlighted by shading. With H the set of 5 lines as shown, $s_0(H,h_0)=17$ and $s_1(H,h_0)=19$. The maximum of $s_k(H,h_0)$, for all sets H of $n+1$ hyperplanes in E^d including h_0, is denoted by $s_k^{(d)}(n)$. It is important to realize that the concept of a zone and of active cells can be defined with respect to any hyperplane of an arrangement, and that the restriction to the special hyperplane h_0 does not imply any loss of generality. By a straightforward perturbation argument, it is easy to see that there is a simple arrangement of $n+1$ hyperplanes in E^d which realizes $s_k^{(d)}(n)$. Notice that $s_k^{(d)}(n)$ is neither an upper bound nor a lower bound on the number of k-faces of the zone Z defined for hyperplane h_0 in any arrangement of $n+1$ hyperplanes in E^d. It fails to be an upper bound since the function accounts for only one side of h_0, while Z contains k-faces on both sides of h_0. This shows that $2s_k^{(d)}(n)$ is an upper bound on the number of k-faces of Z, however. Furthermore, $s_k^{(d)}(n)$ is not a lower bound since it counts active k borders rather than k faces bounding active cells. Since a k face in a simple d-dimensional

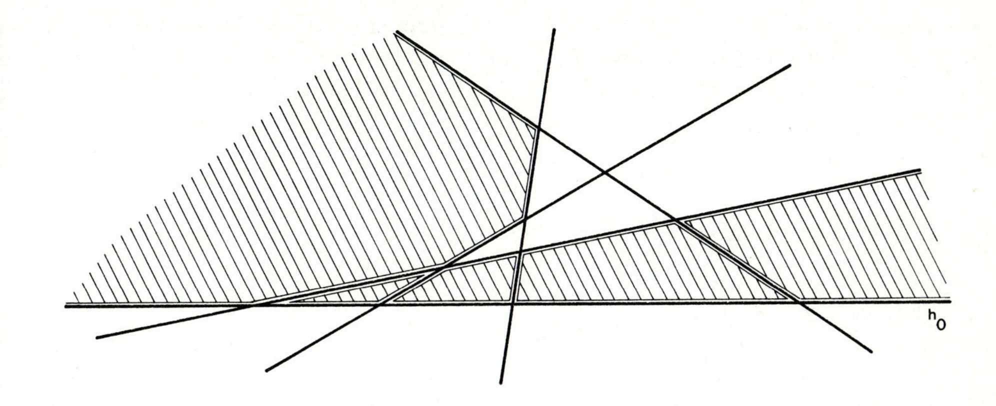

Figure 5.1. Polygons active with respect to h_0.

arrangement bounds 2^{d-k} cells, $s_k^{(d)}(n)/2^{d-k}$ is a lower bound on the number of k-faces of Z. Consequently, the maximum cardinality of any zone in an arrangement of $n+1$ hyperplanes in E^d is in

$$\Theta(\sum_{k=0}^{d-1} s_k^{(d)}(n)).$$

Before we begin with the analysis of $s_k^{(d)}(n)$, let us understand the meaning of supported cells, of zones, and of the function $s_k^{(d)}(n)$ in the dual setting of configurations of points.

As described in further detail in Section 1.4, the set

$$P = \mathcal{D}(H) = \{\mathcal{D}(h) = (\frac{\eta_1}{2}, \frac{\eta_2}{2}, \ldots, \frac{\eta_{d-1}}{2}, -\eta_d) | h\colon x_d = \eta_1 x_1 + \ldots + \eta_{d-1} x_{d-1} + \eta_d \text{ in } H\}$$

is a configuration dual to the arrangement $\mathcal{A}(H)$, provided H contains only non-vertical hyperplanes (as is assumed). Let q be a point in a cell c of $\mathcal{A}(H)$. Then the plane $\mathcal{D}(q)$, defined such that $q = \mathcal{D}(\mathcal{D}(q))$, induces a partition of P into points above and below $\mathcal{D}(q)$. In fact, every point in c induces the same partition since a point p in P is above the hyperplane $\mathcal{D}(q)$ if and only if point q is above the hyperplane $\mathcal{D}(p)$. Any hyperplane h in H that supports c corresponds to a point $p = \mathcal{D}(h)$ in P which is "reachable" from $\mathcal{D}(q)$, that is, $\mathcal{D}(q)$ can be rotated and translated continuously such that no point of P changes from being above $\mathcal{D}(q)$ to below $\mathcal{D}(q)$ or vice versa, and $\mathcal{D}(q)$ ultimately contains p. This leads us to the following conclusions.

Observation 5.1: Let H be a finite set of non-vertical hyperplanes in E^d including h_0: $x_d = 0$, and define $P = \mathcal{D}(H)$.

(i) A cell c in $\mathcal{A}(H)$ is supported by h_0 if and only if $\mathcal{D}(h_0)$ is reachable (in P) from $\mathcal{D}(q)$, for every point q in c.

(ii) A k-face f in $\mathcal{A}(H)$, for $0 \leq k \leq d-1$, is in the zone of h_0 if and only if the points of P contained in $\mathcal{D}(q)$, for any point q of f, are simultaneously reachable from a hyperplane that corresponds to a point of a supported cell in $\mathcal{A}(H)$.

Observation 5.1 implies an obvious interpretation of the cardinality of a zone within the setting of configurations.

5.2. Sweeping a Simple Arrangement

Sections 5.3 and 5.4 will present bounds on $s_k^{(d)}(n)$ using the technique of sweeping an arrangement with a hyperplane that does not belong to the arrangement. To reduce the length and complexity of this discussion, this section appropriately elaborates on the sweep–technique and the phenomena occurring during a sweep. It is sufficient to study the sweep–technique for simple arrangements, since there are simple arrangements of $n+1$ hyperplanes that realize $s_k^{(d)}(n)$ for some hyperplane. We will assume throughout that h_0: $x_d = 0$ is this hyperplane.

Let H be a set of $n \geq d$ hyperplanes in E^d such that $\mathcal{A}(H)$ is simple, and assume without loss of generality that no edge is parallel to the hyperplane $x_d = 0$. Furthermore, let $h(t)$: $x_d = t$ be an additional hyperplane which varies with the (time–) parameter t. Hyperplane $h(t)$ intersects $\mathcal{A}(H)$ in a $(d-1)$-dimensional arrangement of n $(d-2)$–flats in $h(t)$; we denote this arrangement by $\mathcal{A}_t(H)$. $\mathcal{A}_t(H)$ is simple if and only if $h(t)$ contains no vertex of $\mathcal{A}(H)$.

Let $h_i(t)$ be the intersection of $h(t)$ and the hyperplane h_i in H, for $1 \leq i \leq n$. In order to discuss the motion of $(d-2)$–flats within $h(t)$, we use the orthogonal projection of the coordinate system of E^d onto $h(t)$. As $h(t)$ sweeps upwards (that is, as t increases continuously), $h_i(t)$ moves parallel to its initial position in one direction. If h_i is vertical and thus $h_i(t)$ does not change with respect to the coordinate system in $h(t)$, then this is regarded as degenerate motion. Let $t_1 < t_2$ such that $h(t)$ encounters no vertex of $\mathcal{A}(H)$ as t changes from t_1 to t_2. The change from $\mathcal{A}_{t_1}(H)$ to $\mathcal{A}_{t_2}(H)$ is called a 0–*move*. Note that a 0–move does not change $\mathcal{A}_t(H)$ combinatorially, that is $\mathcal{A}_{t_1}(H)$ and $\mathcal{A}_{t_2}(H)$ are combinatorially equivalent. We now assume $t_1 < t_2$ such that $h(t_1)$ and $h(t_2)$ contain no vertex of $\mathcal{A}(H)$ and $h(t)$ contains a vertex u for exactly one point in time t, $t_1 < t < t_2$. Then $\mathcal{A}_{t_1}(H)$ is said to change in a 1–*move* to $\mathcal{A}_{t_2}(H)$. Combinatorially, $\mathcal{A}_{t_2}(H)$ differs from $\mathcal{A}_{t_1}(H)$ as follows.

If $h_{i_1}, h_{i_2}, \ldots, h_{i_d}$ denote the hyperplanes in H that contain u, then $h_{i_1}(t_1), h_{i_2}(t_1), \ldots, h_{i_d}(t_1)$ and $h_{i_1}(t_2), h_{i_2}(t_2), \ldots, h_{i_d}(t_2)$ define two bounded $(d-1)$-dimensional simplices s_1 and s_2 in $\mathcal{A}_{t_1}(H)$ and $\mathcal{A}_{t_2}(H)$, respectively. Since u lies between $h(t_1)$ and $h(t_2)$, s_2 is a negative homothet of s_1 (that is, there is a point p and a negative real number α such that $s_2 = p + \alpha s_1$).

To describe the changes caused by a 1-move appropriately, two relations on the k-borders of $\mathcal{A}(H)$ $(0 \le k \le d)$ are introduced. The first is an equivalence relation and is defined with respect to the k-faces of $\mathcal{A}(H)$ as follows.

Two k-faces f and $\bar{f}$ in $\mathcal{A}(H)$, $0 \le k \le d$, are said to be *equivalent*, denoted as $f \simeq \bar{f}$, if f and $\bar{f}$ satisfy the three conditions below, or if there is a sequence of k-faces commencing with f and ending with $\bar{f}$ such that the conditions below hold for any two consecutive k-faces of this sequence. The conditions are as follows:
(i) f and $\bar{f}$ are contained in a common k-flat,
(ii) f and $\bar{f}$ share a vertex u (whose d^{th} coordinate is v_d), and
(iii) f and $\bar{f}$ lie on different sides of hyperplane $h(t)$ with $t = v_d$.

By transitivity of the equivalence relation $\simeq$, all edges of a 1-flat in $\mathcal{A}(H)$ are equivalent. Furthermore, $h(t)$ intersects exactly one member of each equivalence class unless $h(t)$ contains a vertex of $\mathcal{A}(H)$. It is interesting to note that for $d = 2$ an equivalence class of regions of $\mathcal{A}(H)$ contains exactly the regions between two consecutive levels of $\mathcal{A}(H)$. (To make this statement true, levels have to be defined bottom-up instead of from left to right; see Section 3.2.) There seems to be no relationship between these concepts in dimensions higher than two. To extend the equivalence relation $\simeq$ to k-borders of $\mathcal{A}(H)$, for $k \le d-1$, we introduce the concept of a *cone of* a k-border f_c $(0 \le k \le d-1)$ which is the intersection of all open half-spaces which are bounded by hyperplanes through f and which contain cell c. The above equivalence relation can now be extended to borders as follows:

$$f_c \simeq \bar{f}_{\bar{c}} \text{ if } f \simeq \bar{f} \text{ and the cones of } f_c \text{ and of } \bar{f}_{\bar{c}} \text{ are the same.}$$

Intuitively, two borders are equivalent if they are contained in equivalent faces and if they face the same set of directions. The notion of equivalence is important because equivalent faces and equivalent borders will not be distinguished during a sweep of $h(t)$. In order to illustrate the concept of equivalence between borders, Figure 5.2 displays the crucial portion of a 1-move in E^3. Equivalent 1- and 2-borders are indicated by labeling the intersections with $h(t_1)$ and $h(t_2)$ identically.

The second relation used to describe the actions that are caused by a 1-move associates equivalence classes of k-borders with equivalence classes of $(d-k)$-borders, for $1 \le k \le d-1$. The associated pairs are defined with respect to vertices in $\mathcal{A}(H)$. Intuitively, a class of k-borders and a class of $(d-k)$-borders are associated with respect to a vertex u if they "cross" at u and there is an overlap in the set of directions they face.

Let u be a vertex in $\mathcal{A}(H)$ and let F be an equivalence class of k–borders and G be a equivalence class of $(d-k)$–borders of $\mathcal{A}(H)$, $1 \leq k \leq d-1$. F and G are said *to swap at* u, denoted as $F \sim_u G$, if F contains two k–borders f_c and $\bar{f}_{\tilde{c}}$, $c \neq \tilde{c}$, and G contains two $(d-k)$–borders $g_{\tilde{c}}$ and $\bar{g}_{\bar{c}}$, such that u is the topmost vertex of k–face f, $(d-k)$–face g, and cell c, and u is the bottommost vertex of k–face $\bar{f}$, $(d-k)$–face $\bar{g}$, and cell $\bar{c}$.

If follows that cells c and $\bar{c}$ are equivalent, and that both f and g bound c and both $\bar{f}$ and $\bar{g}$ bound $\bar{c}$. Furthermore, cells c and $\tilde{c}$ share g in their boundaries, and cells $\bar{c}$ and $\tilde{c}$ share $\bar{f}$ in their boundaries. Notice in particular that f and $\bar{g}$ bound two different cells c and $\bar{c}$ which belong to the same equivalence class, and that cell $\tilde{c}$ is common to borders $\bar{f}_{\tilde{c}}$ and $g_{\tilde{c}}$. Intuitively, this means that classes F and G exchange the currently faced equivalence class of cells as $h(t)$ sweeps through vertex u. The intersection of cell c and hyperplane $h(t)$ is a $(d-1)$–simplex s that collapses as $h(t)$ approaches u. As $h(t)$ continues its sweep, a new $(d-1)$–simplex $\bar{s}$ appears in $h(t)$; it is the intersection of $h(t)$ with cell $\bar{c}$. Within the $(d-1)$–dimensional world of $h(t)$, the intersections of faces f and g with $h(t)$ bound the collapsing simplex s. These intersections are such that their closures do not intersect.

To enumerate swapping classes of k– and $(d-k)$–borders, for $k=1,2$, shown in Figure 5.2, let us name a class by its intersection with $h(t_1)$ or $h(t_2)$. Then $F \sim_u G$ for all (F,G) in $\{(A_3,a_2),(B_3,b_2),(C_3,c_2),(a_1,A_1),(b_1,B_1),(c_1,C_1)\}$. Since a $(d-1)$–dimensional simplex is bounded by a total of 2^d-2 $(k-1)$–faces, with k ranging from 1 to $d-1$, we conclude the following result.

Observation 5.2: Let $\mathcal{A}(H)$ be a simple arrangement of at least d hyperplanes in E^d such that, for no real number t, the hyperplane $h(t)$: $x_d = t$ contains an edge of $\mathcal{A}(H)$. A vertex u gives rise to 2^d-2 pairs (F,G) of equivalence classes of borders such that $F \sim_u G$.

Since the two–dimensional case is of particular importance in Section 5.3, we briefly review the definitions for E^2. Each line in an arrangement $\mathcal{A}(H)$ contains the 1–borders of two equivalence classes: the 1–borders of one class face left, and those from the other class face right. For u a vertex and F, G two equivalence classes of 1–borders, we have $F \sim_u G$ if and only if

(i) lines h_i and h_j in H that intersect at u contain F and G, and
(ii) all 1–borders of F and of G face left or all face right.

When $h(t)$ sweeps through a vertex u of $\mathcal{A}(H)$, the equivalence classes of 1–borders on one line through u swap with the equivalence classes of 1–borders on the other line through u.

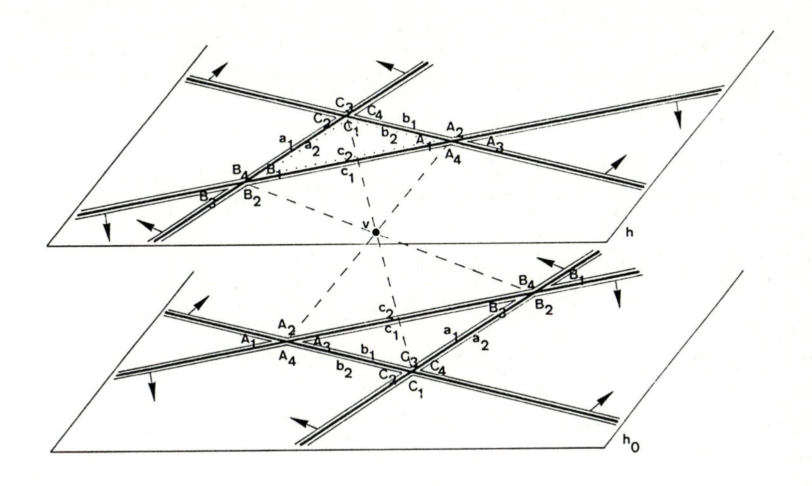

Figure 5.2. A 1–move in three dimensions.

5.3. Tight Bounds in the Plane

Using the sweep–technique already described, this section establishes tight bounds for $s_0^{(2)}(n)$ and for $s_1^{(2)}(n)$. (Recall that $s_k^{(2)}(n)$ denotes the maximum number of k–borders that are active with respect to one of $n+1$ lines in an arrangement in E^2, for $k=0,1$.) By a straightforward perturbation argument, only simple arrangements need be considered. In particular, we prove the following result.

Theorem 5.3: $s_0^{(2)}(n)=5n-3$ and $s_1^{(2)}(n)=5n-1$, for $n\geq 1$.

Proof: Let $H=\{h_0,h_1,...,h_n\}$ be a set of $n+1$ lines in E^2 such that $\mathcal{A}(H)$ is simple. Without loss of generality, we choose h_0: $x_2=0$ and we assume that no two vertices of $\mathcal{A}(H-\{h_0\})$ share the same x_2-coordinate. We will first argue that $s_1^{(2)}(H,h_0)\leq 5n-1$. We will then demonstrate that this bound is tight, and finally, we will show that $s_0^{(2)}(n)=s_1^{(2)}(n)-2$.

Observe that h_0 contains $n+1$ active 1–borders since it intersects n lines. It remains to be shown that there are at most $4n-2$ active 1–borders that are not contained in h_0. To this end, we perform a continuous, upward directed sweep with a horizontal line $h(t)$: $x_2=t$. Initially, at time $t=0$, $h(t)=h_0$, and at each point in time $t>0$, $h(t)$ intersects $\mathcal{A}(H)$ in a one–dimensional arrangement $\mathcal{A}_t(H)$. Let $p_i(t)$ denote the intersection of $h(t)$ with h_i, for $1\leq i\leq n$, and let B_i^L and B_i^R denote the two classes of 1–borders of $\mathcal{A}(H-\{h_0\})$ on h_i facing left and right, respectively. Figure 5.3 provides an illustration without highlighting the equivalence classes of faces and borders.

At each point in time $t>0$, when $h(t)$ contains no vertex of $\mathcal{A}(H)$, an equivalence class B of 1–borders in $\mathcal{A}(H-\{h_0\})$ is in one of three states. Let $b(t)$ be the 1–border in B such that the edge that contains $b(t)$ intersects $h(t)$. Then the states are assigned as follows.

(i) B is *alive* or *live* if $b(t)$ is active,

(ii) B is *dead* if there are two lines h_i, h_j in $H-\{h_0\}$ such that the intersection u of h_i and h_j lies between h_0 and $h(t)$, h_i or h_j contains $b(t)$, and the cone of $b(t)$ contains the sector defined by h_i and h_j that has u as the bottommost vertex; note that death is irreversible in time, and

(iii) B is *sleeping*, otherwise.

Intuitively, B is sleeping when it traverses a "dead sector" of $\mathcal{A}(H)$ and still has a chance to leave it and become alive again. In Figure 5.3, B_3^L and B_1^R are alive, B_3^R, B_2^L, B_2^R, B_4^R, and B_1^L are dead, and B_4^L is sleeping. During the sweep of $h(t)$, the states change only when two points p_i and p_j in $\mathcal{A}_t(H)$ are swapped, that is, when a 1–move occurs. The following rules concerning the initial and ultimate distribution of states can be observed.

R1: Initially, all $2n$ equivalence classes of 1–borders of $\mathcal{A}(H-\{h_0\})$ are alive.

R2: Ultimately, at least two equivalence classes are alive.

Now, let $h(t)$ pass the intersection of h_i and h_j in a 1–move such that p_i is to the left of p_j before they swap. After the move, the states of B_i^L and B_j^L depend only on their states before the 1–move. The same is true for the states of B_i^R and B_j^R,

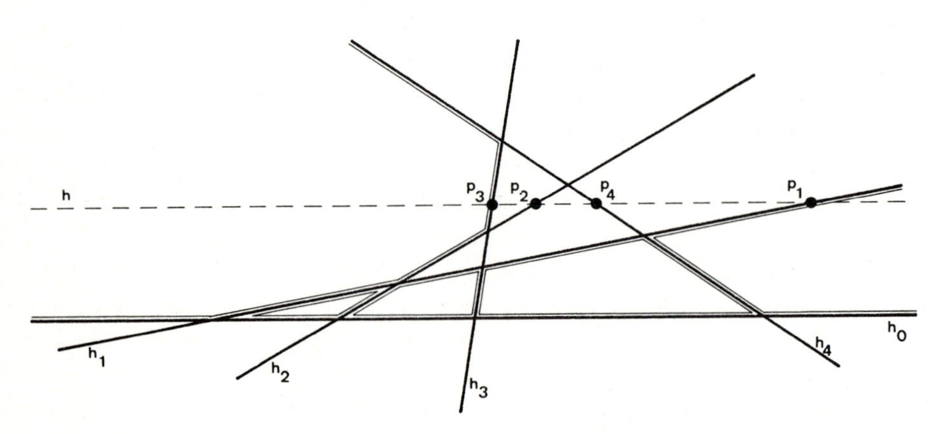

Figure 5.3. Bottom–up sweep.

which are not discussed because the rules they obey are completely symmetric to those for the states of B_i^L and B_j^L. The rules that prescribe the way in which states change are shown in Table 5.4; they follow immediately from the definitions of the states "alive", "dead", and "sleeping". It is important to realize that a class of 1–borders involved in a 1–move which remains or becomes alive causes a live or a sleeping class of 1–borders to die. A refinement of this observation will finally lead us to a counting argument which completes the proof. Note also that there are geometric reasons which prohibit B_i^L being alive or sleeping if B_j^L is dead before the 1–move. Figure 5.5 illustrates the changes of states given in Table 5.4: the ranges during which classes of 1–borders are alive, sleeping, and dead are indicated by solid, dashed, and dotted lines, respectively.

To exploit the rules for an upper bound on the number of active 1–borders that are not contained in h_0, we use four counters to keep track of the sweep as follows:

> ACT designates the current number of active 1–borders which lie entirely below $h(t)$ and which are not contained in h_0, and
> A, S, and D designate the current numbers of equivalence classes of 1–borders which are alive, sleeping, and dead, respectively.

Initially, $ACT=0$, $A=2n$, $S=0$, and $D=0$, by R1, and ultimately, $A \geq 2$, by

Rule	before		after	
R3	alive	alive	dead	alive
R4	dead	alive	dead	sleeping
R5	alive	sleeping	dead	alive
R6	dead	sleeping	dead	sleeping
	dead	dead	dead	dead
R7	sleeping	sleeping	dead	sleeping

Table 5.4. Changes of states in a 1–move.

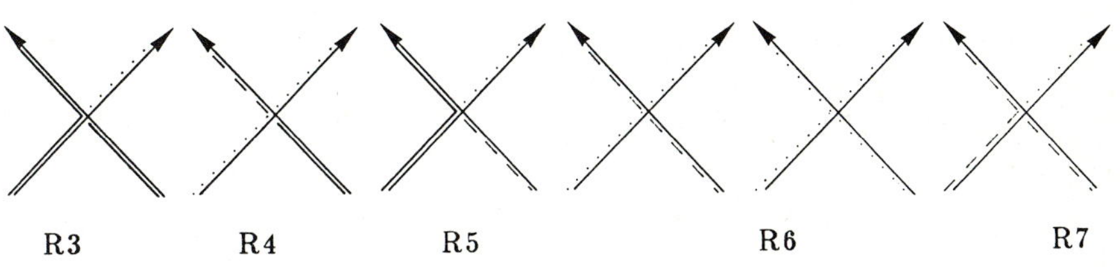

Figure 5.5. Illustration of rules R3 through R7.

R2. Application of rules R3 through R7 causes the changes that are delineated in Table 5.6. The transitions flow only from alive to dead (R3), from alive to sleeping (R4), and from sleeping to dead (R5 and R7). Whenever ACT increases, there is a transition from A to S or D, or from S to D. Both the alive to dead and the alive to sleeping to dead paths increase ACT by at most two. Let U be the ultimate size of A. Since $A=2n$ initially, ACT can grow to at most $4n-2U$. At the end of the sweep when no more vertices are to be passed, ACT denotes the number of active 1-borders below $h(t)$, and U active 1-borders still intersect $h(t)$. The total number of active 1-borders, not counting the $n+1$ 1-borders in h_0, is therefore $(4n-2U)+U$, and this number is maximal if U is as small as possible. By rule R2, $U\geq 2$ which establishes

$$s_1(H,h_0)\leq(n+1)+(4n-2)=5n-1.$$

The arrangement depicted in Figure 5.1 shows that this upper bound can be achieved for $n=4$. In this example, all intersections lie above line h_0, and each line intersects the other lines in the order of their slopes. Evidently, this construction can be generalized to an arbitrary number of n lines. This shows $s_1^{(2)}(n)=5n-1$.

Finally, to establish $s_0(H,h_0)=s_1(H,h_0)-2$, observe that $deg_0(c)=deg_1(c)$, for c a bounded region, and $deg_0(c)=deg_1(c)-1$, for c unbounded. At least the leftmost and the rightmost active regions are unbounded, so $s_0(H,h_0)\leq s_1(H,h_0)-2$, and the obvious generalization of the arrangement in Figure 5.1 shows

$$s_0^{(2)}(n)=s_1^{(2)}(n)-2=5n-3.$$

□

It is useful to note that Theorem 5.3 also holds for arrangements of pseudo-lines (see Section 2.4). The proof remains the same except that the sweep is performed with a pseudo-line.

Rule	ACT	A	S	D
R3	$ACT+2$	$A-1$	S	$D+1$
R4	$ACT+1$	$A-1$	$S+1$	D
R5	$ACT+1$	A	$S-1$	$D+1$
R6	ACT	A	S	D
R7	ACT	A	$S-1$	$D+1$

Table 5.6. The evolution of the counters.

5.4. Asymptotically Tight Bounds in d Dimensions

Bounds on $s_k^{(d)}(n)$, for $0 \le k \le d-1$ and $d \ge 3$, can be established using the ideas applied in the proof of Theorem 5.3. We will not establish tight bounds (although our results will be tight in the asymptotic sense) since the additional number of dimensions increases the complexity of the problem considerably. In particular we show the following result, which is trivial if $d=1$ and which follows from Theorem 5.3 if $d=2$.

Theorem 5.4: $s_k^{(d)}(n)=\Theta(n^{d-1})$, for $d \ge 1$ and $0 \le k \le d-1$.

Proof: Let $H=\{h_0,h_1,...,h_n\}$ be a set of $n+1$ hyperplanes in E^d such that $\mathcal{A}(H)$ is simple, such that h_0 is given by $x_d=0$, and such that no two vertices of $\mathcal{A}(H-\{h_0\})$ share the same x_d-coordinate. Observe that there is no loss of generality as far as an upper bound on $s_k(H,h_0)$, for $0 \le k \le d-1$, is concerned. $\mathcal{A}(H-\{h_0\})$ intersects h_0 in a simple $(d-1)$-dimensional arrangement of n $(d-2)$-flats. This arrangement consists of $\Omega(n^{d-1})$ k-faces, for $0 \le k \le d-1$. Since each such k-face contains at least one active k-border, we have $s_k^{(d)}(n)=\Omega(n^{d-1})$.

For a proof of the upper bound, an upward directed sweep with the hyperplane $h(t)$: $x_d=t$ is performed. At time $t=0$, $h(t)=h_0$, and at each point in time $t>0$, $h(t)$ intersects $\mathcal{A}(H)$ in a $(d-1)$-dimensional arrangement $\mathcal{A}_t(H)$ (see Section 5.2 for further details). If $h(t)$ contains no vertex of $\mathcal{A}(H)$ then $\mathcal{A}_t(H)$ is simple and $h(t)$ intersects a member of each equivalence class of k-borders in $\mathcal{A}(H-\{h_0\})$, for $1 \le k \le d-1$. Depending on the position of $h(t)$, each equivalence class B of k-borders in $\mathcal{A}(H-\{h_0\})$ is in one of three states. Let $b(t)$ be the k-border in B that intersects $h(t)$. Then the states are assigned as follows:

(i) B is *alive* if $b(t)$ is active,

(ii) B is *dead* if there are d hyperplanes in $H-\{h_0\}$ that intersect in a common vertex u between h_0 and $h(t)$, $b(t)$ is contained in the intersection of $d-k$ of these hyperplanes, and the cone of $b(t)$ contains the sector defined by the d hyperplanes which has u as its bottommost vertex (death is thus irreversible in time), and

(iii) B is *sleeping*, otherwise.

The states change only at 1-moves during the sweep, as shown for a 1-move in E^3 in Figure 5.2. Let B and C be two classes of borders which swap at vertex u (consult Section 5.2 for a definition of swapping equivalence classes of borders) and consider the 1-move at which $h(t)$ passes u. It is not hard to see that the states of B and C change according to rules R3 through R7 given in Table 5.4 and illustrated in Figure 5.5. However, rules R1 and R2 no longer apply in $d \ge 3$ dimensions. It will be sufficient for our purposes to replace R1 and R2 by rules

5.7. Bibliographic Notes

Most of the material contained in this section is taken from Edelsbrunner, O'Rourke, Seidel (1986) which includes the two main results (described as Theorems 5.3 and 5.4) as well as Theorem 5.5. Theorem 5.3 was independently derived by Chazelle, Guibas, Lee (1985), who also found Theorem 5.5 restricted to two dimensions. Part of the terminology for proving Theorems 5.3 and 5.4, such as the concept of a 1-move, is taken from Canham (1971). A brief account of arrangements of pseudo-hyperplanes and a definition of this concept that is slightly more general than the one given in the exercise section of this chapter can be found in Goodman, Pollack (1981b). A generalization of the left-facing forest of an arrangement in the plane, defined in Exercise 5.6, has been used in Edelsbrunner, Guibas (1986) as a data structure which supports an algorithmic sweep of an arrangement with a pseudo-line (see also the exercises in Chapter 7).

CHAPTER 6

THE COMPLEXITY OF FAMILIES OF CELLS

It will become apparent in Chapters 7 through 15 of this book that many computational problems in geometry can be solved by constructing arrangements of hyperplanes. Unfortunately, the rather large number of faces of arrangements entails the use of large amounts of storage. It is thus advantageous to construct only part of an arrangement whenever the problem at hand admits it. Examples of useful structures in arrangements are single cells or faces (see Chapter 8), zones (see Chapters 5 and 7), stabbing regions (see Chapter 15), and levels (see Chapters 3, 9, and 13).

To support the analysis of algorithms that construct substructures of an arrangement, this chapter investigates the complexity of collections of cells chosen from some arrangement of hyperplanes. Here, the complexity of a cell is defined as the number of faces contained in its boundary. The complexity of a collection of cells is defined as the sum of the complexities of all cells in the collection. The analysis is given in terms of the number of hyperplanes that define the arrangement and in terms of the cardinality of the collection. Notice that Chapter 5 studies a related problem, namely the complexity of the collection of cells supported by a distinguished hyperplane of the arrangement. It turns out that we can use some results of Chapter 5 to derive an upper bound for the more general problem considered in this chapter.

The results of this chapter include upper and lower bounds on the maximum complexity of a single cell as well as upper and lower bounds on the maximum complexity of collections of non-trivially many cells. In Section 6.2, we consider single cells in E^d and we prove asymptotically tight upper bounds on the number of k-faces of a cell with n facets, for $0 \leq k \leq d-2$. We do this by proving Euler's relation (Section 6.2.2) and the Dehn–Sommerville relations for convex polytopes (Section 6.2.3). To show that the upper bounds are asymptotically tight, we describe so-called cyclic polytopes in E^d, that is, polytopes with particularly many faces (Section 6.2.1). In Sections 6.3 through 6.6, we demonstrate various upper and lower bounds on the maximum number of facets of a collection of m

cells in an arrangement of n hyperplanes in E^d: Section 6.3 gives a tight upper bound for $m \leq (1+\sqrt{2n+1})/2$ regions in E^2, and Sections 6.4 through 6.6 demonstrate bounds for $d \geq 2$ dimensions that cover all values of m.

6.1. Definitions and Preliminary Results

This section presents a few straightforward results on the complexity of families of cells in arrangements of hyperplanes. This complexity is expressed by combinatorial functions which we now define in a formal way.

Let H be a set of n hyperplanes in E^d, and let C be a collection of cells in the arrangement $\mathcal{A}(H)$ defined by H. For $0 \leq k \leq d-1$, we define

$$\mathbf{a}_{C,k}(H) = \sum_{c \in C} deg_k(c),$$

where $deg_k(c)$ denotes the number of k-faces which bound cell c. Thus, $\mathbf{a}_{C,k}(H)$ counts the k-faces that bound cells in C, and a k-face is counted i times if it bounds i cells in C. Furthermore, we define

$$\mathbf{a}_{m,k}(H) = \max\{\mathbf{a}_{B,k}(H) | B \text{ a collection of } m \text{ cells in } \mathcal{A}(H)\}, \text{ and}$$

$$\mathbf{a}_{m,k}^{(d)}(n) = \max\{\mathbf{a}_{m,k}(G) | G \text{ a set of } n \text{ hyperplanes in } E^d\}.$$

For convenience, we write $\mathbf{a}_k(H)$ for $\mathbf{a}_{1,k}(H)$ and $\mathbf{a}_k^{(d)}(n)$ for $\mathbf{a}_{1,k}^{(d)}(n)$.

For example, $\mathbf{a}_{d-1}(H)$ is the largest number of facets of any cell in arrangement $\mathcal{A}(H)$, and $\mathbf{a}_{d-1}^{(d)}(n)$ is the largest number of facets of any cell in any arrangement of n hyperplanes in E^d.

By convexity of the cells in $\mathcal{A}(H)$, each hyperplane can contain at most one facet of any cell c in $\mathcal{A}(H)$. This implies $\mathbf{a}_{d-1}^{(d)}(n) \leq n$. To demonstrate a matching lower bound on $\mathbf{a}_{d-1}^{(d)}(n)$, let each hyperplane of H support the unit sphere in E^d given by the equation

$$x_1^2 + x_2^2 + \ldots + x_d^2 = 1.$$

Each hyperplane in H contains a facet of the cell that contains the origin. The following is thus an immediate result.

Observation 6.1: $\mathbf{a}_{d-1}^{(d)}(n) = n$, for $d \geq 2$.

Now let C be a collection of m cells in an arrangement $\mathcal{A}(H)$ of n hyperplanes in E^d such that

$$\mathbf{a}_{C,k}(H) = \mathbf{a}_{m,k}^{(d)}(n).$$

If $\mathcal{A}(H)$ is not simple, then there are $d-k$ hyperplanes that do not intersect in a common k-flat, for some $-1 \leq k \leq d-2$. The $d-k$ hyperplanes can be perturbed such that their common intersection is a k–flat and such that the number of k–faces of no cell in C decreases. Repeated application of this operation leads to the following result.

Observation 6.2: For any integer numbers k and m, with $0 \leq k \leq d-1$ and $1 \leq m \leq f_d^{(d)}(n)$, there is a set H of n hyperplanes in E^d such that $\mathcal{A}(H)$ is simple and $a_{m,k}(H) = a_{m,k}^{(d)}(n)$.

Recall that $f_k^{(d)}(n)$ denotes the number of k–faces in a simple arrangement of n hyperplanes in E^d, and that $f_k^{(d)}(n) = \Theta(n^d)$.

Observe that a k–face in a simple arrangement in E^d is contained in the intersection of $d-k$ hyperplanes which partition E^d into 2^{d-k} cells. As a consequence, each k–face belongs to the boundary of exactly 2^{d-k} cells. If C is now the collection of all $f_d^{(d)}(n)$ cells of a simple arrangement $\mathcal{A}(H)$ defined by n hyperplanes in E^d, then

$$a_{C,k}(n) = 2^{d-k} f_k^{(d)}(n),$$

for $0 \leq k \leq d-1$. In this case, C trivially achieves the optimum which implies the following result.

Observation 6.3: $a_{m,k}^{(d)}(n) \leq 2^{d-k} f_k^{(d)}(n)$, for $0 \leq k \leq d-1$, and equality holds if and only if $m = f_d^{(d)}(n)$.

We will see in the forthcoming sections of this chapter that the analysis of $a_{m,k}^{(d)}(n)$ is less straightforward for choices of m and k that are different from those taken in this section.

6.2. The Complexity of a Polytope

This section investigates the maximum number of faces of a convex polyhedron defined as the intersection of n closed half–spaces in d–dimensional Euclidean space. Notice that a single face f in an arrangement of hyperplanes in E^d is the relative interior of a polyhedron and that f is the (non–empty) interior of a polyhedron if f is a cell. In order to find upper bounds on the number of faces of a convex polyhedron with a given number of facets, we can restrict our attention to bounded polyhedra, that is, to convex polytopes. The investigations in this section will enhance our knowledge about the combinatorial function $a_k^{(d)}(n) = a_{1,k}^{(d)}(n)$ in a way that will be of particular importance in Chapter 8, where we talk about constructing convex hulls of a finite set of points. Indeed, the convex hull of a finite point set is a convex polytope.

The structure of this section is as follows. In Section 6.2.1, we construct so-called cyclic polytopes and prove some extremum properties. In particular, we prove that every k vertices of a cyclic polytope in E^d form the vertices of a $(k-1)$-face of the polytope, for $0 \le k \le d/2$. In Section 6.2.2, we give a proof of Euler's relation on the number of faces of a convex polytope. This relation is known to be the only linear relation for the number of faces of various dimensions which holds for all convex polytopes. If the polytope is simple (that is, if each k-face is contained in the closures of exactly $d-k$ facets), then the counts of faces of various dimensions also obey the Dehn–Sommerville relations presented in Section 6.2.3. Finally, in Section 6.2.4, we discuss an asymptotic version of the upper bound theorem which gives an upper bound on the number of k-faces of a convex polytope with n facets, for each $0 \le k \le d-2$.

6.2.1. Cyclic Polytopes

This section aims at constructing convex polytopes in E^d that have n vertices and many k-faces, for small integer numbers k. Surprisingly, we will see that there are polytopes such that every k vertices span a $(k-1)$-face if $k \le d/2$. By the duality results of Section 1.6, such polytopes correspond to polytopes with n facets such that the closures of every k facets intersect in a unique $(d-k)$-face. We begin by describing a special class of convex polytopes, the so-called cyclic polytopes.

The *moment curve* M_d *in* E^d is the set of points

$$x(\tau) = (\tau^1, \tau^2, \ldots, \tau^d),$$

for τ ranging over all real numbers. Now, let h be a hyperplane in E^d given by the equation

$$\eta_0 + \eta_1 x_1 + \eta_2 x_2 + \ldots + \eta_d x_d = 0.$$

A point $x(\tau)$ on M_d lies in h if and only if

$$\eta_0 + \eta_1 \tau + \eta_2 \tau^2 + \ldots + \eta_d \tau^d = 0.$$

A polynomial of degree d has at most d roots which implies that the above equation has at most d solutions. This proves the following result.

Lemma 6.4: A hyperplane in E^d intersects the moment curve M_d in at most d points.

Lemma 6.4 implies that any $d+1$ points on M_d are affinely independent. Next, we introduce the central concept of this section:

a convex polytope in E^d is called a *cyclic polytope* if it is the convex hull of a finite set of at least $d+1$ points on M_d.

Let P be the convex hull of set

$$X = \{x(\tau_1), x(\tau_2), \ldots, x(\tau_n)\}$$

and assume without loss of generality that $\tau_1 < \tau_2 < \ldots < \tau_n$. Now let I be a set of $k \leq d/2$ indices between 1 and n, limits included, and define the polynomial

$$p_I(\tau) = \prod_{i \in I} (\tau - \tau_i)^2.$$

The degree of $p_I(\tau)$ is equal to $2k \leq d$ which implies that there is a vector $u = (v_1, v_2, \ldots, v_d)$ and a real number v_0 such that

$$p_I(\tau) = \langle u, t \rangle - v_0,$$

with $t = (\tau, \tau^2, \ldots, \tau^d)$. Obviously, $p_I(\tau) = 0$ if $\tau = \tau_i$, $i \in I$, and $p_I(\tau) > 0$, otherwise. This implies that the moment curve M_d is contained in the closed half-space of points x that satisfy

$$\langle u, x \rangle \geq v_0$$

and that a point $x(\tau)$ of M_d belongs to the hyperplane $\langle u, x \rangle = v_0$ if and only if $\tau = \tau_i$, for some $i \in I$. It is now easy to prove the following result.

Theorem 6.5: A cyclic polytope defined as the convex hull of n points on the moment curve M_d in E^d has $\binom{n}{k}$ $(k-1)$-faces, for $0 \leq k \leq d/2$.

Proof: Let X be the set of n points on M_d that defines the cyclic polytope. For any k points of X, we find a hyperplane through the k points such that all other points of M_d lie on one side of the hyperplane. It follows that the relative interior of the convex hull of these k points is a face of the cyclic polytope. By Lemma 6.4, the k points are affinely independent which implies that they define a $(k-1)$-face. The assertion follows since X has exactly $\binom{n}{k}$ subsets of size k. □

Note that the dual transform D_o, defined in Section 1.6, can be used to map a cyclic polytope P with n vertices to another polytope $D_o(P)$ with n facets. Every k-face of P corresponds to a unique $(d-k-1)$-face of $D_o(P)$ which implies the following result.

Theorem 6.6: For every integer $n \geq d+1$, there is a convex polytope with n facets in E^d that has exactly $\binom{n}{k}$ $(d-k)$-faces, for $0 \leq k \leq d/2$.

maximized by simple polytopes, we have the following upper bound.

Theorem 6.11: A convex polytope with non-empty interior and n facets in E^3 has at most $3n-6$ edges and at most $2n-4$ vertices.

6.2.4. An Asymptotic Version of the Upper Bound Theorem

Using the Dehn-Sommerville relations, stated as Theorem 6.10, we derive asymptotic upper bounds on the number of k-faces of a simple polytope P with n facets. The amount of effort necessary to show these bounds is considerably less than that required to prove tight exact upper bounds, a result known as the upper bound theorem for convex polytopes (see also Exercise 6.3(a)).

The most important ingredients in our development are the Dehn-Sommerville relations which we restate:

$$\sum_{i=0}^{k}(-1)^i\binom{d-i}{d-k}deg_i(P)=deg_k(P),$$

for $0\leq k\leq d$. For the rest of this section, let $r=\lfloor\frac{d-1}{2}\rfloor$. Next, we express the values of $deg_0(P)$ through $deg_r(P)$ in terms of the values of $deg_{r+1}(P)$ through $deg_d(P)$:

$$-deg_k(P)+\sum_{i=0}^{k}(-1)^i\binom{d-i}{d-k}deg_i(P)=0,$$

for $0\leq k\leq r$, and

$$-deg_k(P)+\sum_{i=0}^{r}(-1)^i\binom{d-i}{d-k}deg_i(P)=\sum_{i=r+1}^{k}(-1)^{i+1}\binom{d-i}{d-k}deg_i(P),$$

for $r+1\leq k\leq d$.

It is interesting to note that $\binom{n}{d-k}$ is a trivial upper bound on the number of k-faces of P, and as we have seen in Section 6.2.1, this upper bound is tight for $r+1\leq k\leq d$. As a consequence, the right hand side of the above relations are bounded from above by $c\cdot n^{\lfloor d/2\rfloor}$, for some positive constant number c.

If we define $\binom{i}{j}=0$ for $i<j$, then we can restate the above system of relations as

$$-deg_k(P)+\sum_{i=0}^{r}(-1)^i\binom{d-i}{d-k}deg_i(P)=b_k(P),$$

for $0\leq k\leq d$, where $b_k(P)\leq c\cdot n^{\lfloor d/2\rfloor}$. The system consists of $d+1$ linear relations and contains only $r+1$ unknown variables, namely $deg_0(P)$ through $deg_r(P)$.

Without proof, we state that the system determines the unknown variable, that is, there is a subsystem of $r+1$ independent relations (see also Exercise 6.2(c)).

To obtain an asymptotic upper bound for $deg_i(P)$, $0 \leq i \leq r$, we solve the system of relations under the assumption that the $b_k(P)$ are such that $deg_i(P)$ comes out as a maximum. For every $0 \leq i \leq r$, we get

$$deg_i(P) = \sum_{j=0}^{d} c_{i,j} \cdot b_j(P),$$

where the $c_{i,j}$ are real numbers that depend only on d which is taken as a constant. Thus, we have

$$deg_i(P) \leq c_i \cdot n^{\lfloor d/2 \rfloor},$$

for some constant number c_i, which implies the main result of this section.

Theorem 6.12: $a_k^{(d)}(n) = O(n^{\min\{d-k, \lfloor d/2 \rfloor\}})$, for $d \geq 2$ and $0 \leq k \leq d-1$.

Note that Theorem 6.12 agrees with Theorem 6.11 for the case $d=3$. Since each k-face of a simple polytope is incident upon a constant number of $(k+1)$-faces, for $0 \leq k \leq d$, Theorem 6.12 implies that the number of incident pairs of faces in a simple polytope with n facets is in $O(n^{\lfloor d/2 \rfloor})$. Using a straightforward perturbation argument, it can be shown that the same bound holds for arbitrary convex polytopes with n facets.

6.3. The Complexity of a Few Cells in Two Dimensions

Apart from the case $m=1$, matching upper and lower bounds for $a_{m,k}^{(d)}(n)$ are known if

$$d=2,\ k=1, \text{ and } 4\binom{m}{2} \leq n.$$

In other words, tight upper bounds on the number of edges of m regions in an arrangement of n lines in the plane are known if $4\binom{m}{2}$ does not exceed n. Since the number of vertices of a bounded region is equal to the number of edges, $a_{m,0}^{(2)} = a_{m,1}^{(2)}(n)$, provided $a_{m,1}^{(2)}(n)$ can be obtained with m bounded regions which in fact is the case if $m \geq 3$ or if $n \geq 5$.

Let H be a set of $n \geq 4$ lines in E^2 and let c_1 and c_2 be two regions in $\mathcal{A}(H)$. By convexity of c_1 and c_2, there are at most four lines tangent to both regions (see Figure 6.1); hence $a_{2,1}^{(2)}(n) \leq n+4$ is immediate. The bound is tight since all other lines in H (if there are any) can be chosen to contain an edge either of c_1 or of c_2. This observation can be generalized to the following result.

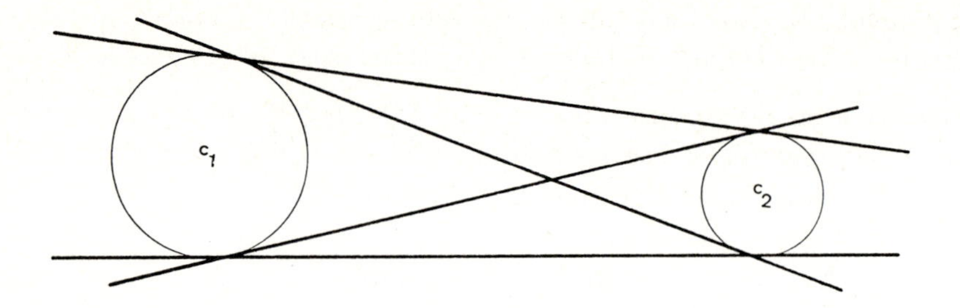

Figure 6.1. Two regions have at most four common tangents.

Theorem 6.13: $\boldsymbol{a}^{(2)}_{m,1}(n) = n + 4\binom{m}{2}$, for $m \geq 2$ and $n \geq 4\binom{m}{2}$.

Proof: Let H be a set of n lines in E^2 and let C be a family of m regions in the arrangement $\mathcal{A}(H)$. We say that a line h *contributes to* a region c if h contains an edge of c. The number of regions in C to which h contributes is called the *contribution* $contr(h)$ *of* h. With this notation we have

$$\boldsymbol{a}_{C,1}(H) = \sum_{h \in H} contr(h).$$

We prove below that this sum is maximized if $contr(h) = 2$, for $4\binom{m}{2}$ lines h, and $contr(h) = 1$, for the remaining $n - 4\binom{m}{2}$ lines. The upper bound of the assertion then follows.

Let $\mathcal{G} = (C, A)$ be a multigraph with the arc $\{c_i, c_j\}$ appearing k times in A if there are k lines in H that contribute to both c_i and c_j. The following observations are straightforward:

1. any two nodes of $\mathcal{G}$ are connected by at most four arcs, so $\text{card}A \leq 4\binom{m}{2}$, and

2. any line h in H gives rise to $\binom{contr(h)}{2}$ arcs in A, so

$$\sum_{h \in H} \binom{contr(h)}{2} \leq 4\binom{m}{2}.$$

The sum of contributions is maximized under the above constraint only if the contributions of two lines does not differ by more than one. By $\text{card}H = n \geq 4\binom{m}{2}$, $contr(h) = 1$ or 2, for each line in H, and without violating $\text{card}A \leq 4\binom{m}{2}$, we have $contr(h) = 2$, for at most $4\binom{m}{2}$ lines of H. An alternate way to finish the proof of the upper bound is to use the Cauchy–Schwarz inequality (see Exercise 6.11(a)).

To prove the lower bound, we place the centers of m sufficiently small discs on a sufficiently large circle and we draw the four common tangents for each pair of discs. This construction gives an arrangement of $4\binom{m}{2}$ lines such that the m regions that contain the interiors of the m discs have a total of $8\binom{m}{2}$ edges. We draw each one of the remaining $n-4\binom{m}{2}$ lines such that it increases the total count by one. □

It is not hard to show

$$a_{m,1}^{(d)}(n) < n+4\binom{m}{2},$$

for $n < 4\binom{m}{2}$ which is equivalent to $m > (1+\sqrt{2n+1})/2$. In fact, as m grows, the upper bound becomes rapidly worse. Also extensions to three and higher dimensions are not without problems. For example, two cells in three dimensions can share an arbitrary number of common tangent planes, which contrasts the two-dimensional result where the number of common tangent lines is bounded from above by four. Consequently, for any positive integer number n there is an arrangement of n planes in E^3 and two cells in the arrangement with n facets each (see also Exercise 6.4).

6.4. Lower Bounds for Moderately Many Cells

This section offers an asymptotic lower bound on $a_{m,d-1}^{(d)}(n)$ which extends the bound of Section 6.3 from two to three and higher dimensions. The lower bound of this section as well as that of the next section are given by constructions of the kind described below.

1. We describe an arrangement $\mathcal{A}(G)$ of at most $n/2$ hyperplanes and pick a set V of m vertices in $\mathcal{A}(G)$.
2. We replace each vertex v in V by the ball $b(v)$ with radius ϵ and center v. Furthermore, we replace each hyperplane g in G that contains a vertex v of V by two parallel hyperplanes that are tangent to ball $b(v)$.

This method yields an arrangement $\mathcal{A}(H)$ which contains a cell for each vertex v in V. Let $deg(v)$ denote the number of hyperplanes in G that contain vertex v in $\mathcal{A}(G)$. If ϵ is chosen small enough then

$$\sum_{c \in C} deg_{d-1}(c) = 2 \sum_{v \in V} deg(v),$$

for C the set of m cells in $\mathcal{A}(H)$ that contain the balls $b(v)$, $v \in V$.

This method of constructing suitable arrangements motivates the introduction of the following notation: let G be a set of n hyperplanes in E^d and let V be a subset of the vertices in $\mathcal{A}(G)$. We define

$$v_V(G) = \sum_{v \in V} deg(v),$$

$$v_m(G) = \max\{v_W(G) \mid W \text{ contains } m \text{ vertices of } \mathcal{A}(G)\}, \text{ and}$$

$$v_m^{(d)}(n) = \max\{v_m(F) \mid F \text{ a set of } n \text{ hyperplanes in } E^d\}.$$

The above implies the following relationship between the combinatorial functions $a_{m,d-1}^{(d)}(n)$ and $v_m^{(d)}(n)$.

Lemma 6.14: $a_{m,d-1}^{(d)}(n) \geq 2v_m^{(d)}(\lfloor n/2 \rfloor)$.

The remainder of this section provides an asymptotic lower bound on $a_{m,d-1}^{(d)}(n)$ using Lemma 6.14.

Since $deg_{d-1}(c) \leq n$, it is obvious that for each cell c in an arrangement of n hyperplanes, $a_{m,d-1}^{(d)}(n) = O(mn)$. For $d \geq 2$ and $m = O(n^{d-2})$, there is a matching asymptotic lower bound.

Theorem 6.15: $a_{m,d-1}^{(d)}(n) = \Omega(mn)$, for $d \geq 2$ and $m = O(n^{d-2})$.

Proof: We show $v_m^{(d)}(n) = \Omega(mn)$, for $m = O(n^{d-2})$. This result is trivial for $d=2$ since $n^{d-2}=1$, in this case. So we assume $d \geq 3$ and we construct a set $G = \{g_0, g_1, \ldots, g_{n-1}\}$ of hyperplanes which implies the claimed lower bound. The construction in three dimensions is essentially a row of p vertices such that all but p hyperplanes in G contain all p vertices, for some integer number p to be determined later. In $d > 3$ dimensions, p^{d-2} vertices are chosen on a $(d-2)$-dimensional orthogonal grid. Below, we give a formal description of the construction.

Define $p = \min\{\lfloor m^{1/(d-2)} \rfloor, \lfloor n/2(d-2) \rfloor + 1\}$ and write $i = (a-1)p + b$, with $0 \leq b \leq p-1$, that is, $a-1$ is the number of times p is contained in i and b is the remainder. For $0 \leq i \leq (d-2)p-1$ we have $1 \leq a \leq d-2$, and we set

$$g_i\colon x_a = b.$$

Observe that all hyperplanes defined so far are normal to the $(d-2)$-flat $f\colon x_{d-1} = x_d = 0$. For the remaining values of i, that is, for $(d-2)p \leq i \leq n-1$, we set

$$g_i\colon x_{d-1} + i x_d = 0.$$

All hyperplanes defined for $i \geq (d-2)p$ contain the $(d-2)$-flat f. Finally, notice that all hyperplanes in G are distinct.

The set $V = \{(v_1, v_2, \ldots, v_{d-2}, 0, 0) \mid v_i = 0, 1, \ldots, p-1 \text{ for } 1 \leq i \leq d-2\}$ contains p^{d-2} vertices of $\mathcal{A}(H)$. By definition of p, we have

$$p^{d-2} \leq m \text{ and } p^{d-2} = \Omega(m).$$

The number of vertices in V is thus not too large and still of the right order of magnitude. The lower bound argument is still valid if the number of vertices in V is less than m. Finally, each vertex v in V is contained in each hyperplane g_i, $(d-2)p \leq i \leq n-1$, and it is contained in exactly $d-2$ hyperplanes g_i, with $0 \leq i \leq (d-2)p-1$. Consequently,

$$deg(v) = n-(d-2)(p-1) \geq \frac{n}{2},$$

for each vertex v in V. This implies $\boldsymbol{v}_V(G) = \Omega(mn)$ and $\boldsymbol{v}_m^{(d)}(n) = \Omega(mn)$. The assertion follows by Lemma 6.14. □

The lower bound for $m = \Theta(n^{d-2})$ stated in Theorem 6.15 also holds as a lower bound if we choose m in $\Omega(n^{d-2})$.

Corollary 6.16: $\boldsymbol{a}_{m,d-1}^{(d)}(n) = \Omega(n^{d-1})$, for $d \geq 2$ and $m = \Omega(n^{d-2})$.

Although the argument for Corollary 6.16 is rather crude, it is interesting that there is no better lower bound known as long as m is in $O(n^{d-3/2})$. In fact, we have seen in Section 6.3 that the result is best possible in two dimensions.

6.5. Lower Bounds for Many Cells

Next, we give non-trivial lower bounds on $\boldsymbol{a}_{m,d-1}^{(d)}(n)$, for m in $\Omega(n^{d-3/2})$. We begin with the case $d=2$, which turns out to be crucial in our development, and then extend the result to higher dimensions.

To do the counting part of the proof below, we need two results from number theory concerning relatively prime pairs of integer numbers, that is, pairs of integer numbers that have no common non-trivial prime factor. For every positive integer number s, let $\phi(s)$ denote the number of integer numbers r, with $1 \leq r \leq s$, such that r and s are relatively prime. For example, we have $\phi(10) = 4$, since 10 is relatively prime with 3, 5, 7, and 9. Function $\phi(s)$ is known as the Euler function. The following bounds on sums of Euler functions are known.

Lemma 6.17: Let n be a positive integer number. Then

$$\text{(i) } \sum_{s=1}^{n} \phi(s) = \Theta(n^2), \text{ and (ii) } \sum_{s=1}^{n} s\phi(s) = \Theta(n^3).$$

The lower bound that we derive for the maximum number of edges in the boundary of m regions in an arrangement of n lines in two dimensions is as follows.

Theorem 6.18: $a_{m,1}^{(2)}(n)=\Omega(m^{2/3}n^{2/3})$.

Proof: We show $v_m^{(2)}(n)=\Omega(m^{2/3}n^{2/3})$ by explicit construction of an arrangement $\mathcal{A}(G)$ of n lines in the plane. Lemma 6.14 then implies the theorem. For the construction, set $p=\lfloor m^{1/2}\rfloor$ and assume $n\geq 4p$, omitting only uninteresting cases.

Define $V=\{(a,b)|1\leq a\leq p, 1\leq b\leq p\}$, a set of p^2 points with integer coordinates. For any line h, we call card$(h\cap V)$ the *contribution of* h, denoted *contr*(h). To make the points of V vertices of $\mathcal{A}(G)$, we define

$$G_1=\{x_1=a\,|1\leq a\leq p\}, \text{ and } G_2=\{x_2=b\,|1\leq b\leq p\},$$

two sets of p lines each, and we let G_1 and G_2 be subsets of G.

To complete the specification of set G, we introduce a few definitions. For two integer numbers r and s, we write $(r,s)=1$ if r and s are relatively prime. To improve the notation, we let $g(i,j,s,r)$ be the line passing through the points with coordinates (i,j) and $(i+s,j+r)$. Finally, we define $f(n,m)=c_0(n/p)^{1/3}$, with c_0 a suitable positive constant to be specified later. Now, we set $G=G_1\cup G_2\cup G_3$ with $G_3=\{g(i,j,s,r)|1\leq j\leq \lfloor p/2\rfloor,\ 1\leq s\leq f(n,m),\ 1\leq i\leq s,$ and $1\leq r\leq s$ with $(r,s)=1\}$. Note that all lines in G_3 are distinct. We have

$$\text{card}\,G_3=\lfloor p/2\rfloor \sum_{s=1}^{f(n,m)} s\phi(s),$$

and by Lemma 6.17(ii), card$G_3=\Theta(n)$. In addition, we can guarantee card$G_3\leq n-2p$ if we choose the constant c_0 appropriately.

Line $g(i,j,s,r)$ contains a point from every s^{th} column starting from the ith column. By the choice of i, the contribution of line $g(i,j,s,r)$ in G_3 is therefore at least $\lfloor \frac{p}{s}\rfloor$. The overall contribution of G_3 is thus at least

$$\lfloor\frac{p}{2}\rfloor \sum_{s=1}^{f(n,m)} \lfloor\frac{p}{s}\rfloor s\phi(s)\geq \lfloor\frac{p}{2}\rfloor^2 \sum_{s=1}^{f(n,m)} \phi(s),$$

which is in $\Theta(p^{4/3}n^{2/3})$, by Lemma 6.17(i). Recall that $p=\Theta(m^{1/2})$, by definition, and consequently,

$$v_m^{(2)}(n)=\Omega(m^{2/3}n^{2/3}),$$

which implies the assertion. □

To generalize the two-dimensional result of Theorem 6.18 to three and higher dimensions, we use a construction similar to that employed in the proof of Theorem 6.15. Intuitively, we choose hyperplanes normal to a given (two-dimensional) plane f such that each plane parallel to f intersects the hyperplanes in a set of lines that define an arrangement of the type described in the proof of Theorem 6.18. Let $m = \Omega(n^{d-2})$, for $d \geq 3$, and set

$$p = \lfloor \frac{m}{(n/(d-1))^{d-2}} \rfloor.$$

We obtain $v_m^{(d)}(n) = \Omega(m^{2/3} n^{d/3})$ from $v_p^{(2)}(n) = \Omega(p^{2/3} n^{2/3})$ by the following construction of a d-dimensional arrangement $\mathcal{A}(G)$, where the set G of hyperplanes is best described as the disjoint union of sets $G_0, G_1, \ldots, G_{d-2}$.

1. The set G_0 contains

$$n-(d-2)\lfloor \frac{n}{d-1} \rfloor \leq \frac{n}{d-1}$$

hyperplanes normal to the (two-dimensional) plane

$$f: x_1 = x_2 = \ldots = x_{d-2} = 0,$$

such that the intersection of the arrangement $\mathcal{A}(G_0)$ with plane f is a two-dimensional arrangement $\mathcal{A}(G_f)$ with $v_p(G_f) = \Omega(p^{2/3} n^{2/3})$ (see the proof of Theorem 6.18). Notice that the two-dimensional arrangement obtained by intersecting $\mathcal{A}(G_0)$ with any plane parallel to f is combinatorially the same as $\mathcal{A}(G_f)$.

2. The remainder of the hyperplanes come in $d-2$ sets, where each set contains $\lfloor \frac{n}{d-1} \rfloor$ parallel hyperplanes. Formally, we define

$$G_i = \{x_i = j \mid j = 0, 1, \ldots, \lfloor \frac{n}{d-1} \rfloor - 1\},$$

for $1 \leq i \leq d-2$.

Note that any $d-2$ hyperplanes, one from each set G_i, $1 \leq i \leq d-2$, intersect in a unique common plane f' parallel to f such that $v_p(G_{f'}) = \Omega(p^{2/3} n^{2/3})$, where $G_{f'} = \{g \cap f' \mid g \in G_0\}$. There are $\lfloor \frac{n}{d-1} \rfloor^{d-2} = \Theta(n^{d-2})$ ways to define such a plane f'. We therefore conclude that

$$v_{m'}(G) = \Omega(p^{2/3} n^{2/3} n^{d-2}),$$

for $m' = p \lfloor \frac{n}{d-1} \rfloor^{d-2} = \Theta(m)$. Since $p^{2/3} n^{d-4/3} = \Theta(m^{2/3} n^{d/3})$, this implies the following result.

Theorem 6.19: $a_{m,d-1}^{(d)} = \Omega(m^{2/3} n^{d/3})$.

The bounds on $a_{m,d-1}^{(d)}(n)$ derived in this section and in Sections 6.3 and 6.4 are illustrated in Figure 6.2 for $d \geq 3$ and in Figure 6.3 for $d = 2$. The figures also indicate the upper bounds of Section 6.6.

6.6. Upper Bounds for Many Cells

We have seen in Section 6.4 that $O(mn)$ is the most obvious asymptotic upper bound for $a_{m,d-1}^{(d)}(n)$, and it is tight for $m = O(n^{d-2})$ (see Theorem 6.15). This section demonstrates two upper bounds which improve on this trivial upper bound. The first result improves the $O(mn)$ upper bound whenever $m = \Omega(n^{d-2})$ while the second result holds only in two dimensions where it improves the trivial upper bound proved in Section 6.3 if $m = \Omega(n^{1/2})$.

The first bound applies to any number d of dimensions and is proved using Theorem 5.5.

Theorem 6.20: $a_{m,d-1}^{(d)}(n) = O(m^{1/2}n^{d/2})$.

Proof: Let C be a collection of m cells in an arrangement $\mathcal{A}(H)$ of n hyperplanes in E^d. Recall that

$$a_{C,d-1}(H) = \sum_{c \in C} deg_{d-1}(c).$$

Next, let C_H be the set of all cells in $\mathcal{A}(H)$. By Theorem 5.5 and since $C \subseteq C_H$, we have

$$\sum_{c \in C} (deg_{d-1}(c))^2 \leq \sum_{c \in C_H} (deg_{d-1}(c))^2 = O(n^d).$$

If we want to maximize $a_{C,d-1}(H)$, this constraint forces all cells to have about equally many facets. That is,

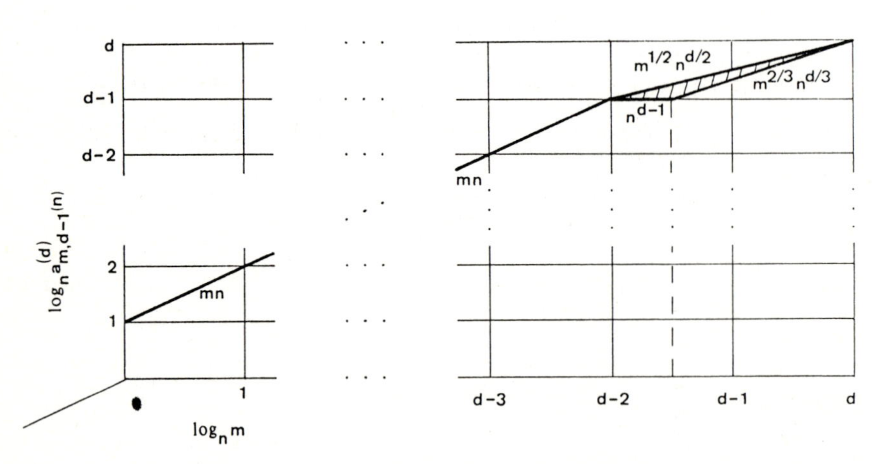

Figure 6.2. Asymptotic results on $a_{m,d-1}^{(d)}(n)$.

$$deg_{d-1}(c) = O((n^d/m)^{1/2}),$$

for each cell c in C. The assertion now follows immediately. Alternatively, we could have used the Cauchy–Schwarz inequality to finish the proof. □

Figure 6.2 illustrates Theorem 6.20 along with all asymptotic bounds on $\boldsymbol{a}_{m,d-1}^{(d)}(n)$ derived in the preceding sections. It thus represents our knowledge about the asymptotic behavior of the combinatorial function and it makes the gap for the range $m = \Omega(n^{d-2})$ obvious. The asymptotic results are displayed using logarithms to the base n of m and of $\boldsymbol{a}_{m,d-1}^{(d)}(n)$.

In two dimensions, the upper bound of Theorem 6.13 can be used to improve Theorem 6.20 if $m = O(n)$. In asymptotic notation, Theorem 6.13 reads

$$\boldsymbol{a}_{m,1}^{(2)}(n) = O(n), \text{ if } m = O(\sqrt{n}).$$

We use this together with a straightforward decomposition argument to show the following result.

Theorem 6.21: $\boldsymbol{a}_{m,1}^{(2)}(n) = O(m \cdot n^{1/2})$, for $m = \Omega(n^{1/2})$.

Proof: Let C be a collection of m regions in an arrangement $\mathcal{A}(H)$ of n lines in E^2. If $m \leq p = \lfloor (1+\sqrt{2n+1})/2 \rfloor$, then Theorem 6.13 implies the assertion. Otherwise, decompose C into $\lceil m/p \rceil$ disjoint subsets of at most p regions each. By Theorem 6.13

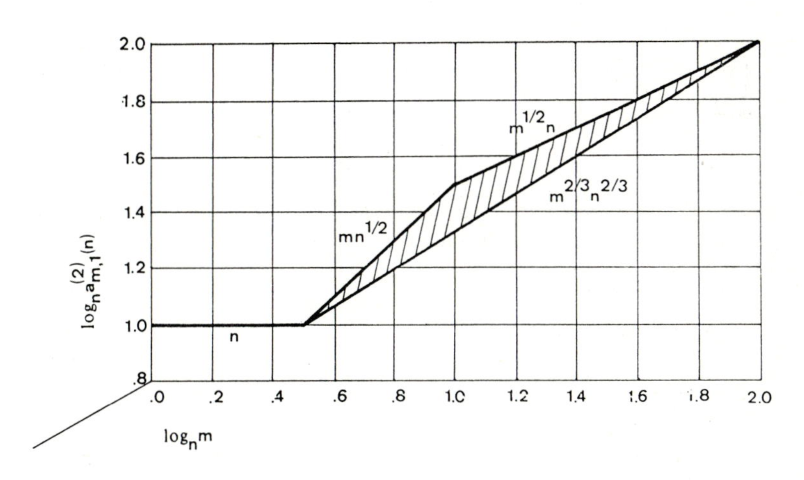

Figure 6.3. Asymptotic behavior of $\boldsymbol{a}_{m,1}^{(2)}(n)$.

$$a_{C',1}(H)=O(n),$$

for each set C' in the decomposition of C. The assertion now follows since $\lceil m/p \rceil = O(m/n^{1/2})$. □

Theorem 6.21 together with all asymptotic results on $a_{m,1}^{(2)}(n)$ demonstrated in Sections 6.3 through 6.6 are illustrated in Figure 6.3. To display the result, we use logarithms to the base n of m and of $a_{m,1}^{(2)}(n)$.

6.7. Exercises and Research Problems

3 **Exercise 6.1:** Prove that Euler's relation (Theorem 6.8) is the only linear relation satisfied by the numbers $deg_{-1}(P)$ through $deg_d(P)$ of every convex polytope P in d dimensions.

Exercise 6.2: Recall that the Dehn–Sommerville relations (Theorem 6.10) for simple convex polytopes in d dimensions form a system of $d+1$ linear relations which are not independent.

3 (a) Prove that the largest independent subsystem consists of at least $\lfloor d/2 \rfloor$ relations. *(Hint: the degrees $deg_0(P)$ through $deg_{d-1}(P)$ of a convex polytope P can be interpreted as the coordinates of a point in d dimensions; now take d cyclic polytopes with $d+1, d+2, \ldots, \lfloor 3d/2 \rfloor + 1$ vertices and prove that their associated points are affinely independent.)*

2 (b) Prove that the largest independent subsystem of the Dehn–Sommerville relations consists of at most $\lfloor d/2 \rfloor$ relations.

3 (c) Prove that the system of relations that is obtained from the Dehn–Sommerville relations by deleting every other relation, starting with the first one, gives an independent system for $deg_0(P)$ through $deg_r(P)$, with $r = \lfloor d-1/2 \rfloor$.

4 **Exercise 6.3:** (a) Prove the following tight upper bounds on the number of k-faces of any convex polytope with n facets in d dimensions: for $2 \le d \le n-1$ and for $0 \le k \le d-2$,

$$a_k^{(d)}(n) = \sum_{i=0}^{\lfloor d/2 \rfloor} \binom{i}{k} \binom{n-d+i-1}{i} + \sum_{i=0}^{\lfloor (d-1)/2 \rfloor} \binom{d-1}{k} \binom{n-d+i-1}{i}.$$

(Comment: the result is known as the upper bound theorem for convex polytopes.)

4 (b) Prove that every simple convex polytope with non-empty interior and n facets in E^d has at least

$$(d-1)n-(d+1)(d-2)$$

vertices and at least

$$\binom{d}{k+1} n - \binom{d+1}{k+1} (d-1-k)$$

k-faces, for $1 \le k \le d-2$. *(Comment: this result is known as the lower bound theorem for convex polytopes.)*

3 **Exercise 6.4:** Prove $a_{m,2}^{(3)}(n) = mn$, for $m \le n+1$ and $n \ge 3$.

Problem 6.5: Define $c_n^{(d)}(n)$ to be the largest number of cells in any arrangement of n hyperplanes in E^d that have a facet in every hyperplane. Notice that $a_{m\ d-1}^{(d)}(n) = mn$, for $m \leq c_n^{(d)}(n)$. It is fairly easy to prove $c_n^{(2)}(n)=1$, for $n \geq 5$. A solution to Exercise 6.4 implies $c_n^{(3)}(n) \geq n+1$.

4 (a) Prove

$$c_n^{(d)}(n) \geq \sum_{i=1}^{\lfloor d/2 \rfloor} \binom{n+1}{d-2i},$$

for $d \geq 2$ and $n \geq d+1$.

5 (b) Determine $c_n^{(d)}(n)$, $d \geq 3$ and $n \geq d+1$.

2 **Exercise 6.6:** Let $c_k^{(d)}(n)$ be the maximum number of cells c with $deg_{d-1}(c)=k$ in a simple arrangement of $n \geq d$ hyperplanes in d-dimensional Euclidean space. Prove that $a_{m,d-1}^{(d)}(n)=2f_{d-1}^{(d)}(n)-d(f_d^{(d)}(n)-m)$, for $f_d^{(d)}(n)-c_d^{(d)} \leq m \leq f_d^{(d)}(n)$.

3 **Exercise 6.7:** (a) Prove $c_2^{(2)}(n)=2\lfloor (n-1)/2 \rfloor+1$, for $n \geq 3$. (See Exercise 6.6 for the definition of $c_k^{(d)}(n)$.)

3 (b) Prove $c_3^{(2)}(n) \geq \frac{n^2}{3}+cn$, for some real constant c. *(Hint: for $n=3^m$ set $k=n/3$ and choose three collections, each of k parallel lines, such that the bounded regions of the arrangement are some number of hexagons and $2k^2-8$ triangles; in a second step, slightly perturb all lines such that each collection of k lines defines an arrangement that is of the same type as the global arrangement.)*

2 (c) Prove $c_d^{(d)}(n)=O(n^{d-1})$.

Exercise 6.8: Prove Lemma 6.17, that is, prove

3 (a) that the number of relatively prime pairs of integer numbers (r,s), with $1 \leq r \leq s \leq n$, is in $\Theta(n^2)$, and

3 (b) that the number of relatively prime pairs of integer numbers (r,s), with $1 \leq r \leq s \leq n$, is in $\Theta(n^3)$ if each such pair is counted s times.

5 **Problem 6.9:** (a) Prove or disprove $a_{m,d-1}^{(d)}(n)=O(n^{d-1})$ if $m=O(n^{d-3/2})$, $d \geq 3$.

4 (b) Prove $v_m^{(2)}(n)=O(m^{2/3}n^{2/3})$, for $m=\Omega(n^{1/2})$.

5 (c) Prove or disprove $v_m^{(d)}(n)=O(m^{2/3}n^{d/3})$, for $d \geq 3$ and $m=\Omega(n^{d-3/2})$.

5 (d) Prove or disprove $a_{m,d-1}^{(d)}(n)=O(v_m^{(d)}(n))$.

5 (e) Prove or disprove $a_{m,d-1}^{(d)}(n)=O(m^{2/3}n^{d/3})$.

5 **Problem 6.10:** Derive non-trivial upper and lower bounds on $a_{m,k}^{(d)}(n)$, for $m>1$ and $k<d-1$. *(Comment: all bounds given in this chapter either consider only one cell or they count only facets of collections of cells.)*

2 **Exercise 6.11:** (a) Prove the Cauchy-Schwarz inequality, that is, show that

$$\left(\sum_{i=1}^{n} a_i b_i\right)^2 \leq \left(\sum_{i=1}^{n} a_i^2\right)\left(\sum_{i=1}^{n} b_i^2\right),$$

where a_i and b_i are arbitrary real numbers.

3 (b) An *r-uniform hypergraph* is a pair (N,A), where N is a finite set of nodes and A is a set of subsets of N such that each subset has cardinality r. The members of A are called *hyperarcs*. We denote (N,A) as $K_r(m)$ if card$N=rm$, and there is a partition

$$N=N_1 \cup N_2 \cup \ldots \cup N_r$$

such that a hyperarc a is in A if and only if it contains a node from each N_i, for $1 \leq i \leq r$. Prove that a hypergraph with n nodes contains at most $O(n^{r-(1/m)^{r-1}})$ hyperarcs unless it contains $K_r(m)$ as a subhypergraph.

6.8. Bibliographic Notes

For general references on combinatorial investigations of convex polytopes we recommend Grünbaum (1967), McMullen, Shepard (1971), and Bronsted (1983). Each of the three books covers the contents of Section 6.2 and gives many additional results. Grünbaum (1967) is an especially broad and deep treatment of the subject and offers detailed historical remarks.

Cyclic polytopes described in Section 6.2.1 were found by Carathéodory (1907, 1911) and were later rediscovered by Gale (1955). The discovery of Euler's relation for the number of k-faces of a convex polytope in three dimensions is usually attributed to Euler (1752/53a, 1752/53b). According to Steinitz (1939) and Steinitz, Rademacher (1934), who discuss convex as well as non-convex polyhedra in three dimensions, René Descartes had a proof of the relation about 100 years before Ludwig Euler did. The generalization of the relation to four and higher dimensions was first performed by Schläfli (1901). The elementary proof of Euler's relation presented in Section 6.2.2 is taken from Nef (1981, 1984). The origins of the Dehn-Sommerville relations, presented in Section 6.2.3, can be found in Dehn (1905), where up to five dimensions are considered, and in Sommerville (1927), where the result is generalized to arbitrary dimensions. Our treatment of the Dehn-Sommerville relations follows the exposition in Bronsted (1983). A celebrated result in the theory of convex polytopes is the upper bound theorem, stated as Exercise 6.3(a), which was first proved by McMullen (1971b). An alternate proof of this theorem can be found in Alon, Kalai (1985). A proof of the so-called lower bound theorem, stated as Exercise 6.3(b), was first given by Barnette (1971, 1973). This theorem gives tight lower bounds on the number of k-faces of a convex polytope with n facets.

The literature on bounds for the maximum number of facets of a collection of cells in an arrangement of hyperplanes is less rich and comparably recent. The tight upper bound on the total number of edges of $m \leq (1+\sqrt{2n+1})/2$ regions in an arrangement of n lines in E^2 is taken from Canham (1969). To prove this bound, we used the Cauchy-Schwarz inequality that can be found in Halmos (1958) and in other textbooks of linear algebra. A more direct proof can be based on an upper bound for the number of hyperarcs of certain hypergraphs given in Erdös (1964); he solves Exercise 6.11(b). Some extensions of this result to three dimensions can be found in White (1939), in Edelsbrunner, Haussler (1986), and in Roudneff (1986) who gives a solution to Exercise 6.4 and Problem 6.5(a). Jean-Pierre Roudneff conjectures that the lower bound on $c_n^{(d)}(n)$ stated in Problem 6.5(a) is tight; a confirmation of this conjecture would settle Problem 6.5(b).

The treatment of asymptotic upper and lower bounds on $a_{m,d-1}^{(d)}(n)$ follows and extends Edelsbrunner, Welzl (1986b) and Edelsbrunner, Haussler (1986). The proofs of Theorems 6.18 and 6.19 use a number-theoretic result (Lemma 6.17) proved for example in Hardy, Wright (1965). A solution to Problem 6.9(b) can be found in Szemerédi, Trotter (1983); the result implies that no improvement of the asymptotic lower bound presented as Theorem 6.18 is possible with the approach taken in Sections 6.4 and 6.5. It is interesting to note that the same problem, that is, counting the maximum number of incidences between a set of lines and a set of points, is applied to proving lower bounds on the complexity of a range search problem (see Fredman (1980)).

Investigations of the maximal number of k-gons in an arrangement of n lines (see Exercises 6.6 and 6.7) are reported in Grünbaum (1972), Strommer (1977), Purdy (1979, 1980), and Füredi, Palásti (1984) who solve Exercise 6.7(b).

PART II

FUNDAMENTAL GEOMETRIC ALGORITHMS

An algorithm can be fundamental because it solves a fundamental problem or because it is the expression of a fundamental technique. We believe that the algorithms of Part II are fundamental for both reasons. We remark that several prerequisites are necessary to appreciate the material of Part II. These prerequisites include some familiarity with basic algorithms for sorting, searching, and for traversing graphs. It is also useful to know the basics about data structures.

CHAPTER 7

CONSTRUCTING ARRANGEMENTS

This chapter investigates the problem of constructing an arrangement of hyperplanes, that is, of creating a structure which represents all faces and all incidences between faces of the arrangement. While it might be true that this problem is interesting in its own right, the applications of an algorithm for this problem, demonstrated in Chapters 12 and 13, make it one of the most fundamental problems in computational geometry. An algorithm that constructs an arrangement $\mathcal{A}(H)$ can be used to answer many questions about the set H of hyperplanes, about the dual configuration $\mathcal{D}(H)$, and about other related concepts.

The algorithm to be described in this chapter constructs an arrangement $\mathcal{A}(H)$ in $O(n^d)$ time, if H is a set of n hyperplanes in E^d. It builds a structure which represents each face and each incidence between two faces explicitly. Notice that $O(n^d)$ is optimal for constructing such a representation of $\mathcal{A}(H)$ since $\mathcal{A}(H)$ consists of $\Omega(n^d)$ faces unless it is highly degenerate. The first step in the description of the algorithm is the specification of a data structure which represents $\mathcal{A}(H)$. This issue is dealt with in Section 7.1. Sections 7.2 through 7.5 give the details of the algorithm. Finally, Section 7.6 analyzes the amount of time needed for the construction, using the combinatorial results of Chapter 5.

7.1. Representing an Arrangement in Storage

Let H be a set of n hyperplanes in E^d. To store the arrangement $\mathcal{A}(H)$, we use the so-called incidence graph $\mathcal{I}(H)$ of $\mathcal{A}(H)$, as defined below. In addition to the regular faces of $\mathcal{A}(H)$ (also called proper faces) we define two *improper faces*: the (-1)-*face* $\emptyset$ and the $(d+1)$-*face* $\mathcal{A}(H)$. For convenience we define the (-1)-face to be incident upon all vertices of $\mathcal{A}(H)$, and the $(d+1)$-face to be incident upon all cells of $\mathcal{A}(H)$. For f a k-face of $\mathcal{A}(H)$ and g an incident $(k+1)$-face, we call f a *subface of* g and g a *superface of* f. The *incidence graph* $\mathcal{I}(H)$ *of*

$\mathcal{A}(H)$ can now be defined as follows:

> for each proper and improper face f of $\mathcal{A}(H)$, $\mathcal{I}(H)$ contains a node $\nu(f)$ that represents f; if two faces f and g are incident upon each other then $\nu(f)$ and $\nu(g)$ are connected by an arc.

Figure 7.1 shows the arrangement of two non-parallel lines and the corresponding incidence graph. By Theorem 1.1 and Observation 6.3, the incidence graph of an arrangement of n hyperplanes in E^d contains $O(n^d)$ nodes and arcs.

When implemented, each node $\nu(f)$ of an incidence graph is a record that contains some auxiliary information to be described, and two lists containing pointers to the subfaces and superfaces of f, respectively. Figure 7.2 shows the implementation of node $\nu(v)$ which represents vertex v shown in Figure 7.1. Note that the pointers to the subfaces and the superfaces of a node are well distinguished, and that both types of incident faces can thus be accessed separately. For the time being, we assume that the auxiliary information stored in a node $\nu(f)$, for $f \neq \emptyset, \mathcal{A}(H)$, consists of the coordinates of a point $p(f)$ in f and a component capable of reflecting one of seven colors. So $p(f)=f$ if f is a vertex. If

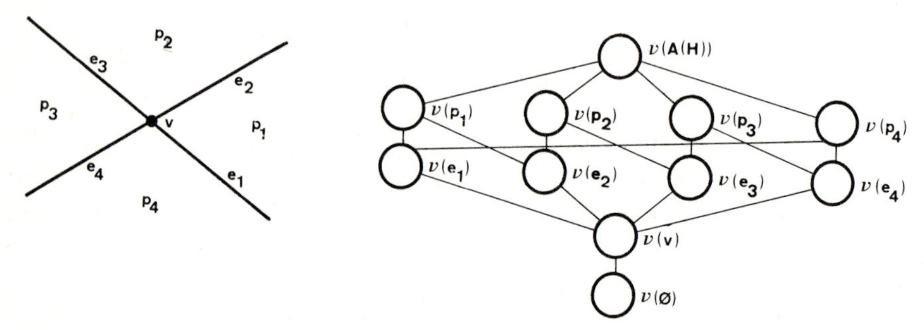

Figure 7.1. Arrangement and incidence graph.

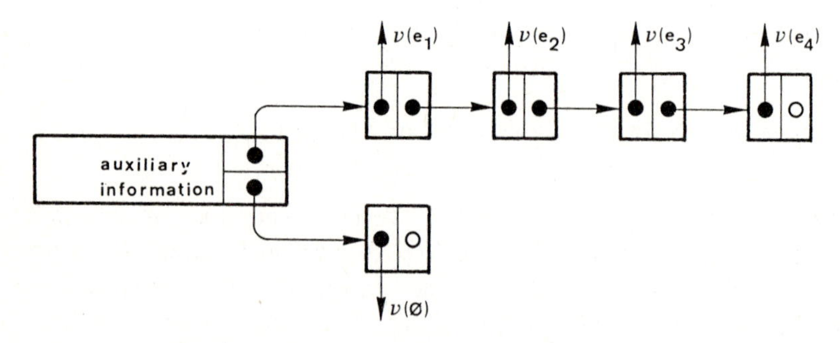

Figure 7.2. Implementing nodes and arcs.

$f_1, f_2, \ldots, f_m$ are the subfaces of f and $m \geq 2$, then

$$p(f) = (\sum_{i=1}^{m} p(f_i))/m$$

is a possible choice. The main function of the auxiliary information is to aid in the manipulation of the incidence graph when hyperplanes are added. Other choices of auxiliary information are conceivable, and we will further augment the incidence graph in particular applications, as in Chapters 12 and 13. Whenever convenient in this chapter, no distinction will be made between the faces of $\mathcal{A}(H)$ and the nodes of $\mathcal{I}(H)$ which store them.

7.2. The Incremental Approach

The algorithm for constructing arrangements that will be described in the forthcoming sections proceeds incrementally, that is, an arrangement is constructed by the addition of one hyperplane at a time to an already existing arrangement. The order in which the hyperplanes are added turns out to be irrelevant. To avoid tedious special cases, we start with a small, carefully chosen subcollection of hyperplanes before using the incremental strategy.

Let H be a set of n hyperplanes denoted as $h_1, h_2, \ldots, h_n$ in E^d. We assume that $\mathcal{A}(H)$ contains at least one vertex. If this is not the case, then each hyperplane intersects the linear subspace of E^d spanned by the normal vectors of all hyperplanes, and the arrangement in this subspace is constructed instead. This lower-dimensional arrangement is guaranteed to contain a vertex, and it captures all of the combinatorial information of $\mathcal{A}(H)$. If $\mathcal{A}(H)$ contains a vertex then there are d hyperplanes which intersect in a vertex; as a consequence, their normal vectors are linearly independent. A formal description of the construction of $\mathcal{A}(H)$, or synonymously of $\mathcal{I}(H)$, follows.

Algorithm 7.1 (Constructing arrangements – global strategy):

Initial step: Rename the hyperplanes in such a way that the normal vectors of the first d hyperplanes $h_1, h_2, \ldots, h_d$ are linearly independent. Construct $\mathcal{A}(\{h_1, h_2, \ldots, h_d\})$.

Main step: For $i = d+1, d+2, \ldots, n$, add h_i to $\mathcal{A}(\{h_1, h_2, \ldots, h_{i-1}\})$ as follows: find an edge e such that cl$e$ intersects h_i; starting from e, mark all faces f whose closures intersect h_i; finally, update all marked faces.

It is an instructive exercise to visualize the addition of a hyperplane for the case of a simple, two-dimensional arrangement. The insertion of a new line h can then be done by "walking along" the boundary of the zone defined by h.

The actual algorithm to be described will not use this intuitively appealing strategy of "walking along" a certain sequence of edges and vertices; rather, it finds these edges and vertices in an unordered fashion, and thus makes the algorithm less intuitive but easier to implement. In fact, for this less intuitive version of the algorithm which lends itself to easy implementation, there is little difference between constructing an arrangement in E^2 and constructing an arrangement in arbitrary dimensions. We will subsequently omit the discussion of the two–dimensional case.

The organization of the description of the algorithm follows. Section 7.3 elaborates on the initial step of Algorithm 7.1 which constructs an arrangement of hyperplanes in arbitrary dimensions and also allows for degeneracies like more than d hyperplanes intersecting in a single point, etc. Section 7.4 presents the geometric fundamentals needed in the specification of the main step, and Section 7.5 outlines the details of the main step.

7.3. Initiating the Construction

To avoid special cases that occur when a hyperplane is added to an empty arrangement or, more generally, to an arrangement without any vertices, we first identify d hyperplanes of H that intersect in a vertex. The arrangement defined by these hyperplanes is then constructed using a trivial method. This section presents the details of this initial step which assumes that these d hyperplanes do indeed exist.

Let $h_1, h_2, \ldots, h_n$ denote the hyperplanes in H, with $n \geq d$. Procedure 7.2 below renames the hyperplanes such that the first d hyperplanes $h_1, h_2, \ldots, h_d$ intersect in a vertex, or equivalently, such that the normal vectors of $h_1, h_2, \ldots, h_d$ span E^d.

Procedure 7.2 (Choosing initial hyperplanes):
Set $i := 1$ and $j := 1$.
while $i < n$ **and** $j < d$ **do**
  if h_1 through h_j and h_{i+1} intersect in a $(d-j-1)$–flat **then**
    Exchange the indices of hyperplanes h_{j+1} and h_{i+1}, and set $j := j+1$.
  endif;
  Set $i := i+1$.
endwhile.

It is worth noting that $j+1 \leq d$ hyperplanes intersect in a common $(d-j-1)$–flat if and only if their normal vectors are linearly independent. This suggests the use of determinants of coordinates of the normal vectors in order to compute the dimensionality of the intersection. Considering d to be a constant, the renaming process is performed in time $O(n)$.

The method of constructing the arrangement of the d hyperplanes which intersect in a vertex is yet to be demonstrated. Let $H'=\{h_1,h_2,\ldots,h_d\}$, where $h_1,h_2,\ldots,h_d$ are the hyperplanes identified by Procedure 7.2. Note that $\mathcal{A}(H')$ is simple. To construct $\mathcal{I}(H')$, we use the following result.

Lemma 7.1: Let H' be a set of d hyperplanes in E^d such that $\mathcal{A}(H')$ is simple. Then,

(i) for each vector $u=(v_1,v_2,\ldots,v_d)$, with $v_i\in\{-1,0,+1\}$ for $1\leq i\leq d$, there is a proper face f in $\mathcal{A}(H')$ with position vector $v(f)=u$, that is, f is above, contained in, or below hyperplane h_i if $v_i=+1$, 0, or -1, respectively, and

(ii) face f is a subface of face g in $\mathcal{A}(H')$ if and only if $f=\emptyset$ and g is a vertex, or if f is a cell and $g=\mathcal{A}(H)$, or if the position vectors $v(f)$ and $v(g)$ of f and g agree up to one component which equals zero in $v(f)$.

Proof: By Theorem 1.3, there are

$$f_i^{(d)}(d)=\sum_{j=0}^{i}\binom{d-j}{i-j}\binom{d}{d-j}$$

i-faces in $\mathcal{A}(H')$, for $0\leq i\leq d$. Since $\mathcal{A}(H')$ is simple, $d-i$ components of $v(f)$ equal 0 if f is an i-face of $\mathcal{A}(H')$. Now, the number of position vectors of length d with $d-i$ 0's is

$$2^i\binom{d}{i}$$

which matches $f_i^{(d)}(d)$ by the combinatorial identity

$$\sum_{j=0}^{i}\frac{1}{j!(i-j)!}=\frac{2^i}{i!}.$$

This proves (i) above since the position vectors of two different faces are necessarily different. (ii) immediately follows from the definition of improper faces (Section 7.1) and proper faces (Section 1.1). □

Lemma 7.1(i) implies that the incidence graph $\mathcal{I}(H')$ contains 3^d+2 nodes. By Lemma 7.1(ii), each i-face has i subfaces, for $1\leq i\leq d$, and $2(d-i)$ superfaces, for $0\leq i\leq d-1$. In addition, the improper $(d+1)$-face has 2^d subfaces and no superface, each d-face has one superface, the one vertex has one subface, and the improper (-1)-face has one superface and no subfaces. It is clear that Lemma 7.1 is a complete description of $\mathcal{I}(H')$, without regard to auxiliary information stored in the nodes. An easy way to compute $\mathcal{I}(H)$ is to produce all 3^d position vectors of length d, to store each position vector in a node by itself, and to create two additional nodes for the two improper faces of $\mathcal{A}(H)$. Finally, the connections between the nodes can be established according to the rules described in Lemma 7.1(ii).

7.4. Geometric Preliminaries

According to Section 7.2, the main step in the construction of an arrangement consists of adding a hyperplane h to an already existing arrangement $\mathcal{A}(H)$. The process which adds h to $\mathcal{A}(H)$ consists of three phases where the second phase marks all faces of $\mathcal{A}(H)$ which interfere with h. Several colors are used to provide a refined handling of the marking process. This section discusses the semantics of the various colors used.

Let f be a proper face of $\mathcal{A}(H)$, and let h be a hyperplane not contained in H. Depending on the relative position of f to h, face f, and synonymously, node $\nu(f)$, is labeled

white, if $\mathrm{cl} f \cap h = \emptyset$,
pink, if $\mathrm{cl} f \cap h \neq \emptyset$ and $f \cap h = \emptyset$,
red, if $f \cap h \neq \emptyset$ and f not contained in h, and
crimson, if $f \subseteq h$.

White faces are considered to be unmarked, and pink, red, and crimson serve as three different kinds of marks. To complete our repertoire of colors, a face in $\mathcal{A}(H \cup \{h\})$ is labeled

white, if $\mathrm{cl} f \cap h = \emptyset$,
grey, if $\mathrm{cl} f \cap h \neq \emptyset$ and $f \cap h = \emptyset$, and
black, if $f \subseteq h$.

Note that white, pink, and crimson faces in $\mathcal{A}(H)$ turn into white, grey, and black faces in $\mathcal{A}(H \cup \{h\})$, respectively. Every red face is replaced by two grey faces and one black face.

The following geometric facts provide the foundations for the algorithm outlined in Section 7.5, which adds a hyperplane h to the arrangement $\mathcal{A}(H)$. All facts follow from the definition of a proper face (Section 1.1) and from the rules which assign colors to faces according to their positions relative to h. The first result concerns the possible combinations of colors assigned to pairs of incident faces in $\mathcal{A}(H)$. For convenience, the results are presented in a matrix with four rows and four columns shown in Table 7.3. A "1" in the matrix means that a proper face with the color of the row can have a proper superface with the color

subface	white	pink	red	crimson
	superface			
white	1	1	1	0
pink	0	1	1	0
red	0	0	1	0
crimson	0	1	0	1

Table 7.3. Colors of pairs of incident faces.

of the column. A "0" indicates the impossibility of such a combination.

In addition to the facts expressed in Table 7.3, we observe the following rules obeyed by the faces in $\mathcal{A}(H)$.

Observation 7.2: (i) A vertex can only be white or crimson, and a cell cannot be crimson.

(ii) Each proper non-white k-face, $k \geq 2$, is bounded by at least one non-white edge.

(iii) A superface of a crimson face is crimson if all of its subfaces are crimson. Otherwise, it is pink.

(iv) A superface of a pink k-face, $k \leq d-1$, is red if it has a red subface or pink subfaces on both sides of the added hyperplane h. Otherwise, it is pink.

As suggested by Observation 7.2(ii), the algorithm in Section 7.5 first colors the edges and then colors higher-dimensional faces. To find all non-white edges, we exploit the straightforward fact that the skeleton of edges and vertices in an arrangement of hyperplanes is connected unless the normal vectors of the hyperplanes are linearly dependent. The intersection of an arrangement in E^d with a new hyperplane h yields a $(d-1)$-dimensional arrangement, and the normal vectors of the $(d-2)$-flats in h which determine this arrangement do not span h only if all normal vectors of the hyperplanes of the original arrangement do not span E^d. This produces the following result.

Observation 7.3: Let $\mathcal{A}(H)$ be an arrangement in E^d, $d \geq 2$, with at least one vertex, and let h be a hyperplane not contained in H. The intersection of h with the union of 2-faces, edges, and vertices in $\mathcal{A}(H)$ is connected.

The next result concerns the changes that a red face in $\mathcal{A}(H)$ undergoes.

Recall that if h is a non-vertical hyperplane then h^+ denotes the open half-space bounded by h from below and h^- is the open half-space bounded by h from above.

Observation 7.4: Let $\mathcal{A}(H)$ be an arrangement in E^d, h a non-vertical hyperplane not in H, and g a red k-face in $\mathcal{A}(H)$, for $1 \leq k \leq d$, that is, g intersects h but is not contained in h.

(i) $g \cap h$ is a black $(k-1)$-face, and $g \cap h^+$ and $g \cap h^-$ are grey k-faces in $\mathcal{A}(H \cup \{h\})$.

(ii) A $(k-1)$-face f in $\mathcal{A}(H \cup \{h\})$ is a subface of $g \cap h^+$ $(g \cap h^-)$ if and only if (1) f is a white or pink subface of g in $\mathcal{A}(H)$ that lies above (below) h, (2) $f = f' \cap h^+$ $(f' \cap h^-)$, for f' a red subface of g in $\mathcal{A}(H)$, or (3) $f = g \cap h$.

(iii) A $(k-2)$-face e in $\mathcal{A}(H \cup \{h\})$ is a subface of the black $(k-1)$-face $g \cap h$ if and only if (1) $k-2=-1$, (2) $e = f' \cap h$, for f' a red subface of g in $\mathcal{A}(H)$, or (3) e is a crimson subface of a pink subface of g in $\mathcal{A}(H)$.

The reader is encouraged to take the time to verify the facts expressed in Table 7.3 and in Observations 7.2 through 7.4, at least for $d=2$ and $d=3$.

7.5. Incrementing the Arrangement

This section concentrates on the process that adds a hyperplane h to an already existing arrangement $\mathcal{A}(H)$ in E^d, $d \geq 2$. $\mathcal{A}(H)$ is represented by the incidence graph $\mathcal{I}(H)$. Hyperplane h and all hyperplanes in H are assumed to be non-vertical. We also assume that h differs from any hyperplane in H, and that $\mathcal{A}(H)$ contains at least one vertex. The latter assumption is justified by the initial step of Algorithm 7.1 which guarantees the existence of a vertex before any hyperplane is added. The existence of a vertex in $\mathcal{A}(H)$ is sufficient to make Observation 7.3 true which is essential for finding all non-white edges of $\mathcal{A}(H)$. According to the description of the overall strategy (Algorithm 7.1), a hyperplane is added in three phases:

Phase 1: An edge e_0 in $\mathcal{A}(H)$ with $\text{cl}\, e_0 \cap h \neq \emptyset$ is determined.
Phase 2: Starting with e_0, all faces f with $\text{cl}\, f \cap h \neq \emptyset$ are marked pink, red, or crimson, as appropriate.
Phase 3: The marked faces are updated.

The remainder of this section describes the details of the actions taken in each phase. The amount of time needed is analyzed in Section 7.6.

To find a suitable edge e_0, the following strategy is applied in Phase 1:

Procedure 7.3 (Phase 1):

Let u be an arbitrary vertex in $\mathcal{A}(H)$ and let e be an incident edge with the line aff$e$ not parallel to h.

while cl$e \cap h = \emptyset$ **do**

Let v be the vertex of e that is closer to h, and let $e' \neq e$ be the edge incident upon v with affe'=aff$e$. Set $e := e'$.

endwhile;

Set $e_0 := e$.

Phase 2 marks all k-faces f of $\mathcal{A}(H)$ for which cl$f \cap h \neq \emptyset$ is true and stores them in a list $\mathcal{L}_k$ which contains all non-white k-faces of $\mathcal{A}(H)$, for $0 \leq k \leq d$. This is performed in two stages. First, only vertices, edges, and 2-faces of $\mathcal{A}(H)$ are considered (see Observation 7.3), and the vertices and edges are appropriately colored. The 2-faces are temporarily stored in a queue $\mathcal{Q}$. No attempt is made to find the correct color of a 2-face at this stage of the algorithm, and we assign the temporary color green to every non-white 2-face in order to distinguish them from the white 2-faces. Second, working in the direction of increasing dimensionality of the faces, the k-faces are marked on the basis of the colors assigned to their subfaces, for $2 \leq k \leq d$. The algorithm now follows.

Procedure 7.4 (Phase 2):

Place an arbitrary 2-face incident upon e_0 into $\mathcal{Q}$ and mark it green.

while $\mathcal{Q}$ is not empty **do**

Delete the first 2-face r from $\mathcal{Q}$.

for each incident white edge e of r **do**

if cl$e \cap h \neq \emptyset$ **then**

Mark each white incident vertex of e which is contained in h crimson and place it into list $\mathcal{L}_0$. Mark e pink if $e \cap h = \emptyset$, crimson if $e \subseteq h$, and red, otherwise. In either case place e into list $\mathcal{L}_1$. Mark all white 2-faces incident upon e green and put them into $\mathcal{Q}$.

endif

endfor

endwhile;

for $k := 2$ **to** d **do**

for each face f in $\mathcal{L}_{k-1}$ **do**

for each white or green superface g of f **do**

Case 1: f is pink.

Mark g red if it has a red subface or pink subfaces on both sides of h, and mark g pink, otherwise (see Table 7.3, second row, and Observation 7.2(iv)).

Case 2: f is red.

Mark g red (see Table 7.3, third row).

Case 3: f is crimson.
Mark g crimson if all subfaces of g are crimson, and pink, otherwise (see Table 7.3, fourth row, and Observation 7.2(iii)).
In any case, place g into list $\mathcal{L}_k$.
endfor
endfor
endfor.

Finally, Phase 3 exploits the colors assigned to the faces and manipulates all marked faces stored in lists $\mathcal{L}_0, \mathcal{L}_1, \ldots, \mathcal{L}_d$. In this phase, each red face is split into two grey faces and one black face, and all incidences of these faces are established (see Observation 7.4).

Procedure 7.5 (Phase 3):
for $k := 0$ to d do
for each face g in $\mathcal{L}_k$ do
Case 1: g is pink.
Change the color of g to grey.
Case 2: g is crimson.
Change the color of g to black.
Case 3: g is red.
Step 1: Replace g by the grey faces $g_a = g \cap h^+$ and $g_b = g \cap h^-$ (see Observation 7.4(i) and Figure 7.4). Reflect this change in $\mathcal{I}(H)$ and $\mathcal{L}_k$, that is, $\mathcal{I}(H)$ and $\mathcal{L}_k$ should now contain g_a and g_b instead of g.
Step 2: Create the black face $f = g \cap h$, connect it to g_a and g_b, and place f into $\mathcal{L}_{k-1}$ (see Observation 7.4(ii) and Figure 7.4).
Step 3: Connect each superface of g with g_a and g_b (see Figure 7.4).
Step 4: Connect each white or grey subface of g with g_a if it is in h^+, and with g_b, otherwise (see Observation 7.4(ii)).
Step 5: If $k=1$ then connect f with the (-1)-face, and connect f with the black subfaces of the grey subfaces of g, otherwise (see Observation 7.4(iii) and Figure 7.4).
endfor
endfor;
Finally, color all grey and black faces white and empty the lists $\mathcal{L}_0, \mathcal{L}_1, \ldots, \mathcal{L}_d$.

Notice that Step 3 above does not connect faces g_a and g_b to their correct superfaces, unless g_a and g_b are cells. In fact, these superfaces will be created only during the next execution of the outermost for–loop. To resolve this dilemma, Step 3 connects g_a and g_b to the red superfaces of g. Step 4 will correct this connection when these red faces are split into the correct superfaces of g_a and g_b. The connection of the $(k-1)$–face f to its correct superfaces is done in a similar fashion: during the next execution of the for–loop, the superfaces of f which are contained in hyperplane h are created and connected to f in Step 5. Figure 7.4 illustrates the actions taken in Steps 1 through 5. Red, pink, and grey nodes are shaded, and crimson and black nodes are cross–hatched. Observe that no mention of the auxiliary information $p(f)$ in node $\nu(f)$ is made during the description of the algorithm. The remainder of this section briefly addresses this issue.

Recall that $p(f)$ is not defined for $f=\emptyset, \mathcal{A}(H)$, and that $p(f)$ is a point in f, otherwise. All tests that determine whether f is above or below a non–vertical hyperplane h (provided $f \cap h = \emptyset$) can be conveniently decided using $p(f)$. When a new face f is created, $p(f)$ can be computed from the points of its subfaces unless f is a vertex or an unbounded edge. In the former exception, $p(f)$ can be obtained by intersecting an edge with h. In the latter exception, we suggest the following strategy.

> Let g be the 2–face in $\mathcal{A}(H)$ with $f = g \cap h$, and let $a \neq b$ denote two parallel lines in the plane aff$g$ that are not parallel to h. Lines a and b intersect h in two distinct points p and q which span the line containing f. Notice that at most one of p and q can coincide with the one end-point of edge f. Using the vertex of f and one of p and q which differs from this vertex, it is easy to determine a point $p(f)$ in f.

The reader should note that the algorithm for hyperplanes can be slightly altered to be used in the construction of arrangements of pseudo–hyperplanes. The only assumption needed to prove the validity of this statement is that the

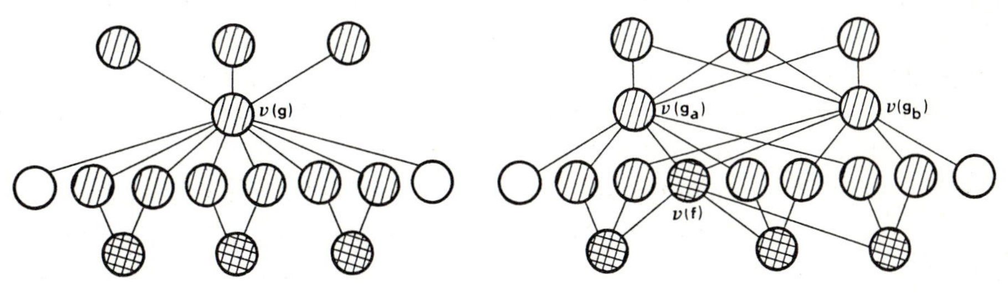

Figure 7.4. Updating a red node $\nu(g)$.

pseudo–hyperplanes should allow for certain primitive operations, such as intersections.

7.6. The Analysis of the Algorithm

The analysis of the time–complexity of Algorithm 7.1 which constructs arrangements of hyperplanes is based on Theorem 5.4. This theorem ensures that the zone of a hyperplane in an arrangement of $n+1$ hyperplanes in E^d consists of at most $O(n^{d-1})$ faces. As we will see, the time needed to insert a hyperplane h into an arrangement $\mathcal{A}(H)$ is proportional to the cardinality of the zone of h in $\mathcal{A}(H)$.

Lemma 7.5: Let H be a set of n non–vertical hyperplanes in E^d, $d \geq 2$, and let h be a non–vertical hyperplane not in H. Procedures 7.3 through 7.5 construct $\mathcal{A}(H \cap \{h\})$ in time $O(n^{d-1})$ from $\mathcal{A}(H)$.

Proof: It is trivial to implement Procedure 7.3 so that it takes time proportional to the number of edges incident upon the vertices on the line chosen. By Exercise 1.2, this number is in $O(n^{d-1})$. Consequently, Phase 1 of the insertion of hyperplane h into $\mathcal{A}(H)$ can be performed in time $O(n^{d-1})$.

The time required by Procedure 7.4 is proportional to the total number of incidences of the marked faces in $\mathcal{A}(H)$. Since each proper superface of a marked face is also marked (see Table 7.3, first column), we can restrict our attention to the analysis of

$$\sum_{k=1}^{d} \sum_{g \in M_k} deg_{k-1}(g),$$

where M_k is the set of marked k–faces in $\mathcal{A}(H)$. By Theorem 5.4,

$$\sum_{g \in M_d} deg_{d-1}(g) = O(n^{d-1}).$$

The restriction that Theorem 5.4 considers only red cells is insignificant since each cell in $\mathcal{A}(H)$ that is pink with respect to h is red with respect to one of the hyperplanes $h+(0,\ldots,0,\epsilon)$ and $h-(0,\ldots,0,\epsilon)$, if $\epsilon > 0$ is sufficiently small. For counting purposes, face g in M_k $(1 \leq k \leq d-1)$ is attributed to the unique k–flat p that contains g. With $H_p = \{h^* | h^* = h' \cap p,\ h' \text{ in } H \text{ with } p \text{ not contained in } h'\}$, $\mathcal{A}(H_p)$ is the k–dimensional subarrangement of $\mathcal{A}(H)$ in p. If we identify the k–flat p with the k–dimensional Euclidean space E^k, then $\mathcal{A}(H_p)$ is simply a k–dimensional arrangement of at most n $(k-1)$–flats, or synonymously, of at most n hyperplanes in E^k. Since the marked faces in M_p are exactly the faces in $\mathcal{A}(H_p)$ that would be marked if $h \cap p$ was added to $\mathcal{A}(H_p)$, we can use Theorem 5.4 to get an upper bound on the number of $(k-1)$ faces bounding marked faces in $\mathcal{A}(H_p)$. Let M_p be the set of k-faces in M_k contained in p. By Theorem 5.4,

$$\sum_{g \in M_p} deg_{k-1}(g) = O(n^{k-1}).$$

In order to obtain an upper bound for all marked k-faces, we sum over all k-dimensional subarrangements of $\mathcal{A}(H)$. Since the hyperplanes in H define at most $\binom{n}{d-k} = O(n^{d-k})$ k-flats, we have

$$\sum_{g \in M_k} deg_{k-1}(g) = O(n^{d-1}),$$

and the same asymptotic bound is true if we compute the sum with k varying from 1 to d. This proves that Phase 2 takes only time $O(n^{d-1})$.

With the exception of Step 5, all actions taken in Procedure 7.5 require time proportional to the number of incidences of the grey and black faces in $\mathcal{A}(H \cup \{h\})$. The former number is in $O(n^{d-1})$ as shown above, and the latter is in $O(n^{d-1})$, since h cuts $\mathcal{A}(H)$ in a $(d-1)$-dimensional arrangement defined by at most n $(d-2)$-flats. We now concentrate on Step 5. The time needed to process a red k-face g in Step 5 is proportional to the number of subfaces of the grey subfaces $\overline{g}$ of g. An upper bound can be obtained if we count the subfaces of each grey $(k-1)$-face $\overline{g}$ $super(\overline{g})$ times, where $super(\overline{g})$ is the number of superfaces of $\overline{g}$. We therefore consider

$$\sum_{k=2}^{d} \sum_{\overline{g} \in G_{k-1}} super(\overline{g}) deg_{k-2}(\overline{g}),$$

where G_{k-1} is the set of grey $(k-1)$-faces in $\mathcal{A}(H \cup \{h\})$. This sum is a maximum only if there are maximally many $(k-1)$-faces which implies that $super(\overline{g}) = 2(d-k+1)$, for each grey $(k-1)$-face $\overline{g}$. In this case, however, $super(\overline{g})$ is only a constant factor in the above sum, and the analysis presented earlier in this proof shows that the sum is in $O(n^{d-1})$. □

Since the amount of time needed for actions not covered by Lemma 7.5 is only nominal, we conclude with the following result.

Theorem 7.6: Let H be a set of n non-vertical hyperplanes in E^d, $d \geq 2$. The incidence graph $\mathcal{I}(H)$ of $\mathcal{A}(H)$ can be constructed in time $O(n^d)$, and this is optimal.

It is interesting to note that Theorem 7.6 implies Theorem 1.4, which asserts that there are at most 2^{cn^d} combinatorially non-equivalent arrangements of n hyperplanes in E^d, for some positive real number c. It turns out that Theorem 1.4 also holds for arrangements of pseudo-hyperplanes in two and three dimensions. The generalization to arrangements of pseudo-hyperplanes in $d \geq 4$ dimensions is not straightforward since it is not known whether or not every such arrangement can be swept by a pseudo-hyperplane (see also Problem 5.7). The argument which finally proves Theorem 1.4 is given below.

Each decision in Algorithm 7.1 is of bounded degree, that is, one of at most a constant number of possibilities is chosen. Two arrangements $\mathcal{A}(H_1)$ and $\mathcal{A}(H_2)$, with $\mathrm{card}H_1=\mathrm{card}H_2=n$, are combinatorially equivalent if the control flow in the construction of $\mathcal{A}(H_1)$ equals that of $\mathcal{A}(H_2)$. Theorem 1.4 easily follows, since Theorem 7.6 limits the capacity of the algorithm to $2^{O(n^d)}$ different flows of control.

7.7. Exercises and Research Problems

2 **Exercise 7.1:** Extend Procedure 7.2 in such a way that it determines the linear subspace of E^d spanned by the normal vectors of hyperplanes $h_1, h_2, \ldots, h_n$.

2 **Exercise 7.2:** (a) Implement the construction of $\mathcal{I}(H')$, for H' a set of d hyperplanes in E^d and $\mathcal{A}(H')$ simple.

3 (b) Implement Algorithm 7.1 which constructs the incidence graph of the arrangement of n hyperplanes in E^d.

3 (c) Modify Algorithm 7.1 so that it works on-line, that is, it works strictly incrementally adding one hyperplane at a time. *(Comment: notice that the current formulation of Algorithm 7.1 is not strictly incremental since it chooses the first d hyperplanes at its convenience.)*

2 **Exercise 7.3:** Prove the following combinatorial identity:

$$\sum_{j=0}^{i} \frac{1}{j!(i-j)!} = \frac{2^i}{i!},$$

for $i \geq 0$.

1 **Exercise 7.4:** Verify the facts expressed by Table 7.3.

1 **Exercise 7.5:** Verify Observation 7.2.

Exercise 7.6: Let $\mathcal{I}(H)$ be the incidence graph of an arrangement $\mathcal{A}(H)$, where H is a set of n hyperplanes in E^d. Design an algorithm that augments, in time $O(n^d)$, each node $\nu(f)$, $f \neq \emptyset, \mathcal{A}(H)$,

3 (a) with the list of hyperplanes that contain f, and

3 (b) with the numbers $a(f)$, $o(f)$, $b(f)$ of hyperplanes h in H such that f is in h^-, h, h^+, respectively.

3 **Exercise 7.7:** Describe an algorithm that constructs the zonotope defined by n line segments in E^d in time $O(n^{d-1})$. *(Hint: use the transformation given in Section 1.7).*

4 **Exercise 7.8:** Design a sweep algorithm that constructs an arrangement of n lines in E^2 in $O(n^2)$ time and $O(n)$ storage if the amount of storage needed to represent the arrangement is not counted. *(Comment: in applications where there is no need to store the entire arrangement but where all of its faces need to be inspected, this algorithm avoids the high storage costs.)*

3 **Exercise 7.9:** Design a sweep algorithm that constructs an arrangement of n hyperplanes in E^d in time $O(n^d \log n)$.

3 **Exercise 7.10:** Implement Algorithm 7.1.

7.8. Bibliographic Notes

The material contained in this chapter is taken from Edelsbrunner, O'Rourke, Seidel (1986). The concept of an incidence graph appeared in Grünbaum (1967), who used it to represent convex polytopes. For the two-dimensional case, the incremental algorithm described in this chapter was independently developed in Chazelle, Guibas, Lee (1985). Algorithms that follow the plane-sweep technique in the construction of arrangements appear in Edelsbrunner, Welzl (1986a) for the case $d=2$, and in Bieri, Nef (1982) for d in general. The former settles Exercise 7.9 for the case $d=2$. A "topological" sweep algorithm which constructs a two-dimensional arrangement in optimal time and only linear working storage can be found in Edelsbrunner, Guibas (1986); this algorithm solves Exercise 7.8. Their method uses a pseudo-line to sweep the plane - in contrast to the standard sweep method which sweeps the plane with a uni-directed line.

CHAPTER 8

CONSTRUCTING CONVEX HULLS

This chapter investigates the problem of constructing the convex hull of a finite set of points in E^d, that is, of producing a meaningful representation of the convex hull. If P is a finite set of points in E^d, then we write conv$P$ for the convex hull of P. By the definitions given in Appendix A, conv$P$ is the set of convex combinations of P. Equivalently, conv$P$ can be defined as

the smallest convex set that contains P, or
the intersection of all convex sets that contain P, or
the intersection of all half-spaces that contain P.

In particular, the last of the definitions makes it obvious that conv$P$ is a convex polytope in E^d. It therefore seems natural to use the incidence graph of its faces to represent conv$P$. Recall that a face of a convex polytope is the relative interior of the intersection with a hyperplane that avoids the interior of the polytope. The issue of storing conv$P$ will be discussed in Section 8.2.

It is interesting to note that convex hulls of finite point sets relate in at least two rather different ways to arrangements of hyperplanes. Convex hulls in E^d relate by duality to intersections of half-spaces in E^d and therefore to cells in d-dimensional arrangements. This suggests that there is strictly less structure to convex hulls than to arrangements. Section 8.1 will present the details of this correspondence. Another geometric transformation implies that for each arrangement of hyperplanes in E^{d-1} there is a zonotope in E^d with essentially the same combinatorial structure as the arrangement (see Section 1.7). Now, a zonotope is a special kind of convex polytope and therefore the convex hull of a finite set of points in E^d. However, it so happens that a zonotope that corresponds to an arrangement of n hyperplanes has on the order of n^{d-1} vertices.

As was already mentioned, the convex hull of every finite point set is a convex polytope, and, by definition, every convex polytope is the convex hull of a finite set of points. As a consequence, the results of Chapter 6 imply bounds on the number of faces in the boundary of the convex hull of a set of n points in E^d. Section 8.1 will be more specific about these implications. Section 8.2 discusses the issue of representing convex hulls, and Sections 8.3 through 8.5 present two methods for efficiently constructing such representations. Section 8.3 is tutorial and introduces both methods in two dimensions. Section 8.4 generalizes the so-called beneath-beyond method from two to $d>2$ dimensions. The other algorithm, which follows the divide-and-conquer paradigm, is generalized to three dimensions in Section 8.5.

8.1. Convex Hulls and Duality

This section reviews the close relationship between convex hulls of point sets and intersections of half-spaces described in Section 1.6. The relationship is best explained using a geometric transform that maps a set P of points in E^d to a set S of half-spaces in E^d, and vice versa, such that $\mathcal{P}=\text{conv}P$ and $\mathcal{Q}=\bigcap_{s\in S} s$ are dual to each other. We first recall the transformation and then show the details of the correspondence between $\mathcal{P}$ and $\mathcal{Q}$.

Let P be a finite set of points in E^d. We assume that the origin o is contained in the interior of $\mathcal{P}=\text{conv}P$ and that no point of P coincides with o. The dual transform $\mathcal{D}_o$, introduced in Section 1.6, maps a point p to the hyperplane $\mathcal{D}_o(p)$: $\langle x,p\rangle=1$, and vice versa. Notice that the vector p is thus a normal vector of $\mathcal{D}_o(p)$ and that the distance of o from $\mathcal{D}_o(p)$ is the reciprocal of its distance from point p. Hyperplane h: $\langle x,v\rangle=1$ avoids the origin if it is the image of a point in $E^d-\{o\}$. We exploit this fact when we define h^{pos} as the open half-space bounded by h that contains o, and h^{neg} as the other open half-space. Thus,

$$h^{pos}: \langle x,v\rangle<1 \text{ and } h^{neg}: \langle x,v\rangle>1.$$

To prove that $\mathcal{P}=\text{conv}P$ and $\mathcal{Q}=\bigcap_{p\in P}\mathcal{D}_o(p)^{pos}$ are dual to each other, we need the following straightforward result which can also be found in Section 1.6. It is an implication of the observation that the formula $\langle x,v\rangle=1$ can be interpreted in two ways.

Observation 8.1: Let $p\neq o$ be a point in E^d and let h be a hyperplane that avoids the origin o. Then point p belongs to half-space h^{pos}, hyperplane h, or half-space h^{neg} if and only if point $\mathcal{D}_o(h)$ belongs to half-space $\mathcal{D}_o(p)^{pos}$, hyperplane $\mathcal{D}_o(p)$, or half-space $\mathcal{D}_o(p)^{neg}$, respectively.

As an immediate consequence of Observation 8.1, the hyperplanes that correspond to the points of a k-flat intersect in a common $(d-1-k)$-flat. Furthermore, there is an incidence preserving one-to-one correspondence between the k-faces of P and the $(d-1-k)$-faces of Q, for $0 \leq k \leq d-1$, such that a hyperplane h supports P in a k-face if and only if $D_o(h)$ is a point of the corresponding $(d-1-k)$-face of Q. This observation allows us to rewrite Theorem 6.12 in such a way that it gives upper bounds on the number of faces and the number of incidences between faces of polytope P in terms of the number of vertices of P.

Corollary 8.2: Let P be the convex hull of n points in E^d. The number of k-faces and the number of incidences between k-faces and $(k+1)$-faces of P is in $O(n^{\min\{\lfloor d/2 \rfloor, k+1\}})$, for $0 \leq k \leq d-1$. These bounds are asymptotically tight.

Corollary 8.2 thus gives a lower bound on the worst-case complexity of every algorithm that computes all faces and incidences of the convex hull of a set of points in E^d.

Since there is a one-to-one correspondence between the faces of P and Q which preserves incidences, an algorithm that constructs the convex hull of a set of points can also be used to compute the intersection of a given set of half-spaces if a point in the interior of this intersection is known. This point can be used to map the half-spaces to points and the intersection problem to a convex hull problem. We refer to Chapter 10 for an algorithm that finds such a point efficiently.

8.2. The Incidence Graph of a Convex Polytope

The incidence graph of a convex polytope is defined in the same way as the incidence graph of an arrangement, and it serves the same purpose: it is used to represent the polytope. Let P be a convex polytope with non-empty interior in E^d. In the same manner as for arrangements, we use "vertex", "edge", and "facet" as synonyms for 0-face, 1-face, and $(d-1)$-face of P, respectively. In addition, we call a $(d-2)$-face of P a *ridge*. For convenience, we define int$P$ to be the only d-face of P, unless int$P = \emptyset$, and we define the empty set as the only (-1)-face of P. For $-1 \leq k \leq d-1$, a k-face f and a $(k+1)$-face g are *incident* (*upon* each other) if f belongs to the boundary of g; in this case, f is called a *subface of* g and g is called a *superface of* f. Two facets of P are *adjacent* (*to* each other) if they are incident upon a common ridge, and two vertices are *adjacent* (*to* each other) if they are incident upon a common edge. The *incidence graph* $I(P)$ *of* P is an undirected graph whose nodes are in one-to-one correspondence with the faces of P, and $I(P)$ contains an arc between two nodes if their corresponding faces are incident. Figure 8.1 shows the incidence graph of

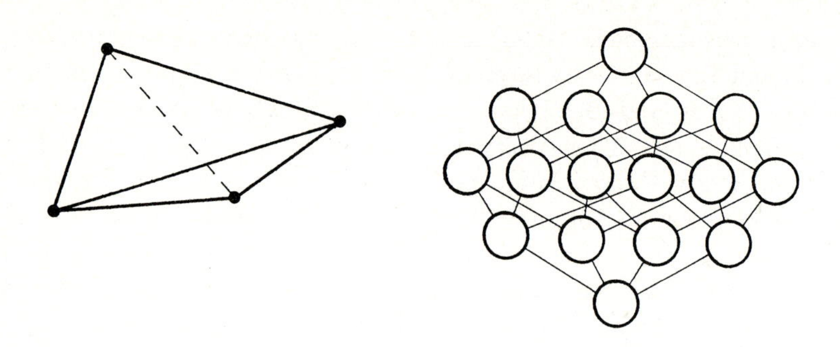

Figure 8.1. A tetrahedron and its incidence graph.

a tetrahedron in E^3; the nodes drawn in the $(4-k)^{\text{th}}$ row from the top correspond to k-faces of the tetrahedron.

In order to store P using its incidence graph, we equip $I(P)$ with some additional information which affixes P in space. Such additional information may include the coordinates of the vertices, the hyperplanes that contain the facets, and the dimension of P.

It is worthwhile to note that the incidence graph $I(P)$ neglects all one– or higher–dimensional order structures among the faces of P, at least as it is currently defined. In Section 8.5, where we treat three–dimensional polytopes, we will incorporate the linear order of edges incident upon a common vertex or upon a common facet into the incidence graph. This piece of additional structure will be crucial in the efficient construction of convex hulls of three–dimensional point sets.

8.3. Two Algorithms in Two Dimensions

This chapter presents two different approaches to computing convex hulls: the first method works incrementally and the second method follows the divide–and–conquer paradigm. Both methods are introduced in this section which is intended to be tutorial. It therefore concentrates on the algorithmic principles of the methods rather than on the geometry of the problem. In particular, both methods are applied to computing convex hulls in two dimensions. In this relatively easy case, the convex hull of a finite set of points is a convex polygon. Furthermore, we assume that there are no degeneracies in the point sets treated; as such, there are no three collinear points and no two points on a common vertical line.

Before we get to the algorithms, we review the incidence graph of a convex

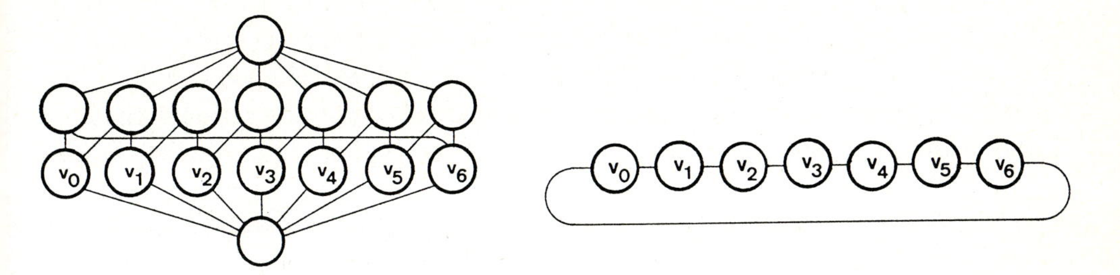

Figure 8.2. Compressing the incidence graph of a polygon.

polygon in E^2 and modify it to our convenience. Figure 8.2 shows the incidence graph of a convex heptagon. Notice that each vertex of a polygon is incident upon exactly two edges, and that each edge is incident upon exactly two vertices. Hence, the incidence graph of a polygon can be compressed to a cyclic sequence of nodes corresponding to the vertices of the polygon: two nodes are connected by an arc if the two corresponding vertices are incident upon a common edge. The data structure used by the algorithms described in the following two sections is thus a doubly-linked list of nodes, where each node stores the coordinates of a point. Section 8.3.1 presents the beneath-beyond method and Section 8.3.2 discusses the divide-and-conquer approach; both methods will be extended to higher dimensions in Sections 8.4 and 8.5.

8.3.1. The Beneath-Beyond Method

The beneath-beyond method works incrementally, that is, it adds one point at a time to an already constructed convex hull. It turns out to be advantageous to presort the points with respect to some direction in order to guarantee that a point lies outside the current convex hull when it is added. Algorithm 8.1 below offers a formal description of the method.

Algorithm 8.1 (Beneath-beyond method in the plane – global strategy):

Initial step: Sort the n points with respect to their x_1-coordinates and relabel them such that $(p_1, p_2, \ldots, p_n)$ is the sorted sequence of points. Construct $P_3 = \mathrm{conv}\{p_1, p_2, p_3\}$.

Iteration: Complete the construction point by point:

for $i := 4$ **to** n **do**

Add point p_i to the current convex hull, that is, update the representation of $P_{i-1} = \mathrm{conv}\{p_1, p_2, \ldots, p_{i-1}\}$ so that it represents $P_i = \mathrm{conv}\{p_1, p_2, \ldots, p_{i-1}, p_i\}$.

endfor.

The remainder of this section shows how a new point p_i can be added to the already existing convex hull $P_{i-1}=\text{conv}\{p_1,p_2,\ldots,p_{i-1}\}$. The procedure that performs this task takes advantage of the fact that p_i lies outside polygon P_{i-1}, that point p_{i-1} is a vertex of P_{i-1}, and that the relatively open line segment that connects points p_{i-1} and p_i avoids P_{i-1}. It also assumes that the cyclic list of nodes that represents P_{i-1} stores the vertices of P_{i-1} in counterclockwise order. That is, if v is a vertex of P_{i-1}, then $succ(v)$ gives the next vertex in the counterclockwise direction, and $pred(v)$ gives the next vertex in the clockwise direction; therefore, $pred(succ(v))=succ(pred(v))=v$. Intuitively, the procedure scans the vertices of P_{i-1} beginning at p_{i-1} in a clockwise and in a counterclockwise direction until it finds the two vertices v_t and v_b such that the lines $\text{aff}\{p_i,v_t\}$ and $\text{aff}\{p_i,v_b\}$ are tangent to P_{i-1} (see Figure 8.3). We assume that w is the node that stores p_{i-1} and we draw no distinction between a node and the represented vertex when we describe the algorithm.

Procedure 8.2 (Adding a point in the plane):

Step 1: Find vertex v_t as follows:

Set $v:=w$, so v represents point p_{i-1}.

while point p_i lies above the line through v and $succ(v)$ **do**

Set $v:=succ(v)$.

endwhile;

Save v, that is, set $v_t:=v$.

Step 2: Find vertex v_b analogously.

Step 3: Remove the nodes between v_b and v_t and add p_i as follows:

Let u be the new node that stores point p_i, and set $pred(v_t):=u$, $succ(v_b):=u$, $pred(u):=v_b$, and $succ(u):=v_t$.

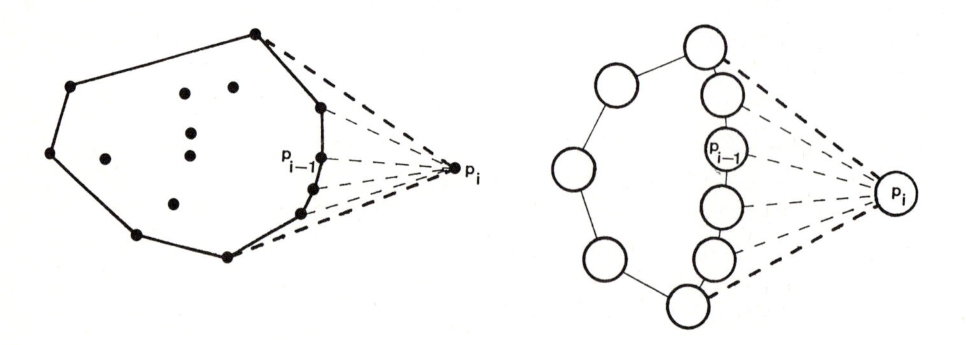

Figure 8.3. Adding a point to the current convex hull.

Although one single execution of Procedure 8.2 can take time linear in the number of points, there is a charging scheme that proves that the iteration part of Algorithm 8.1 altogether takes only time $O(n)$.

Theorem 8.3: Algorithm 8.1 takes time $O(n\log n)$ to construct the convex hull of n points in the plane.

Proof: Using an asymptotically optimal sorting algorithm for the initialization part of Algorithm 8.1 takes time $O(n\log n)$. The iteration takes time $O(n)$ since each step taken within an execution of Procedure 8.2 can be attributed to either a node deleted from the list that represents the convex hull, or to the node that stores a new vertex and is added to the list. Altogether, there are only $n-3$ nodes added to the list. Since each node is created only once and deleted only once, if at all, at most constant time can be spent per node. □

8.3.2. Using Divide–and–Conquer

The idea of the divide–and–conquer paradigm is to fragment a problem into subproblems of the same kind, to solve the subproblems recursively, and, finally, to combine the solutions of the subproblems into a solution of the original problem. In our particular case, this can mean that we partition the given set P of points in the plane into two subsets, that we compute the convex hulls of both subsets, and that we "merge" the two convex hulls into a single polygon. For efficiency reasons, we partition P into two sets P_1 and P_2 of approximately the same size so that there is a vertical line that separates P_1 and P_2. Furthermore, we presort P with respect to the x_1-coordinates of the points, and we relabel the points so that $(p_1,p_2,\ldots,p_n)$ is the sorted sequence. Below, we give a more formal description of the global strategy.

Algorithm 8.3 (Divide–and–conquer in the plane):
if $n \leq 3$ **then**
Construct the convex hull of P using a trivial algorithm.
else
DIVIDE: Set $k := \lfloor n/2 \rfloor$ and define sets $P_1 := \{p_1,p_2,\ldots,p_k\}$ and $P_2 := \{p_{k+1},p_{k+2},\ldots,p_n\}$.
RECUR: Compute $\mathrm{conv}P_1$ and $\mathrm{conv}P_2$ recursively.
MERGE: Combine the two convex hulls to form $\mathrm{conv}P$.
endif.

The only non–trivial part of Algorithm 8.3 is the MERGE step which combines the convex hulls of sets P_1 and P_2. To do so, we need to find the edges in the boundary of $P=\text{conv}P$ that are not edges of $P_1=\text{conv}P_1$ or of $P_2=\text{conv}P_2$. Since P is convex and P_1 and P_2 are separated by a vertical line l, there are exactly two such edges and both intersect line l. Let b_1 and b_2 be the two edges of P that intersect line l such that b_1 intersects l above b_2. We call b_1 the *upper bridge of* P_1 and P_2 and b_2 the *lower bridge of* P_1 and P_2. To find both bridges, we take advantage of the fact that point p_k is the rightmost point of set P_1, and that point p_{k+1} is the leftmost point of P_2. It follows that the line segment $s=\text{ri}\,\text{conv}\{p_k,p_{k+1}\}$ avoids P_1 and P_2. To find the upper bridge b_1, we repeatedly advance the endpoint of s on the boundary of P_1 in a counterclockwise direction and the endpoint of s on the boundary of P_2 in a clockwise direction. The lower bridge b_2 of P_1 and P_2 can be found analogously. Figure 8.4 illustrates all concepts introduced in this paragraph, and it shows possible intermediate steps of line segment s, as its endpoints are advanced on the boundaries of P_1 and P_2.

We give a more formal description of the MERGE step below. The procedure that implements it accepts as input the cyclic doubly-linked lists representing P_1 and P_2 together with pointers to nodes u_1 and u_2 that store points p_k and p_{k+1}. For convenience, we do not distinguish between a node and the point it stores.

Procedure 8.4 (Find bridges in the plane):

Step 1: To find the upper bridge set $v:=u_1$ and $w:=u_2$ and perform the following iteration which assumes that h is the line through the current points v and w:

while one of the points $succ(v)$ and $pred(w)$ lies above h **do**
    if $succ(v)$ lies above h **then** set $v:=succ(v)$
                        else set $w:=pred(w)$
    endif
endwhile;

Step 2: Find the lower bridge analogously.

Note that at the end of the while loop we have vertices v and w such that $succ(v)$ and $pred(w)$ lie below the line h through v and w. The relatively open line segment between v and w cannot intersect P_1 since the process starts with v the rightmost vertex of P_1 and the algorithm maintains the non-intersection property. If v is advanced in counterclockwise direction around P_1, then the position of $succ(v)$ guarantees that the new line segment does not intersect P_1. If the right endpoint, w, of the line segment is moved, then the angle between the line segment and the edge connecting v and $pred(v)$ strictly increases. This implies that the line segment does not intersect P_1 because v lies on the upper part of the boundary of P_1, that is, the vertical upward directed ray emanating from v avoids P_1. The symmetric argument shows that the line segment between v and

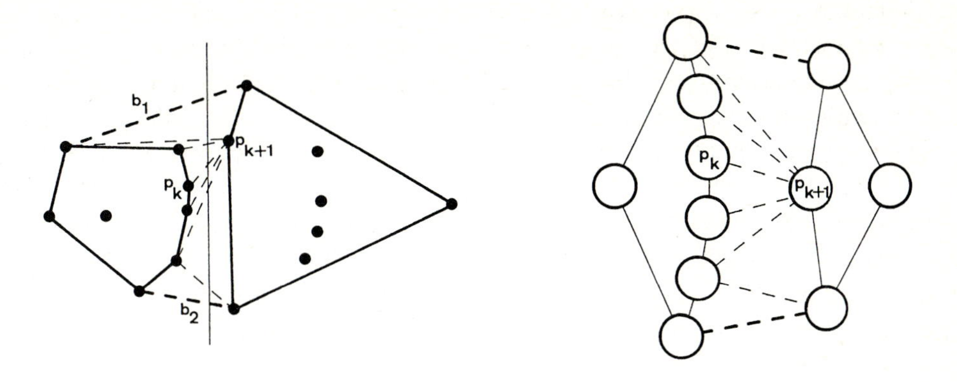

Figure 8.4. Finding the lower bridge and the upper bridge.

w cannot intersect $\mathcal{P}_2$. It follows that Procedure 8.4 is correct.

To assess the amount of time needed by Algorithm 8.3, we observe that each step taken in "merging" two convex hulls either deletes an edge of the two convex hulls to be "merged" or it creates a new edge which is either the lower or the upper bridge of the two convex hulls. Since each invocation of the MERGE step creates only two new edges and since each edge is deleted at most once, we conclude that the total amount of time spent to "merge" convex hulls is linear in the number of points. Every other step in Algorithm 8.3 takes constant time. The amount of time needed to presort the points for Algorithm 8.3 is in $O(n\log n)$ which implies the following result.

Theorem 8.4: Algorithm 8.3 takes time $O(n\log n)$ to construct the convex hull of n points in the plane.

8.4. The Beneath-Beyond Method in d Dimensions

This section describes an algorithm that constructs the convex hull of a finite set P of points in E^d, for arbitrary positive integer numbers d. The global strategy is the same as the one used in Section 8.3.1 for sets of points in two dimensions; the only difference is that we no longer assume that the points are in non-degenerate position. This entails that we distinguish between two fundamentally different cases: in the first case, the point added to $\mathcal{P} = \text{conv}P$ is not contained in the affine hull of $\mathcal{P}$; in the second case, the new point belongs to the affine hull of $\mathcal{P}$. Notice that the dimension of aff$\mathcal{P}$ increases by one each time the first case occurs, which implies that it can occur at most d times. The global strategy can now be described as follows.

Algorithm 8.5 (Beneath–beyond method in E^d – global strategy):

Initial step: Sort the n points lexicographically and let $(p_1,p_2,\ldots,p_n)$ be the sorted sequence of points. Let the initial convex hull be the empty set represented by a single node.

Iteration: Complete the construction adding one point at a time:

for $i:=1$ **to** n **do**

Add point p_i to the convex hull of $\{p_1,p_2,\ldots,p_{i-1}\}$. To do so, distinguish between the case that $\text{aff}\{p_1,p_2,\ldots,p_{i-1}\}$ contains point p_i and the case that it does not contain p_i.

endfor.

The organization of this section is as follows. Section 8.4.1 describes how the incidence graph of the convex hull changes when a point inside or outside the affine hull is added. Sections 8.4.2 through 8.4.4 use the results of this study and outline procedures which actually add a new point to the convex hull. Finally, Section 8.4.5 addresses the analysis of these procedures and of Algorithm 8.5 above.

8.4.1. Geometric Preliminaries

This section gives a precise characterization for how the incidence graph of a polytope $\mathcal{P}'=\text{conv}(P\cup\{p\})$ differs from the incidence graph of $\mathcal{P}=\text{conv}P$. Obviously, $\mathcal{P}=\mathcal{P}'$ if point p belongs to the convex hull of P. So we assume that p lies outside $\mathcal{P}$. Depending on whether or not p belongs to $\text{aff}\mathcal{P}$, there are two different cases to consider. We treat the easier case first.

Let h be a hyperplane that contains $\mathcal{P}$ and avoids point p. If $f\neq\text{int}\mathcal{P}$ is a face of $\mathcal{P}$ then there is a hyperplane h^* with

$$f=\text{ri}(\mathcal{P}\cap h^*)$$

and h^* supports $\mathcal{P}$, that is, $h^*\cap\text{ri}\mathcal{P}=\emptyset$. Define $g=h\cap h^*$ and note that g contains face f. We can rotate h^* about g such that it avoids $\text{ri}\hat{\mathcal{P}}$ which implies that f is also a face of $\hat{\mathcal{P}}=\text{conv}(P\cup\{p\})$. If we rotate h^* into a position where it contains point p, then

$$\hat{f}=\text{ri}\,\text{conv}(f\cup\{p\})=\text{ri}(\hat{\mathcal{P}}\cap h^*),$$

and consequently, $\hat{f}$ is a face of $\hat{\mathcal{P}}$. All supporting hyperplanes that cannot be reached from a supporting hyperplane of $\mathcal{P}$ as explained, cut $\hat{\mathcal{P}}$ in point p only. This shows that the description of faces of $\hat{\mathcal{P}}$ given below is exhaustive. Thus, we have the following result.

Lemma 8.5: Let $\mathcal{P}$ be a convex polytope in E^d such that $k=\dim\text{aff}\mathcal{P}<d$,

and let p be a point not in aff$P$. Define $\hat{P} = \text{conv}(P \cup \{p\})$. Then for every face f of P, f is also a face of $\hat{P}$ and $\hat{f} = \text{ri conv}(f \cup \{p\})$ is a face of $\hat{P}$. Moreover, $\hat{P}$ has no other faces.

The polytope $\hat{P}$ in Lemma 8.5 is usually called a *pyramid* with *base* P and *apex* p. Lemma 8.5 gives a characterization of the faces of $\hat{P}$ in terms of the faces of P. Thus, we know all nodes of the incidence graph $I(\hat{P})$. To determine the arcs of $I(\hat{P})$, we also need a characterization of the incident pairs of faces of $\hat{P}$. To simplify the notation, we write $f \vdash g$ if f is a subface of g.

Observation 8.6: Let $\hat{P}$ be a pyramid with base P and apex p. For faces f and g of P, define $\hat{f} = \text{ri conv}(f \cup \{p\})$ and $\hat{g} = \text{ri conv}(g \cup \{p\})$.
(i) $f \vdash g$ in $\hat{P}$ if and only if $f \vdash g$ in P,
(ii) $f \vdash \hat{g}$ if and only if $f = g$, and
(iii) $\hat{f} \vdash \hat{g}$ if and only if $f \vdash g$.

Lemma 8.5 and Observation 8.6 imply that $I(\hat{P})$ basically consists of $I(P)$ and a copy of $I(P)$, with an additional arc between each node of $I(P)$ and its copy. Refer back to Figure 8.1 for an appropriate illustration. The incidence graph of the tetrahedron shown consists of 16 nodes, and it can be constructed by four copy-and-connect steps that begin with a single node representing the empty set. Let us now turn to the more complicated case when the new point p is contained in aff$P$. We can assume without loss of generality that d is the dimension of aff$P$.

First, we introduce a classification of the faces of P which depends on their relative position to point p. For f a facet of P, this classification is defined as follows:

f is *red* if the hyperplane $h = \text{aff} f$ avoids p and separates p from the interior of P,
f is *yellow* if h contains p, and
f is *blue* if p and the interior of P lie on the same side of hyperplane h.

Intuitively, point p sends out intense red light. This light colors a facet red if the facet is reached by the light, and it colors a facet yellow it this facet is parallel to the direction of the light. The facets not reached or passed by the light emanating from p remain cold blue. If there is no point p that casts red light onto the polytope then we assume that all faces are without color, that is, their color is not defined. No color is assigned to the only d-face int$P$ of P, even if a point p sheds light onto P. If e is a k-face of P, with $k < d-1$, then we assign to e the mixture of the colors assigned to all facets which contain e in their boundaries:

e is *orange* if it belongs to the boundary of red and yellow facets,

e is *green* if it belongs to the boundary of yellow and blue facets,
e is *purple* if it belongs to the boundary of red and blue facets, and
e is *brown* if it belongs to the boundary of red, yellow, and blue facets.

We say that face e has a *blue component* if it is blue, green, purple, or brown, e has a *red component* if it is red, orange, purple, or brown, and e has a *yellow component* if it is yellow, orange, green, or brown. Notice that the color of e can also be defined in terms of the colors of its superfaces: e has exactly all components that occur in its superfaces. The empty set is therefore purple if no facet of P is parallel to the direction of the light, and it is brown, otherwise. With this notation, the following result is analogous to Lemma 8.5.

Lemma 8.7: Let P be a polytope with non–empty interior in E^d, let p be a point outside P, and let $\bar{P}=\text{conv}(P\cup\{p\})$. Let f be a face of P. Then
(i) f is also a face of $\bar{P}$ if f has a blue component,
(ii) $\hat{f}=\text{ri conv}(f\cup\{p\})$ is a face of $\bar{P}$ if f has a blue and a red component, and
(iii) $\bar{f}=\text{ri conv}(f\cup\{p\})$ is a face of $\bar{P}$ if f is yellow (or if $f=\text{int}P$).
Furthermore, for every face g of $\bar{P}$, there is a face f of P such that (i), (ii), or (iii) implies the existence of g.

Proof: Parts (i), (ii), and (iii) of Lemma 8.7 follow by the reasoning used to prove Lemma 8.5; we therefore omit the argument. The claim that this enumeration exhausts all faces of $\bar{P}$ is not completely trivial, since there can be a green face f of P such that $\bar{f}=\text{ri conv}(f\cup\{p\})$ is a face of $\bar{P}$. In this case, however, $\bar{f}=\text{ri conv}(g\cup\{p\})$, for some yellow face g. In fact, g is the relative interior of $h\cap P$, where h is a supporting hyperplane of $\bar{P}$ such that $\bar{f}=\text{ri conv}(h\cap\bar{P})$. □

We remark that Lemma 8.7 implies that no yellow face g of P can have a brown subface f. Both $\bar{g}=\text{ri conv}(g\cup\{p\})$ and $\hat{f}=\text{ri conv}(f\cup\{p\})$ are faces of $\hat{P}$, and each vertex of $\hat{f}$ is also a vertex of $\bar{g}$. This implies that $\hat{f}$ is a subface of $\bar{g}$ which contradicts the fact that $\hat{f}$ and $\bar{g}$ have the same dimension.

As Lemma 8.5 does for the pyramidal case, Lemma 8.7 characterizes the faces of $\bar{P}$ in terms of the faces of P, that is, the nodes of $I(\bar{P})$ in terms of the nodes of $I(P)$. To specify $I(\bar{P})$ completely, we need a characterization of the arcs of $I(\bar{P})$.

Lemma 8.8: Let P and $\bar{P}$ be as in Lemma 8.7, and let f and g be two faces of P, both different from $\text{int}P$. We write f' for f if f has a blue component, $\hat{f}$ for $\text{ri conv}(f\cup\{p\})$ if f has a red and a blue component, and $\bar{f}$ for $\text{ri conv}(f\cup\{p\})$ if f is yellow. Analogous conventions are used for face g. Then

(i) $f' \vdash g'$ if and only if $f \vdash g$,
(ii) $f' \vdash \hat{g}$ if and only if $f = g$,
(iii) $f' \vdash \bar{g}$ if and only if $f \vdash g$,
(iv) $\hat{f} \vdash \hat{g}$ if and only if $f \vdash g$,
(v) $\hat{f} \vdash \bar{g}$ if and only if $f \vdash x \vdash g$, for some subface x of g, and
(vi) $\bar{f} \vdash \bar{g}$ if and only if $f \vdash g$.

Moreover, the d–face $\mathrm{int}\bar{P}$ of $\bar{P}$ is a superface of all facets of $\bar{P}$, and this exhausts the pairs of incident faces of $\bar{P}$.

Proof: Note first that by Lemma 8.7 every face of $\bar{P}$ is of the form f', $\hat{f}$, or $\bar{f}$ for some face f of P. Now, parts (i), (iii), and (vi) are trivial. Parts (ii) and (iv) follow from Lemma 8.5 applied to the pyramid $\hat{g}$. The "if" direction of part (v) follows from $\dim \mathrm{aff}\hat{f} = \dim \mathrm{aff}\bar{g} - 1$ (so the pyramid $\hat{f}$ has the right dimensionality) and from the fact that the set of vertices of $\hat{f}$ is a subset of the set of vertices of $\bar{g}$. For the "only if" direction, note that if f is not a subface of a subface of g then $\hat{f}$ either has the wrong dimensionality to be a subface of $\bar{g}$, or one of its vertices is not in the boundary of $\bar{g}$. For the impossibility of $\hat{f} \vdash g'$ and $\bar{f} \vdash g'$ note that point p lies in the boundaries of faces $\hat{f}$ and $\bar{f}$, but p does not belong to the boundary of g'. Furthermore, $\bar{f} \vdash \hat{g}$ is impossible, by application of Lemma 8.5 to pyramid $\hat{g}$. □

For the efficiency of the algorithms to be presented, we need one more lemma. It sharpens the conditions in point (v) of Lemma 8.8.

Lemma 8.9: Let $f \vdash x \vdash g$ be three faces of a polytope P, with f brown and g yellow. Face x is uniquely determined by faces f and g and the requirement that it be orange.

Proof: Since $\dim \mathrm{aff} f = \dim \mathrm{aff} g - 2$, f is subface of exactly two subfaces of g; call them x and y. Both x and y have a yellow component. We showed above that no yellow face has a brown subface, which implies that x and y are neither brown nor yellow. Thus, x and y can only be green or orange, and the proof is complete if we show that they cannot have the same color. Since face f is brown, $\hat{f} = \mathrm{ri}\,\mathrm{conv}(f \cup \{p\})$ is a face of $\bar{P} = \mathrm{conv}(P \cup \{p\})$, and $\hat{f}$ is a subface of $\bar{g}$, by Lemma 8.8(v). As a consequence, there is a hyperplane h_f through p that contains $\hat{f}$ and avoids $\bar{g}$. Let k be the dimension of f and intersect g with a $(d-k)$–flat h^* that contains p and intersects f in a point f^*. Thus, h^* intersects g in a 2–face f^*, and h^* intersects x and y in edges x^* and y^* incident upon point f^*. Furthermore, h^* intersects $h_f \cap \mathrm{aff} g$ in a line through p that touches the boundary of g^* in point f^*. Thus, the affine hull of one of x^* and y^* separates point p from $\mathrm{ri} g^*$ and the other does not, which implies that one of x and y has a red component and the other has a blue component. □

What completes this section is a short note as to why Lemma 8.9 is really needed. Lemma 8.8 characterizes the incident pairs of faces of $\overline{P}$ in terms of the incident pairs of faces of P. Each of the six possible conditions involves just one incident pair of faces of P, except for condition (v) which involves a chain of three faces, $f \vdash x \vdash g$. The natural way to detect such chains is to look for orange faces x, and to look for brown subfaces f of x and for yellow superfaces g of x. Each such pair (f,g) will induce an incidence relation between $\hat{f} = \text{ri conv}(f \cup \{p\})$ and $\overline{g} = \text{ri conv}(g \cup \{p\})$ on $\overline{P}$. Lemma 8.9 ensures that such a relationship is detected only once.

8.4.2. Towards the Incrementation of the Convex Hull

Before writing out algorithms, we want to specify how our representation of the incidence graph $I(P)$ of a polytope P looks. For each face f of P, we need a record with the following fields:

> $super_f$ and sub_f, which point at lists of superfaces and subfaces of f, respectively,
>
> $color_f$, which stores one out of eight colors, and
>
> $induced_f$, which can store a pointer to a new node; $induced_f$ is also used to mark and unmark face f.

In the case when the dimension of aff$f$ is one lower than the dimension of P there are also fields

> a_f and α_f, with the property that f belongs to the hyperplane $\langle x, a_f \rangle = \alpha_f$, and $P - f$ is contained in the half-space $\langle x, a_f \rangle < \alpha_f$.

Further information such as the coordinates of f if f is a vertex could also be stored. However, this information turns out to be unessential for our algorithms. The global information that we need to store includes $k = \dim \text{aff} P$, a description of aff$P$ in terms of $d-k$ hyperplanes with common intersection equal to aff$P$, and pointers to the nodes that store the only k-face and the only (-1)-face of P. In the description of the algorithms, we will draw no distinction between a face f of P and a node in $I(P)$ that represents f. We assume the following four basic operations to work in constant time:

> create a new face f (this creates a record representing f),
>
> delete face f (this removes f along with all incidences it is involved in),
>
> establish face f as a subface of face g (this inserts f in the list sub_g and g in $super_f$), and
>
> remove the incidence between faces f and g (here we assume that a pointer to f in list sub_g or to g in list $super_f$ is given).

In order to perform the last operation in constant time, the lists $super_f$ and sub_g need to be represented by appropriate pointer structures. One possibility is to have the lists doubly-linked and to have a connection between f in sub_g and g in $super_f$.

After this fairly rigorous treatment we should recall that the problem at hand is to build up the incidence graph of a polytope incrementally by adding one point at a time. In order to get this algorithm off the ground, we specify what the annotated incidence graph of the empty polytope looks like. It has only one face f, namely the empty set $\emptyset$. The information associated with f is as follows:

$$super_f = \textbf{nil},\ sub_f = \textbf{nil},\ color_f \text{ is not defined, and}$$
$$induced_f = \textbf{nil}.$$

We now have everything ready to present the procedures for updating the incidence graph of $I(P)$ to $I(\overline{P})$, $\overline{P}=\text{conv}(P\cup\{p\})$, for some arbitrary point p not in P. To decide whether or not p belongs to aff$P$, we check the $d-k$ hyperplanes which determine aff$P$.

8.4.3. Pyramidal Updates

If point p does not belong to the affine hull of P, then there is a hyperplane h in the set of hyperplanes that determine aff$P$ which avoids p. This hyperplane will be used later when the annotated information is updated. The new polytope $\hat{P} = \text{conv}(P\cup\{p\})$ can be constructed by the following recursive procedure which duplicates the input face g. Initially, g is ri$P$, the highest-dimensional face of P.

Procedure 8.6 (Duplicate the incidence graph):
Create a new face f and set $induced_g := f$.
for all subfaces e of g with $induced_e = \textbf{nil}$ **do**
Recur for face e. Then establish $induced_e$ as a subface of $induced_g$.
endfor;
Establish f as a subface of g.

After executing Procedure 8.6, every node of the old incidence graph points to its copy via its induced field. Define $k = \dim \text{aff} P$, and, therefore, $\dim \text{aff} \hat{P} = k+1$. To update the hyperplanes stored in the k-faces of $\hat{P}$, we need to distinguish two cases. If k-face f is the pyramid ri conv$(e\cup\{p\})$ for e a $(k-1)$-face of P (in this case, the node which stores f is the same node which represented e before point p was added), then the hyperplane of e must be rotated about its intersection with hyperplane h until it contains point p. If $f = \text{ri}P$ then hyperplane h contains f. Next, the hyperplanes that determine the affine hull of P are adjusted, that is, h is removed and the other hyperplanes are rotated about their intersection with h until they contain p. Finally, the induced fields of all faces are reset to **nil**.

8.4.4. Non-Pyramidal Updates

Next, we deal with the more complicated case where the new point p belongs to the affine hull of P. We assume without loss of generality that $\dim \operatorname{aff} P = d$. The new polytope $\overline{P} = \operatorname{conv}(P \cup \{p\})$ is constructed in three phases:

Phase 1: For each face f of P, determine the color of f with respect to point p.
Phase 2: Change the incidence graph of P to the incidence graph of $\overline{P}$.
Phase 3: Clean up the incidence graph.

From Lemmas 8.7 and 8.8, we know that only the non–blue faces are of importance in Phase 2. Thus, it is desirable that Phase 1 discover these faces without considering many blue faces. A way of achieving this is as follows. First, discover some red facet. Next, discover all other red or yellow facets by performing a depth–first–search through the graph whose nodes are the facets of P and whose arcs are the ridges of P. (In this graph, two nodes are connected if the two corresponding facets are adjacent, that is, if there is a ridge incident upon both facets. Recall that every ridge is a subface of exactly two facets; thus, this graph is well defined.) The depth–first–search backtracks whenever it reaches a blue facet. Finally, having discovered all red and yellow facets, determine the colors of all faces in their boundaries; those are the ones that are not blue. Notice that this strategy makes use of the fact that the subgraph searched is connected.

A more formal description of this strategy is given below. It produces lists L_i containing the non–blue i–faces of P, for $-1 \leq i < d$. List L_{d-1} will also contain a number of blue facets.

Procedure 8.7 (Coloring – Phase 1):
 Step 1: Find one red facet r. (How this step can be done quickly will be explained later.)
 Step 2: Find all non–blue facets and ridges using depth–first–search as follows. This search works recursively and takes a facet f as input. Initially, $f := r$.
  Depth–first–search: Add facet f to list L_{d-1} and determine its color using the hyperplane stored with f.
   if f is not blue **then**

```
            for all subfaces e of f do
                Let g ≠ f be the other superface of ridge e.
                if color_g is not defined then
                    Recur with facet g.
                endif;
                if color_e is not defined then
                    Add ridge e to list L_{d-2} and compute color_e
                    from the colors of facets f and g.
                endif
            endfor
        endif;
Step 3: Find and color all remaining non–blue faces:
    for i := d-3 downto -1 do
        for each face f in L_{i+1} do
            for each subface e of f do
                if color_e is not defined then
                    Add face e to L_i.
                endif;
                If necessary, change the color of e such that it con-
                tains its old components and the components of f.
            endfor
        endfor
    endfor.
```

Phases 2 and 3 are quite straightforward. Phase 2, the actual updating of the incidence graph, utilizes the characterizations given in Lemmas 8.7 through 8.9. It uses the lists $\mathcal{L}_i$ generated by Procedure 8.7. The third phase just removes obsolete information. We offer more detailed descriptions of both phases below.

```
Procedure 8.8 (Phase 2):
    for i := -1 to d-1 do
        for each face f in list L_i do
            Case 1: face f is purple or brown (see Lemma 8.7(ii)).
                Create a new face f̂ and set induced_f := f̂.
                if i = d-2 then
                    Let g and g' be the superfaces of f, and compute aff f̂
                    as the hyperplane through point p and the intersec-
                    tion of aff g and aff g'. Establish f̂ as a subface of ri P̄.
                endif;
                Make f a subface of f̂ (see Lemma 8.8(ii)).
                for all subfaces e of f do
```

```
                Make f̂ a superface of induced_e (if induced_e ≠ nil).
            endfor (see Lemma 8.8(iv));
        Case 2: face f is orange.
            for all superfaces g of f do
                if g is yellow then
                    for all brown subfaces e of f do
                        Establish g as a superface of induced_e.
                    endfor
                endif
            endfor (see Lemma 8.5(v));
            Delete face f from list L_i and from the incidence graph.
        Case 3: face f is red.
            Delete face f from list L_i and from the incidence graph.
        Default: Do nothing.
    endfor
endfor.
```

After completion of Procedure 8.8, the lists $\mathcal{L}_{-1}$ through $\mathcal{L}_{d-1}$ store all manipulated faces which are not to be deleted from the incidence graph. We now empty these lists and reset the colors and induced fields of all their faces to undefined and to **nil**, respectively.

The remainder of this section describes how the first red facet can be detected without checking too many facets of $\mathcal{P}$. The strategy we propose takes advantage of the fact that the points are sorted in lexicographical order. Thus, the new point is guaranteed to lie outside $\mathcal{P}$, and if q is the last point added before p, then q is a vertex of $\mathcal{P}$ and, most importantly, at least one of the facets of $\mathcal{P}$ that have q as a vertex is red. Since all these facets were created at the previous update, they can all be saved until p is added. To find one red facet, it thus suffices to check all saved faces.

8.4.5. The Analysis of the Algorithm

To assess the running time of Algorithm 8.5, we do an analysis which is amortized over each step, where a step is defined as all actions taken to add one point to the current convex hull. To this end, we will attribute the time to the faces manipulated, and we will then use a counting argument to give an upper bound on the amount of time needed.

Let $P=\{p_1,p_2,...,p_n\}$ be a set of n points in E^d, and assume that point p_i is lexicographically less than p_{i+1}, for $1 \leq i \leq n-1$. It is not hard to see that the amount of time $T(P)$ needed to construct $\mathcal{P}_n = \text{conv}P$ is proportional to the change in the incidence graph that occurred during the process. That is, if $I(p_i)$ denotes the number of new faces and new incidences that appear when $P_i = \text{conv}(P_{i-1} \cup \{p_i\})$ is formed from $P_{i-1} = \text{conv}\{p_1, p_2, \ldots, p_{i-1}\}$, and if $D(p_i)$ is the number of old faces and incidences that disappear, then the running time $T(p_i)$ needed to form P_i from P_{i-1} is proportional to

$$I(p_i) + D(p_i) + \rho(p_i),$$

where $\rho(p_i)$ denotes the amount of time necessary to find one red facet of P_{i-1}, in the case when a non-pyramidal update is performed. Hence,

$$T(P) = \sum_{i=1}^{n} T(p_i) + O(1) = O(\sum_{i=1}^{n} (I(p_i) + D(p_i) + \rho(p_i))).$$

To simplify this formula, notice that $\rho(p_i) \leq I(p_{i-1})$, as the procedure which finds a first red facet checks only facets which have point p_{i-1} as a common vertex, and these facets are created during the $(i-1)^{\text{st}}$ step. Furthermore, each facet and each incidence between two facets can be deleted at most once, and if it has been, then it must have been created first. Consequently,

$$\sum_{i=1}^{n} D(p_i) \leq \sum_{i=1}^{n} I(p_i).$$

Thus, the running time of Algorithm 8.5 can be expressed as

$$T(P) = O(\sum_{i=1}^{n} I(p_i)).$$

To get a handle on this sum, we recall that the number of faces and incidences of the convex hull of n points in E^d is in $O(n^{\lfloor d/2 \rfloor})$, see Corollary 8.2 and Chapter 6 for further details. We use this fact to prove the following result.

Lemma 8.10: Let P be a convex polytope with n vertices in E^d, and let p be a vertex of P. The number of faces of P which contain p in their boundary plus the number of incidences among them is in $O(n^{\lfloor (d-1)/2 \rfloor})$.

Proof: Since p is a vertex of polytope P, there is a hyperplane h that avoids all vertices of P and separates p from the other vertices of P. Hyperplane h cuts P in a $(d-1)$-dimensional polytope P_h with at most $n-1$ vertices, since vertex p is endpoint of at most $n-1$ edges. Moreover, h cuts each face of P which has p in its boundary (except for p itself), and two such faces of P are incident only if the corresponding faces of P_h are incident. □

Lemma 8.10 is of importance for us since, when obtaining P_i from P_{i-1}, all the new faces contain point p_i in their boundary. Furthermore, every new face has one old face as subface, unless it is not a pyramid, that is, it is not the extension of a yellow face of P_{i-1}, in which case it can have more than one old subface.

However, the worst case occurs when no yellow faces exist, since it is always possible to perturb p_i such that all yellow facets of P_{i-1} become red and the colors of the other facets remain unchanged. This perturbation does not decrease the number of faces or incidences of P_i in question. It follows immediately that

$$I(p_i) = O((i-1)^{\lfloor (d-1)/2 \rfloor}),$$

and thus,

$$\sum_{i=1}^{n} I(p_i) = O(\sum_{i=1}^{n} (i-1)^{\lfloor (d-1)/2 \rfloor}) = O(n^{\lfloor (d+1)/2 \rfloor}).$$

This completes the analysis of the amount of time needed by Algorithm 8.5. The amount of storage needed is easily analyzed since it is proportional to the size of the incidence graph, which is in $O(n^{\lfloor d/2 \rfloor})$ by Corollary 8.2. We therefore conclude with the main result of Section 8.4.

Theorem 8.11: Let P be a set of n points in E^d. Algorithm 8.5 constructs the convex hull of P in $O(n\log n + n^{\lfloor (d+1)/2 \rfloor})$ time and $O(n^{\lfloor d/2 \rfloor})$ storage.

Notice that Corollary 8.2 implies that the above result is optimal in the worst case if the number of dimensions is larger than two and even. For $d=2$, the optimality of the result follows from the fact that an algorithm that computes the convex hull of a set of n points in E^2 can be used to sort n real numbers. Now, for sorting n real numbers there is an $\Omega(n\log n)$ lower bound. The next section improves Theorem 8.11 for three dimensions, where it fails to be optimal for the first time.

8.5. Divide-and-Conquer in Three Dimensions

This section generalizes the divide-and-conquer algorithm outlined in Section 8.3.2 from two to three dimensions. The global strategy is still the same, that is, the problem is broken up into two smaller problems of the same kind, both subproblems are solved recursively, and, finally, we "wrap" the two resulting convex polytopes together into one polytope. There are some additional complications introduced by raising the problem from E^2 to E^3, however. First, the number of special cases that occur due to degenerate positions of points in E^3 is considerably larger than in E^2, and each is a challenge to the programmer. Section 8.5.1 will present a general method which allows us to neglect degeneracies altogether. However, this method entails that we "wrap" only polytopes with non-empty interiors and thus forces us to use a different method if there are seven or fewer points in the set. Second, the incidence graph of a convex polytope in E^3 can no longer be mapped to a simple cyclic list of vertices. Even worse, we need to modify the specification of the incidence graph so that it captures more structural information in order to guarantee efficient behavior of the algorithm to be described. We will return to this issue in Section 8.5.2.

The remainder of this section offers a formal outline of the global strategy for computing the convex hull of a set $P=\{p_1,p_2,...,p_n\}$ of points in E^3. The details will be provided by the forthcoming sections. The algorithm assumes that the points are presorted, that is, the x_1-coordinate of point p_i is smaller than the x_1-coordinate of p_{i+1}, for $1 \le i \le n-1$. By the method to be introduced in Section 8.5.1, we can neglect the case of coincident first coordinates.

Algorithm 8.9 (Convex hulls in E^3 - global strategy):
if $n \le 7$ **then**
Construct $\mathcal{P}=\text{conv}P$ using a straightforward method.
else
DIVIDE: Set $k:=\lfloor n/2 \rfloor$, and define sets $P_1:=\{p_1,p_2,...,p_k\}$ and $P_2:=\{p_{k+1},p_{k+2},...,p_n\}$.
RECUR: Compute polytopes $\mathcal{P}_1:=\text{conv}P_1$ and $\mathcal{P}_2:=\text{conv}P_2$ recursively.
MERGE: "Wrap" $\mathcal{P}_1$ and $\mathcal{P}_2$ into one single polytope $\mathcal{P}$, that is, compute $\mathcal{P}:=\text{conv}(\mathcal{P}_1 \cup \mathcal{P}_2)$.
endif.

Note that polytopes $\mathcal{P}_1$ and $\mathcal{P}_2$ can be separated from each other by a plane h_0 which can be chosen normal to the x_1-axis. Clearly, h_0 intersects $\mathcal{P}=\text{conv}(\mathcal{P}_1 \cup \mathcal{P}_2)$ in a two-dimensional convex polygon, and each facet and edge of $\mathcal{P}$ which is not a facet or edge of $\mathcal{P}_1$ or $\mathcal{P}_2$ must intersect plane h_0. It follows that the facets and edges to be constructed in the MERGE step come in a cyclic sequence. How this sequence of facets and edges is constructed is outlined in Section 8.5.4.

8.5.1. Coping with Degenerate Cases

The method that we propose to take care of special cases caused by degeneracies of the point set $P=\{p_1,p_2,...,p_n\}$ simulates a perturbation of the points. It eliminates degeneracies, such as four coplanar points, or three collinear points. The perturbation is performed on a conceptual level only, and the simulation of the perturbation is based on a symbolic representation of the coordinates by polynomials in one variable. More specifically, the perturbation is a mapping depending on the real number ϵ defined as follows:

for every point $p_i=(\pi_{i,1},\pi_{i,2},\pi_{i,3})$ in P, we define $\pi_{i,j}(\epsilon)=\pi_{i,j}+\epsilon^{2^{3i-j}}$, for $j=1,2,3$, and we set $p_i(\epsilon)=(\pi_{i,1}(\epsilon),\pi_{i,2}(\epsilon),\pi_{i,3}(\epsilon))$, for $1 \le i \le n$, and

$$P(\epsilon)=\{p_i(\epsilon)|1\le i\le n\}.$$

For sufficiently small $\epsilon>0$, no two points of $P(\epsilon)$ can be on a common plane normal to any coordinate-axis, no three points can be on a common line, and no four points can be on a common plane. Notice also that the perturbed set $P(\epsilon)$ equals P for $\epsilon=0$, and that the distance between points p_i and $p_i(\epsilon)$ goes to zero as ϵ approaches 0.

Throughout the algorithm, decisions about the location of a point p_i relative to other points of P are drawn on the basis of the perturbed points and under the assumption that ϵ is positive but sufficiently small. It follows that we "lose" no extreme points by perturbing. Notice that an efficient simulation of the perturbation prohibits the evaluation of the polynomials, even if a sufficiently small value of ϵ were known. Already the evaluation of the exponents of ϵ is too expensive. Nevertheless, we can simulate the perturbation by carefully isolating the primitives needed in our computations, and by designing them such that they compare sums and products of polynomials by comparing the indices of the points that give rise to the polynomials involved. Further details about the simulation of the introduced perturbation are offered in Chapter 9, where the perturbation is a crucial part of the algorithm, unlike in this section, where the perturbation is used for convenience.

There are two prices to be paid for the effective elimination of special cases from the algorithm. One is the careful design of the primitive operations mentioned above. The other is a side effect of the perturbation on the result of the computation. Suppose that point p_i lies on a facet or edge of polytope $\mathcal{P}=\text{conv}P$. Depending on its index i and on the indices of the points that span the facet or edge which contains p_i, the perturbation decides whether or not p_i is considered an extreme point in P. If $p_i(\epsilon)$ is extreme in $P(\epsilon)$, then point p_i is considered to be extreme, and, therefore, p_i shows up as a (degenerate) vertex of the computed convex hull. To remedy this deficiency, we remove the following types of degenerate faces from the convex hull after its construction:

(i) edges that separate coplanar facets,
(ii) vertices that separate collinear edges, and
(iii) vertices that are incident upon coplanar edges only.

It goes without saying that we merge facets as we remove edges, and that we merge edges as we remove vertices.

8.5.2. The Upgraded Incidence Graph

Recall the specification of the incidence graph of a convex polytope in E^3 (see Section 8.2), and notice that it does not particularly aid operations which make crucial use of the linear order relation of the edges around a facet or around a vertex. It is exactly this kind of operation that needs to be performed efficiently if one wants to obtain an efficient algorithm for "wrapping" two separated convex polytopes. We therefore modify the representation of the incidence graph such that it explicitly represents this order information. It will come naturally that this modification puts more emphasis on the role of the edges than of the vertices and facets.

It is convenient to impose a direction on the representation of each edge; in fact, we store each edge e of a polytope in two nodes, each one representing one of the two directed variants of e. Thus, let $\vec{e}$ be a directed edge with origin u and destination v. If we view $\vec{e}$ from the outside of P, then there is a unique facet f to the left of $\vec{e}$, and another facet g to the right of $\vec{e}$. The record storing (the node which represents) edge $\vec{e}$ is equipped with three pointer fields:

$orig(\vec{e})$ to vertex u, the origin of $\vec{e}$,
$left(\vec{e})$ to f, the facet to the left of $\vec{e}$, and
$sym(\vec{e})$ to the symmetric copy of $\vec{e}$ which leads from the destination v of $\vec{e}$ to the origin u of $\vec{e}$.

Since $sym(\vec{e})$ is represented by the same kind of record as $\vec{e}$ itself, we have constant time access to the destination v of $\vec{e}$ and to the facet g to the right of $\vec{e}$ as follows:

$orig(sym(\vec{e}))$ gives a pointer to v, and
$left(sym(\vec{e}))$ gives a pointer to g.

In addition, the record of $\vec{e}$ stores pointers

$onext(\vec{e})$ to edge $\vec{e}_o$ such that $\vec{e}$ and $\vec{e}_o$ have the same origin and $left(sym(\vec{e}_o)) = left(\vec{e})$, and
$lnext(\vec{e})$ to edge $\vec{e}_l$ such that $\vec{e}$ and $\vec{e}_l$ have the same facet to their left and $orig(\vec{e}_l) = orig(sym(\vec{e}))$ (see Figure 8.5).

Using $lnext$-pointers, we can traverse the edges of a facet in counterclockwise order and constant time per edge. Similarly, we can visit all edges with common origin in counterclockwise order and constant time per edge if we use $onext$-pointers. To traverse edges in clockwise order around a facet or a vertex, we alternate between sym- and $onext$-pointers or between sym- and $lnext$-pointers.

Apparently, there is no need for a vertex v to store pointers to all its incident edges. Instead, we store one pointer

$edge(v)$ to an arbitrary edge with origin v.

This allows us to visit all edges incident upon v in time proportional to their number. Similarly, the record that represents a facet f stores a pointer

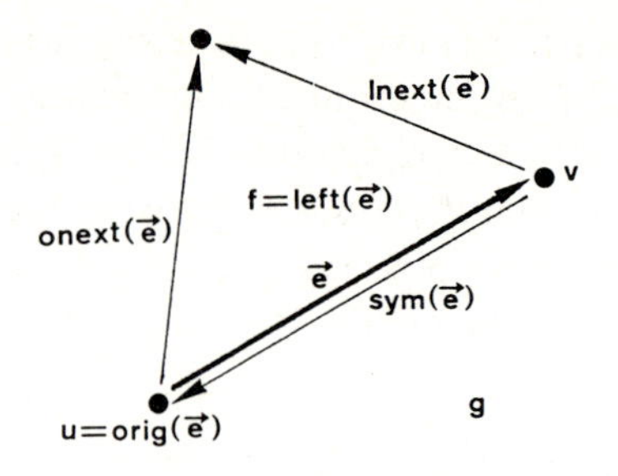

Figure 8.5. Representing a directed edge.

edge(f) to an arbitrary edge $\vec{e}$ with $left(\vec{e})=f$.

This allows us to traverse all edges in the boundary of f in time proportional to their number. In addition, each record has a field for marking the represented vertex, edge, or facet using one of three colors.

The following straightforward operations are assumed to take only constant time:

1. add a facet f (where all directed edges $\vec{e}$ with $left(\vec{e})=f$ are already in the incidence graph and a pointer to one of them is given),
2. add a directed edge $\vec{e}$ (here we assume that the edges which are to become $onext(\vec{e})$ and $lnext(\vec{e})$ are known),
3. remove a facet, and
4. remove a directed edge.

8.5.3. Geometric Preliminaries

Before we go into the details of the algorithm, we investigate the geometry of "wrapping" two separated convex polytopes in E^3. This study gains simplicity from our assumption that both polytopes have non-empty interiors and that no four vertices are coplanar.

Let P_1 and P_2 be two convex polytopes, let h_0 be a plane normal to the x_1-axis that separates P_1 and P_2, and define $P=\text{conv}(P_1 \cup P_2)$. Since h_0 separates all vertices of P_1 from all vertices of P_2, every edge and facet of P is either an edge or facet of P_1 or P_2, or it intersects plane h_0. In the latter case, we say that it is a *new edge* or a *new facet of* P. Conversely, every face of P_1 and P_2 is either a face of P or it belongs to the interior of P, in which case we say that it is a *red face of* P_1 or P_2. We use two colors to classify a face f of P_1 or P_2 which is also a face of P:

f is *blue* if f is a facet, or if f is an edge and it is incident upon two blue facets, or if f is a vertex and it is incident upon blue edges only, and

f is *purple* if f is an edge and it is incident upon at least one red facet, or if f is a vertex and it is incident upon at least one purple or red edge.

To memorize the meaning of the colors, imagine that both P_1 and P_2 send out intense red light. Pieces of the surface that receive red light from the other polytope are burned red, while the pieces that do not receive light stay cold blue. The edges and vertices that are only passed by rays of light form the border between blue and red and are therefore purple. To allow for refined arguments on directed edges, we say that a directed version $\vec{e}$ of a purple undirected edge e of P_1 or P_2 is

purple, if the facet $left(\vec{e})$ to the left of $\vec{e}$ is red, and
blue, if $left(\vec{e})$ is blue.

A directed version of a red or a blue undirected edge is also red or blue, respectively. It turns out that only the purple directed versions of a purple undirected edge are directly involved in the computations. It is tempting to conjecture that the purple edges and vertices of P_1, and similarly of P_2, define a graph which is equal to a simple cycle; but this is not so. For example, there are cases where a purple edge is incident upon two red facets and therefore upon no blue facet. Similarly, a purple vertex v can be endpoint of an arbitrary non–negative number of purple edges. If the number is zero, however, then v is the only non–red face of its polytope. To see how a purple edge can be incident upon two red facets we observe that each edge e of P_1 defines two planes that contain e and support P_2. If both planes intersect the interior of P_1, then e is either blue or red, and e is purple if at least one of the planes also supports P_1. It can happen that both planes support P_1 and P_2. In this case, both planes contain facets of P, and the two facets f' and f'' of P_1 incident upon e are red. To construct such a case, choose e such that its line $\text{aff}\,e$ misses P_2 and choose f' and f'' such that their planes $\text{aff}\,f'$ and $\text{aff}\,f''$ intersect the interior of P_2.

Below, we offer straightforward characterizations of red facets and red vertices that will be used when we remove obsolete parts of P_1 and P_2 after they are "wrapped" into one package.

Observation 8.12: Let P_1 and P_2 be two separated convex polytopes with non–empty interior in E^3.

(i) A facet f of P_i is red if and only if $f = left(\vec{e})$, for at least one purple or red directed edge $\vec{e}$ of P_i, for $i = 1,2$, and

(ii) a vertex is red if and only if all its incident edges are red.

To prove some properties of the collection of purple edges, we formally define the graph that they induce. For $i=1,2$, let $\mathcal{G}_i$ be the directed graph formed by the purple vertices and the purple directed edges of P_i. Below, we prove that the graphs $\mathcal{G}_i$, $i=1,2$, introduced are rather special. All arguments will be given for $\mathcal{G}_1$ only, since the arguments for $\mathcal{G}_2$ are symmetric.

Assume for convenience that the plane h_0: $x_1=0$ separates polytopes P_1 and P_2, and that all points of P_1 have negative x_1-coordinates; so all points of P_2 have positive x_1-coordinates. For a directed plane g, non-parallel to h_0, whose positive half-space we denote by g^{pos}, we define its *angle* $ang(g)$ as follows:

> let line $l=g\cap h_0$ be directed such that $g^{pos}\cap h_0$ is the half-plane to the right of l (where h_0 and l are viewed from P_1), and define $ang(g)$ as the clockwise angle between the positive x_2-axis and line l (see Figure 8.6).

For a purple directed edge $\vec{e}$ of P_1, let $f(\vec{e})$ be the new facet of P to the left of $\vec{e}$, and define $ang(\vec{e})=ang(g)$, for $g=\text{aff}\,f(\vec{e})$ and g^{pos} the half-space bounded by g that contains P_1 and P_2 (see Figure 8.6). We call $ang(\vec{e})$ the *angle of* $\vec{e}$, and we define $ang(a)=ang(\vec{e})$, for $a=(v,w)$ the corresponding arc of $\vec{e}$ in $\mathcal{G}_1$. Now we prove the following result on the structure of $\mathcal{G}_1$.

Lemma 8.13: Let P_1 and P_2 be two convex polytopes with non-empty interior in E^3, and let $\mathcal{G}_1$ be the directed graph of P_1 defined as above. The sequence $(a_1,a_2,\ldots,a_k)$ of all arcs of $\mathcal{G}_1$, with $ang(a_i)<ang(a_{i+1})$, for $1\leq i\leq k-1$, is a eulerian tour of $\mathcal{G}_1$.

Proof: For every angle α in $[0,2\pi)$, we define a directed plane $g(\alpha)$ with positive closed half-space $g(\alpha)^{pos}$, such that

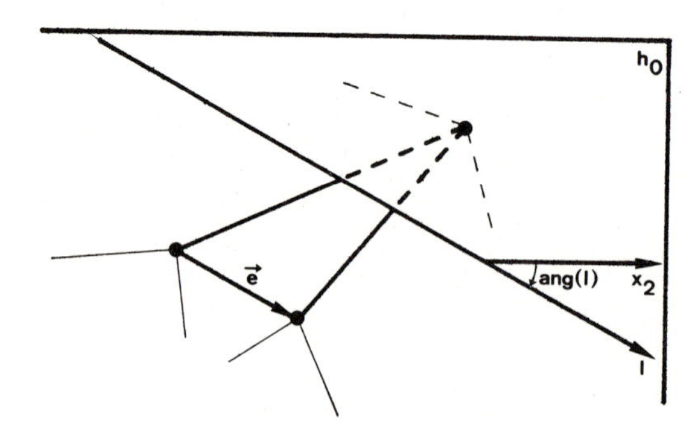

Figure 8.6. The angle of a purple directed edge.

(i) $ang(g(\alpha))=\alpha$,

(ii) $g(\alpha)^{pos}$ contains P_1 and P_2, and

(iii) $g(\alpha)$ is tangent to both P_1 and P_2, that is, $g(\alpha)\cap P_1\neq\emptyset$ and $g(\alpha)\cap P_2\neq\emptyset$.

Clearly, $g(\alpha)$ is unique, and it contains edge $\vec{e}$ and facet $f(\vec{e})$ if and only if $ang(\vec{e})=ang(g(\alpha))$, where $f(\vec{e})$ is the facet of P to the left of $\vec{e}$. As α runs from 0 to 2π, plane $g(\alpha)$ rotates around P_1 and P_2. Obviously, plane $g(\alpha)$ visits the purple directed edges of P_1 in the same order as they correspond to the arcs in the sequence $(a_1,a_2,\ldots,a_k)$. To complete the proof, we need to show that the origin of arc a_i equals the destination of a_{i+1}, for $1\leq i\leq k$ and for the indices taken modulo k. This follows, however, from the observation that plane $g(\alpha)$ pivots about the destination of arc a_i while α belongs to the open interval $(ang(a_i),ang(a_{i+1}))$, for $1\leq i\leq k-1$. If $i=k$, then $g(\alpha)$ pivots about the destination of a_k while α belongs to the open interval $(ang(a_k),ang(a_1)+2\pi)$. □

The algorithm of Section 8.5.4, which "wraps" polytopes P_1 and P_2, maintains a plane rotating around both polytopes and thus computes eulerian tours of graphs $\mathcal{G}_1$ and $\mathcal{G}_2$. For efficiency reasons, we need two monotonicity results that hold for all edges incident upon a common vertex.

Lemma 8.14: Let P_1 and P_2 be two convex polytopes with non-empty interior in E^3, and let $(\vec{e}_1,\vec{e}_2,\ldots,\vec{e}_k)$ be the sequence of purple directed edges of P_1 such that $ang(\vec{e}_i)<ang(\vec{e}_{i+1})$, for $1\leq i\leq k-1$. For each vertex u, let $1\leq i_1<i_2<\ldots<i_{k_u}\leq k$ be the indices i_j such that u is either the origin or the destination of edge $\vec{e}_{i_j}$.

(i) k_u is even.

(ii) Vertex u is the destination of edge $\vec{e}_{i_j}$ if j is odd, and u is the origin of $\vec{e}_{i_j}$ if j is even (unless u is the origin of $\vec{e}_1$ in which case it is the other way round).

(iii) The sequence of edges $\vec{e}_{i_j}$, for $j=1,2,\ldots,k_u$, is ordered in a clockwise direction around vertex u. If two consecutive edges $\vec{e}_{i_j}$ and $\vec{e}_{i_{j+1}}$ are the directed versions of the same common undirected edge, then u is the origin of $\vec{e}_{i_j}$ and the destination of $\vec{e}_{i_{j+1}}$ (unless u is the origin of $\vec{e}_1$ in which case it is the other way round).

Proof: Parts (i) and (ii) are trivial since the sequence of edges specified in the premise of the lemma corresponds to a eulerian tour of $\mathcal{G}_1$. To show (iii), we investigate the edges of polytope P incident upon vertex u and the superfaces of these edges. There is a new facet of P to the left of an edge $\vec{e}_m$ of P_1 with origin or destination u if and only if $m = i_j$, for some $1 \leq j \leq k_u$. All new facets of P intersect the plane h_0 which separates P_1 from P_2. Since h_0 intersects polytope P in a convex polygon, and since each new facet of P intersects h_0 in an edge of this polygon, we conclude that the new facets with common vertex u intersect h_0 in the order of their angles. As we move a copy h of plane h_0 parallel to itself and continuously towards vertex u, the intersection of h with polytope P changes but it always remains a convex polygon. Furthermore, the facets that we are interested in intersect h in edges of this polygon until h reaches u. It follows that the order of these edges around the polygon of intersection remains the same during the sweep and is therefore the same around vertex u. Figure 8.7 illustrates this argument and shades new facets of P. An incoming edge of u cannot precede its outgoing symmetric copy, for this implies that both of their superfaces are blue facets of P_1, which is a contradiction. The only exception to this rule is when u is the origin of edge $\vec{e}_1$, and it is an exception for trivial reasons. The order of the directed edges given in (iii) follows. □

To formulate the other monotonicity property, we need a few definitions. Let u be a vertex of polytope P_1, and let g be a directed plane such that $u = g \cap P_1$ and g^{pos} contains P_1. We call g a *u-tangent plane of P_1*. A line t in plane g is termed a *pivot of g* if it goes through u and if it is not parallel to plane h_0 separating polytopes P_1 and P_2. Vertex u is termed the *anchor of t in P_1*. We define line $l = g \cap h_0$, and we direct l such that $g^{pos} \cap h_0$ is the half-plane to the right of line l in h_0 (see Figures 8.6 and 8.8). If $t' \neq t$ is another pivot of plane g, then we say that t *precedes* t' and t' *succeeds* t if the point $t \cap h_0$ precedes the point $t' \cap h_0$ on the directed line l. If plane g pivots about t in the direction of increasing angle $ang(g)$, g will eventually run into another vertex $w_1(t)$ of P_1. Since P_1 is a convex polytope, vertices u and $w_1(t)$ are neighbors, that is, they

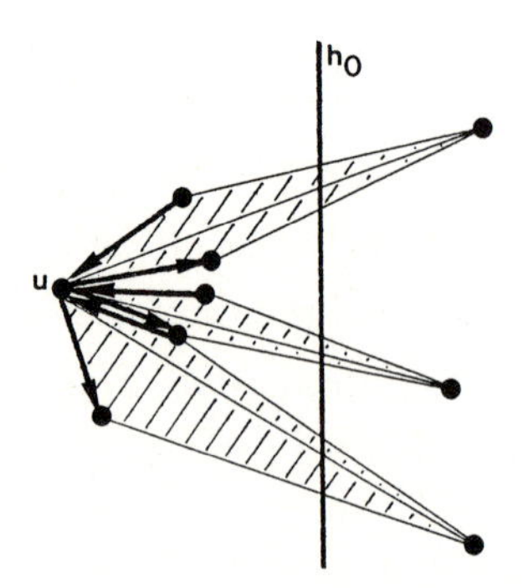

Figure 8.7. New and old facets around a vertex of P_1.

are incident upon a common edge. As a matter of fact, if $w_1(t)$ is unique, then it is the only neighbor w of vertex u such that both the successor and the predecessor of w in the ordered sequence of neighbors of u are contained in $\hat{g}^{pos}$, where $\hat{g}$ is g rotated such that it contains vertex w. We will only be interested in cases when $w_1(t)$ is unique, and we call $w_1(t)$ the *winning vertex of* pivot t and polytope P_1. The following observation is a consequence of the convexity of P_1, and it is illustrated by Figure 8.8.

Observation 8.15: Let P_1 and P_2 be two convex polytopes with non-empty interiors, and let h_0: $x_1=0$ separate P_1 and P_2. Let u be a vertex of P_1, and let g_1 and g_2 be two u-tangent planes such that $ang(g_1) < ang(g_2)$ (to avoid the discontinuity of this order relation, we assume that there is no u-tangent plane with vanishing angle). Furthermore, let t_0 and t_1 be pivots of g_1 and let t_1 and t_2 be pivots of g_2 (therefore, $t_1 = g_1 \cap g_2$), such that t_0 precedes t_1 in g_1 and t_2 succeeds t_1 in g_2. Then the winning vertex $w_1(t_1)$ belongs to the clockwise enumeration of all neighbors of vertex u from $w_1(t_0)$ to $w_1(t_2)$.

Intuitively, Observation 8.15 claims that the winning vertex of the current pivot t moves in clockwise order around its anchor in P_1 if the rotating plane g runs into a vertex of P_2 and therefore advances the pivot. Symmetrically, the winning vertex of t moves in counterclockwise order around the anchor of t in P_2 if g runs into a vertex of P_1.

8.5.4. Wrapping Two Convex Polytopes

We are now ready to outline the algorithm which "wraps" two convex polytopes,

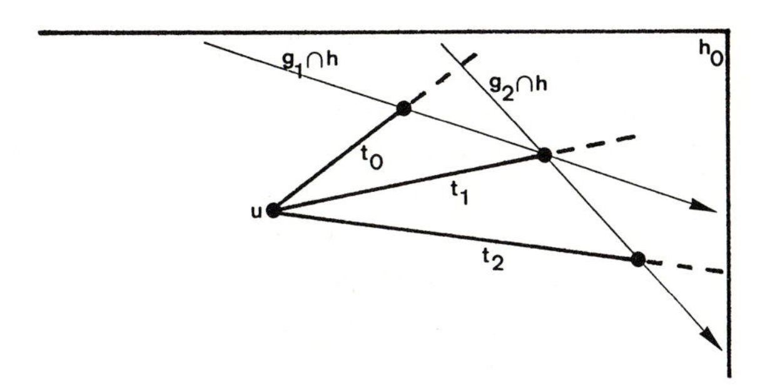

Figure 8.8. Three consecutive pivots.

that is, it computes $P = \text{conv}(P_1 \cup P_2)$, where P_1 and P_2 are two convex polytopes in E^3 which are separated by a plane h_0 normal to the x_1-axis. By the perturbation method of Section 8.5.1, we can assume that both P_1 and P_2 have non-empty interiors, and that no four vertices of P_1 and P_2 are coplanar. Both assumptions allow us to neglect tedious special cases. The algorithm works in three phases:

Phase 1: Find an edge of P that intersects plane h_0.
Phase 2: Perform the rotation of a directed plane g around polytopes P_1 and P_2. During this process, compute all purple directed edges of P_1 and P_2 in the order of their angles, record the color of each purple vertex, and record the color of each red edge of P_1 and P_2 that is incident upon a purple vertex.
Phase 3: Find and remove the red faces from P_1 and P_2, and merge P_1 and P_2 by adding the new faces of P.

One of the easiest ways to implement Phase 1 in time linear in the number of vertices is to mimic the two-dimensional construction of a bridge described in Procedure 8.4. This computation uses the the orthogonal projections $P_1(h)$ of P_1 and $P_2(h)$ of P_2 onto the plane h: $x_2=0$. Both $P_1(h)$ and $P_2(h)$ are convex polygons in h. In order to define the concepts "left" and "right", and "clockwise" and "counterclockwise", we let the x_3-axis be vertical and upwards directed, and we let the x_1-axis be horizontal and directed from left to right. Let u be the rightmost vertex of P_1, that is, u has the largest x_1-coordinate of all vertices of P_1, and let v be the leftmost vertex of P_2. Both u and v project onto vertices of $P_1(h)$ and $P_2(h)$, and the projection of the relatively open segment with endpoints u and v avoids both $P_1(h)$ and $P_2(h)$. Let $h(u,v)$ be the plane through vertices u and v normal to plane h. Furthermore, let $succ(u)$ be the vertex of P_1 that projects onto the successor of u's projection in the counterclockwise sequence of vertices around polygon $P_1(h)$. Symmetrically, we define $pred(v)$ as the vertex of P_2 that projects onto the predecessor of v's projection in the counterclockwise order of the vertices around $P_2(h)$. Vertex $succ(u)$ is a neighbor of vertex u and can be found in time proportional to the number of neighbors of u. The same is true for vertices v and $pred(v)$. To mimic Procedure 8.4 for $P_1(h)$ and $P_2(h)$, we set u to $succ(u)$ whenever $succ(u)$ lies above plane $h(u,v)$, and we set v to $pred(v)$ whenever $pred(v)$ lies above $h(u,v)$. Notice that the above strategy computes the topmost new edge of P, that is, no vertex of P_1 or P_2 lies above plane $h(u,v)$, for u and v the endpoints of the computed edge.

To implement Phase 2, we need a procedure that determines, for a given pivot t_1 through vertices u of P_1 and v of P_2, the winning vertices $w_1(t_1)$ of P_1 and $w_2(t_1)$ of P_2. For a neighbor w of u, let $g(w)$ be the directed plane through vertices u, v, and w, and let $g(w)^{pos}$ be the closed half-space bounded by $g(w)$ such that (u,v,w) is the clockwise enumeration of the vertices of the triangle $\text{conv}\{u,v,w\}$ viewed from outside $g(w)^{pos}$. Notice that $w = w_1(t_1)$ if and only if $g(w)^{pos}$ contains P_1. To find $w_1(t_1)$, we scan the neighbors of vertex u in clockwise order beginning at the winning vertex of the previous pivot t_0 if t_0 contains u, and at the anchor of t_0 in P_1, otherwise. We assume that this vertex is the destination of the directed edge $edge(u)$. The scan continues until $w_1(t_1)$ is found. By Lemmas 8.13 and 8.14, the edge $\text{ri}\,\text{conv}\{u,w\}$ is red if w is scanned and not found to be the winning vertex, unless w is the first vertex of the scan and the directed edge from w to u is purple. A more formal description of the procedure is given below.

Procedure 8.10 (Winning vertex):

Initial step: Initialize $\vec{e}$ to $edge(u)$, set $\vec{e}_p$ and $\vec{e}_s$ to the predecessor and the successor of $\vec{e}$ in the clockwise order of edges with common origin u, and let w_p, w, and w_s be the destinations of $\vec{e}_p$, $\vec{e}$, and $\vec{e}_s$.
unless $sym(\vec{e})$ is purple **do**
Mark $\vec{e}$ and $sym(\vec{e})$ red.
endunless;

while $g(w)^{pos}$ does not contain both vertices w_p and w_s **do**
Advance all edges and vertices by one position, that is, set $\vec{e}_p := \vec{e}$, $\vec{e} := \vec{e}_s$, and $\vec{e}_s := lnext(sym(\vec{e}_s))$, and set $w_p := w$, $w := w_s$, and $w_s := orig(sym(\vec{e}_s))$. Mark $\vec{e}$ and $sym(\vec{e})$ red.
endwhile;

Reset the color of $\vec{e}$ and $sym(\vec{e})$ to blue and set $edge(u) := \vec{e}$.

If Procedure 8.10 is used for polytope P_2, then we need to advance edges and vertices in a counterclockwise direction.

To compute all purple directed edges of P_1 and P_2, we iterate Procedure 8.10, beginning at the topmost new edge e_{top} of P until e_{top} is reached a second time. The purple directed edges of P_1 are stored in a list $\mathcal{L}_1$ in the order in which they are encountered by the rotating plane g, and the purple directed edges of P_2 are stored in a list $\mathcal{L}_2$ in the same fashion. Notice that the sequence of directed edges which is ultimately stored in $\mathcal{L}_1$ corresponds to a eulerian tour of $\mathcal{G}_1$ (see Lemma 8.13), while the sequence in $\mathcal{L}_2$ corresponds to the reverse of a eulerian tour of $\mathcal{G}_2$. For a purple edge $\vec{e}$ of P_1, $\mathcal{L}_1$ stores a pointer $match(\vec{e})$ to the vertex v of P_2 such that g touches P_2 in v at the same time when it encounters the destination of $\vec{e}$. Symmetric information is provided for each purple edge of P_2 in $\mathcal{L}_2$. The lists $\mathcal{L}_1$ and $\mathcal{L}_2$ will be used to construct the incidence graph $\mathcal{I}(P)$ from $\mathcal{I}(P_1)$ and $\mathcal{I}(P_2)$. Below, we give a formal description of the rotation process which constructs the lists $\mathcal{L}_1$ and $\mathcal{L}_2$.

Procedure 8.11 (Wrapping P_1 and P_2):

Initial step: Set $e := e_{top}$, where e_{top} is the undirected edge of P computed in Phase 1, and let u be its endpoint in P_1 and let v be its endpoint in P_2. Initialize the current pivot t to aff$\{u,v\}$, and use Procedure 8.10 without its coloring features to initialize the winning vertices in P_1 and P_2.

Iteration: Find all purple edges as follows:

repeat

Step 1: Determine the winning vertices $w_1(t)$ and $w_2(t)$ of pivot t in P_1 and P_2, and assume without loss of generality that $w_2(t)$ belongs to half-space $g(w_1(t))^{pos}$, that is, $w_1(t)$ is hit first by a tangent plane g pivoting about line t in the direction of increasing angle $ang(g)$.

Step 2: Set $w := w_1(t)$ and mark vertex w purple. Set $edge(w) := sym(edge(u))$, where $edge(u)$ is the directed edge from u to w.

Step 3: Mark edge $\vec{e} := edge(u)$ purple, add it to list L_1, and set $match(\vec{e}) := v$.

Step 4: To prepare the next iteration, set $t :=$ aff$\{w,v\}$ and $u := w$.

until ri conv$\{u,v\} = e_{top}$;

Final step: Invoke Procedure 8.10 with its coloring features for the pivot spanned by edge e_{top}.

After executing Procedure 8.11, we have two lists of purple directed edges: list L_1 for polytope P_1 and L_2 for P_2. The edges in L_1 are sorted by angle, and for each edge $\vec{e}$, we have a pointer $match(\vec{e})$ to the vertex v of P_2 such that $\vec{e}$ and v span a new facet of P which is to the left of $\vec{e}$. The symmetric statement can be made for list L_2 and polytope P_2. In addition to these lists, Procedure 8.11 records the correct color of each purple vertex, of each purple directed edge, and of each red directed edge with purple origin or destination. We perform Phase 3 in three stages: the first stage finds all red faces and marks them red, the second stage removes all nodes of red faces from the incidence graphs $I(P_1)$ and $I(P_2)$, and the third stage creates the nodes for the new faces of P and connects $I(P_1)$ and $I(P_2)$ by adding these nodes.

The first stage can be implemented as a depth-first-search of the graph whose nodes are the red directed edges and whose arcs are the interconnections of these nodes via *sym*–, *onext*–, and *lnext*–pointers. Here, we use the fact that every red edge is either incident upon a purple vertex, or there is a path of red edges which connects it with such a red edge. The search backtracks when it encounters an edge that is already marked red: purple edges stand as guards at the border between red and blue facets and edges. To identify red facets and vertices, we make use of Observation 8.12. After the coloring process, it is easy to remove all red faces.

For the third stage, we scan lists $\mathcal{L}_1$ and $\mathcal{L}_2$ once. Notice, however, that either list could be empty. To formalize this process, we write $next_i(\vec{e})$ for the edge stored in the element of $\mathcal{L}_i$ that succeeds the element that stores $\vec{e}$, for $i=1,2$. Recall that the origin of $next_1(\vec{e}_1)$ coincides with the destination of $\vec{e}_1$ for every edge $\vec{e}_1$ in $\mathcal{L}_1$, and that the destination of $next_2(\vec{e}_2)$ is the origin of $\vec{e}_2$ for every edge $\vec{e}_2$ in $\mathcal{L}_2$. Furthermore, we write $\vec{e}_{u,v}$ for the directed edge with origin u and destination v. For two directed edges $\vec{e}$ and $\vec{e}'$ with common endpoint x, we say that $\vec{e}'$ *comes after* $\vec{e}$ if $\vec{e}'$ immediately succeeds $\vec{e}$ in a clockwise order around x. Symmetrically, we say that $\vec{e}$ *comes before* $\vec{e}'$. At each step of the algorithm, a new facet and two new directed edges are created and added to the incidence graph of $\mathcal{P}$. A formal description of the process follows.

Procedure 8.12 (Adding bridging faces):

Initial step: Let $\vec{e}_1$ be the first edge of $\mathcal{L}_1$, unless $\mathcal{L}_1$ is empty, in which case $\vec{e}_1$ is **nil**. Symmetrically, define $\vec{e}_2$ as the first edge in $\mathcal{L}_2$, unless $\mathcal{L}_2$ is empty. Set $u := orig(\vec{e}_1)$ and $v := orig(sym(\vec{e}_2))$. (If $\vec{e}_1$ is **nil** then set $u := match(\vec{e}_2)$, and set $v := match(\vec{e}_1)$ if $\vec{e}_2$ is **nil**.) Add edge $\vec{e}_{u,v}$ before $\vec{e}_1$ at vertex u and after $sym(\vec{e}_2)$ at vertex v. (If the list of edges at u is empty then we set $onext(\vec{e}_{u,v}) := \vec{e}_{u,v}$, and if the list of edges at v is empty then we set $lnext(\vec{e}_{u,v}) := \vec{e}_{u,v}$.)

Iteration: Complete the "wrapping" as follows:

while one of $\vec{e}_1$ and $\vec{e}_2$ is not **nil do**

Without loss of generality assume that $match(\vec{e}_1)$ is equal to $orig(sym(\vec{e}_2))$, so $\vec{e}_1$ must be processed prior to $\vec{e}_2$.

Step 1: Set $w := orig(sym(\vec{e}_1))$. Add $\vec{e}_{v,u}$ as the symmetric copy of $\vec{e}_{u,v}$. Add the directed edge $\vec{e}_{w,v}$ after $sym(\vec{e}_1)$ at w and before $\vec{e}_{v,u}$ at v.

Step 2: Create a new facet f, set $edge(f) := \vec{e}_1$, and set $left(\vec{e}_1) := f$, $left(\vec{e}_{w,v}) := f$, and $left(\vec{e}_{v,u}) := f$.

Step 3: Prepare the next iteration, that is, set $u := w$ and $\vec{e}_1 := next(\vec{e}_1)$.

endwhile;

Final step: Merge the node of the last directed edge $\vec{e}_{u,v}$ with the node of the edge $\vec{e}_{u,v}$ created in the initial step. Furthermore, mark all non-blue faces of $\mathcal{P}$ blue and empty lists $\mathcal{L}_1$ and $\mathcal{L}_2$.

Figure 8.9 illustrates the process of adding new faces to $\mathcal{P}$. The above procedure does not mention that the improper faces int$\mathcal{P}_1$ and int$\mathcal{P}_2$ are to be merged the above convex hull algorithm as the first step in producing a tetrahedrization of a given finite point set in three dimensions. Their results solve Exercise 8.6(c) and (d).

The beneath–beyond method presented in Sections 8.3.1 and 8.4 can be used to construct convex hulls of finite point sets in arbitrary dimensions. An earlier publication of the algorithm is Seidel (1981), where the algorithm is described in dual space, however. We acknowledge that Raimund Seidel rewrote his algorithm in the primal space and that he provided his write–up as a manuscript of which Section 8.4 is a minor revision. Different convex hull algorithms that work in arbitrary dimensions can be found in Chand, Kapur (1970), with the analysis of their algorithm given in Swart (1985), and in Seidel (1986).

The algorithms presented in this chapter have the best worst–case behavior of the currently known algorithms. Nevertheless, there are algorithms which are preferable to them in cases where the convex hull turns out to consist of considerably fewer faces than is possible in the worst case for the same number of given points. Rényi, Sulanke (1963) and Raynaud (1970) show that the number of vertices of the convex hull of n points in $\boldsymbol{E}^d$ is sublinear for very reasonable distributions of the points; in fact, they offer solutions to Exercise 8.8. Jarvis (1973) was the first to describe an algorithm that takes advantage of this observation; his algorithm constructs the convex hull of n points in $\boldsymbol{E}^2$ in time $O(nh)$, if h of the n points are extreme (see Exercise 8.9). This result was improved to $O(n\log h)$ by Kirkpatrick, Seidel (1986) who also offer a matching lower bound. An algorithm that constructs the convex hull of n points in $\boldsymbol{E}^2$ in expected time $O(n)$, where the points are drawn from several reasonable distributions, can be found in Bentley, Shamos (1978). Exercise 8.12 gives an extension of their method to three dimensions using a twist communicated by Raimund Seidel.

Algorithms which construct convex hulls can be used to construct the onion of a set of points often referred to as the "iterated hull" of the set (see Exercise 8.10). Overmars, van Leeuwen (1981) gave the first algorithm which is more efficient than this trivial approach. They show how to maintain the convex hull of a current number of n points in $\boldsymbol{E}^2$ in time $O(\log^2 n)$ per insertion or deletion of a point. By repeated deletion of extreme points, the onion of n points in $\boldsymbol{E}^2$ can thus be constructed in time $O(n\log^2 n)$. An optimal $O(n\log n)$ time algorithm can be found in Chazelle (1985a). An application of onions to range search in two dimensions is given in Chazelle, Guibas, Lee (1985).

CHAPTER 9

SKELETONS IN ARRANGEMENTS

We have seen in Chapters 7 and 8 that there are optimal algorithms known for constructing a complete arrangement of hyperplanes (Chapter 7) and for constructing a single cell in an arrangement provided the number of dimensions is even or equal to three (Chapter 8). Less satisfying methods are available if a structure in an arrangement is to be computed that is more complicated than a single cell. There are two obvious strategies for such computations, which are not without disadvantages, however.

(i) Compute the complete arrangement first and select the required faces then. This method is expensive in time and space which is not justified if a substantial number of the faces in the arrangement is not needed.

(ii) Construct each required face individually using the algorithms of Chapter 8. The amount of time required for this approach is prohibitive unless there are only few faces to be determined.

This chapter presents yet another method for constructing a structure in an arrangement. The method is fairly general in the sense that it can be used to construct any connected structure that consists of edges and vertices of the arrangement. Such a structure is called a skeleton. For example, all edges and vertices of an arrangement form a skeleton and all edges and vertices in the boundary of a cell form a skeleton. Another example is the set of edges and vertices contained in a level of an arrangement (see Chapter 3 for a definition of levels), and we will see that such skeletons can be used to compute Voronoi diagrams and power diagrams (see Chapter 13).

Intuitively, the new method constructs a skeleton edge by edge, and it maintains data structures that require little work per edge if the edges are processed in a suitable order. It is fairly obvious that strategies (i) and (ii) sketched above are far from optimal when it comes to constructing the skeleton defined by the edges of a level in an arrangement or other skeletons described in Exercises 9.10 and

9.11. The method developed in this chapter is at its best for this type of skeleton since it is able to reuse information when it proceeds from edge to edge.

The organization of this chapter is as follows: Section 9.1 formally introduces the notion of a skeleton and describes the determination of a sequence of edges that favors the efficient construction of a skeleton. Data types supporting the necessary computations are addressed in Section 9.2, and Section 9.3 outlines the overall algorithm which assumes that the underlying arrangement is simple. Section 9.4 describes a general programming technique, called the "simulation of simplicity", that copes with all kinds of degeneracies in arrangements. With this technique, we are able to use the algorithm that assumes simplicity also for arrangements that are not free of degeneracies. Finally, Section 9.5 develops efficient data structures that support so-called penetration queries needed in the construction of skeletons. We concentrate on the data structure for the three-dimensional case, since skeletons in three-dimensional arrangements turn out to have more applications than skeletons in any other dimension.

9.1. Skeletons and Eulerian Tours

In this section, we formally define what we mean by a skeleton in an arrangement and we demonstrate a method that traverses the edges of a skeleton in a connected sequence such that each edge is traversed exactly twice, once in each direction. It will be important that the method uses only local information when it decides on which edge it leaves a vertex.

Let H be a finite set of hyperplanes in E^d, and let E be a subset of the edges in $\mathcal{A}(H)$. We call the set

$$S_E = \bigcup_{e \in E} \mathrm{cl}\, e$$

a *skeleton in* $\mathcal{A}(H)$ if S_E is connected. The edges in E are the *edges of* S_E, and the endpoints of edges in E are the *vertices of* S_E. Examples of skeletons are the *complete skeleton* that contains all edges of $\mathcal{A}(H)$, and the *k-level skeleton* defined by the set of edges of the k-level of $\mathcal{A}(H)$ (see Chapter 3). Figure 9.1 shows the complete skeleton of an arrangement of six lines in the plane, and if we ignore non-bold edges, it shows the 3-level skeleton of the arrangement.

An algorithm that constructs the complete skeleton of an arrangement efficiently follows from the methods given in Chapter 7. Notice that after constructing the complete skeleton, it is easy to construct any other skeleton by removing all edges and vertices that do not belong to the skeleton. When the number of edges of the desired skeleton is considerably smaller than the number of edges of the complete skeleton, then other methods that avoid the construction of all edges of the arrangement are to be preferred. Such an algorithm is described in

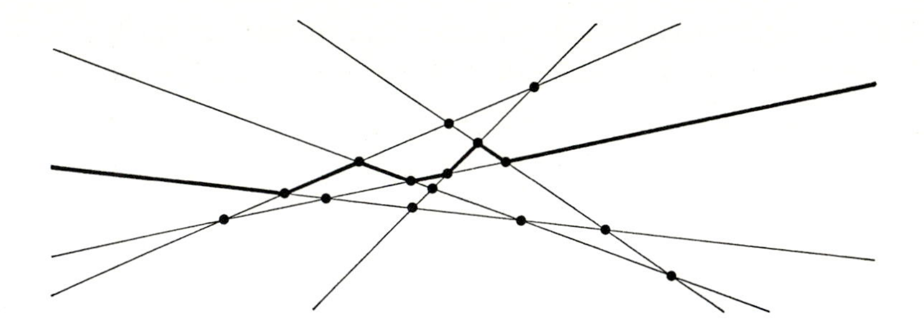

Figure 9.1. Skeletons in the plane.

this chapter. It rests on the possibility of traversing the edges of a skeleton in a certain kind of sequence. Such sequences are best explained for digraphs naturally defined for skeletons.

Let S_E be a skeleton in an arrangement $\mathcal{A}(H)$. S_E defines a digraph $\mathcal{G}_E = (N_E, A_E)$, termed the *skeleton graph of* S_E, as follows:

(i) N_E is the set of vertices of S_E, where we adopt the convention that an unbounded edge has a unique vertex at infinity, and

(ii) a directed arc $a = (v, w)$ is in A_E if E contains an edge incident upon v and w.

By definition, the directed arc $a = (v, w)$ is in A_E if and only if $-a = (w, v)$ is in A_E. Consequently, the number m of arcs of $\mathcal{G}_E$ is necessarily even. If we denote the in-degree of a node v by $deg^-(v)$ and the out-degree of v by $deg^+(v)$, then we also have

$$deg^-(v) = deg^+(v),$$

for each node v in N_E. Below, we show that the latter property is necessary and sufficient for the existence of a *eulerian tour*, that is, a sequence

$$a_0, a_1, \ldots, a_{m-1}$$

of all arcs of $\mathcal{G}_E$ such that the destination of arc a_i is the origin of arc a_{i+1}, for $0 \leq i \leq m-1$ and $a_m = a_0$.

Lemma 9.1: Let $\mathcal{G}$ be a weakly connected digraph with $deg^-(v) = deg^+(v)$, for each node v of $\mathcal{G}$. Then there is a eulerian tour of $\mathcal{G}$.

Proof: Notice that a eulerian tour of $\mathcal{G}$ is uniquely determined if we list the nodes in the sequence they are visited. Each time a node v is visited, it is entered via an incoming arc and it is left via an outgoing arc. Consequently, the number of incoming arcs of any node v of $\mathcal{G}$ in any eulerian tour is the same as the number of outgoing arcs of v. This proves the necessity of the condition.

To see the sufficiency of the condition, assume inductively that there is a eulerian tour if the digraph contains at most k arcs. Now let $\mathcal{G}$ contain $k+2$ arcs. We construct an arbitrary non-empty tour $\mathcal{T}$ in $\mathcal{G}$ and remove all arcs of $\mathcal{T}$ from $\mathcal{G}$. Notice that we still have $deg^-(v) = deg^+(v)$, for each node v. Each weakly connected component of the new digraph has at most k arcs which implies that there is a eulerian tour of each component. To obtain a eulerian tour of the original digraph, we concatenate $\mathcal{T}$ with the eulerian tour $\mathcal{T}'$ of each connected component. To this end, we cut the cyclic sequences $\mathcal{T}$ and $\mathcal{T}'$ at a common node and glue the loose ends of $\mathcal{T}$ and $\mathcal{T}'$ together. □

The existence of a eulerian tour in a digraph implies of course that the digraph is strongly connected. We infer that every weakly connected digraph with the property that the in-degree of each node is the same as the out-degree is strongly connected.

Lemma 9.1 is actually stronger than what we need, namely the existence of eulerian tours in skeleton graphs. The remainder of this section offers an algorithm that finds a eulerian tour of a skeleton graph. The algorithm is a variant of the well-known depth-first-search method. We write the algorithm as a recursive procedure which accepts a node v and an integer number i as its input. Initially, we assume that all nodes of the skeleton graph are unmarked, and we call the procedure with an arbitrary node v and with $i=1$. The mark of a node v is denoted by $mark(v)$.

Procedure 9.1 (Depth-first-search):
    (This procedure takes as input a node v and an integer number i).
    Set $mark(v) := i$.
    for each outgoing arc (v,w) **do**
        unless w is marked and $mark(w) < mark(v)$ **do**
            Traverse arc (v,w).
            if w is unmarked **then**
                Call Procedure 9.1 recursively with parameters w and $i+1$.
            endif;
            Finally, traverse arc (w,v).
        endunless
    endfor.

To see that Procedure 9.1 traverses a eulerian tour note that it traverses a tour and that it traverses all incoming and outgoing arcs of each marked node exactly once. Now, assume that there is an unmarked node left. Since $\mathcal{G}$ is connected, there must be an unmarked node u adjacent to a marked node v which contradicts the control flow of the algorithm.

9.2. Towards the Construction of a Skeleton

On a high level of understanding, the construction of a skeleton is done during the traversal of a eulerian tour of the associated skeleton graph. This traversal is accompanied by a number of primitive actions supported by two abstract data types,

(i) an augmentable representation of (the currently known parts of) the skeleton graph, and

(ii) a representation of a set of hyperplanes that supports so-called penetration queries and can be maintained through a sequence of insertions and deletions.

A data structure that implements the first data type is treated in detail in this section. Efficient implementations of the second data type are considerably more involved. Therefore, we delay the development of such implementations until Section 9.5. Only a minimal amount of information about such implementations that is necessary to understand the algorithm of the next section is presented at the end of this section.

We begin with the data structure that represents a skeleton graph. Let $\mathcal{G}=(N,A)$ be a skeleton graph. At any point in time t, we let N_t denote the set of nodes visited by the algorithm, and we let A_t denote the set of arcs with origins in N_t. Only nodes in N_t and arcs in A_t are stored in the representation of the skeleton graph. Note that not every arc in A_t is also traversed by the algorithm. An arc that leads from a node in N_t to a node in $N-N_t$ is necessarily not traversed yet, whereas an arc leading from a node in N_t to another node in N_t can be of either type.

We proceed with the detailed specification of a data structure ADJ that stores the skeleton graph in a collection of lists (see Figure 9.2).

(i) Each node v in N_t is stored in a record that contains a field $mark(v)$ and a pointer to a list $A(v)$ that stores the outgoing arcs of v.

(ii) Each arc a in A_t with origin v is stored in a record in $A(v)$; if a is already traversed then this record contains a pointer to the destination of a, otherwise, it contains the empty pointer.

(iii) Each record representing a node or an arc x stores the set $H(x)$ of hyperplanes that contain the corresponding vertex or edge.

(iv) All records representing nodes in N_t are organized in a dictionary that discriminates by sets of containing hyperplanes.

Figure 9.2(b) illustrates the data structure ADJ that represents the incomplete skeleton graph shown in Figure 9.2(a): u is the first vertex of the tour and v is the vertex currently considered. Non-traversed arcs are indicated by broken

Algorithm 9.2 which assumes the simplicity of the arrangement. Sections 9.4.2 and 9.4.3 describe how such a simulation can be put into use without sacrificing the efficiency of Algorithm 9.2. We remark that the method of simulating simplicity is also used in Section 8.5 to construct the convex hull of point sets in three dimensions.

There is, however, one side-effect of the simulation of simplicity that needs to be mentioned. If some k hyperplanes lie in special position, then the technique pretends that they do not. For example, if two hyperplanes are parallel, then the technique pretends that they intersect in a common $(d-2)$-flat, and if $d+1$ hyperplanes intersect in a common vertex, then the simulation of simplicity pretends that the hyperplanes determine $d+1$ vertices. The technique thus gives rise to the construction of faces that lie at infinity, of vertices that are connected by zero-length edges, etc. Depending on the application we have in mind, we may or may not want to remove these vacuous faces in a postprocessing step.

9.4.1. A Conceptual Perturbation

To define a perturbation of a set of hyperplanes, we add to each coefficient of a hyperplane a term of the form ϵ^{ℓ}, with ℓ a positive constant and ϵ an arbitrarily small positive real number which remains without further specification. The terms will be defined such that the perturbed hyperplanes approach the original hyperplanes when ϵ goes to zero.

Let $H=\{h_0,h_1,\ldots,h_{n-1}\}$ be a set of n non-vertical hyperplanes in E^d, with $\mathcal{A}(H)$ not necessarily simple. For each hyperplane

$$h_i\colon\ x_d=\eta_{i,1}x_1+\eta_{i,2}x_2+\ldots+\eta_{i,d-1}x_{d-1}+\eta_{i,d}$$

in H and for every non-negative real number ϵ, we define

$$h_i(\epsilon)\colon\ x_d=\eta_{i,1}(\epsilon)x_1+\eta_{i,2}(\epsilon)x_2+\ldots+\eta_{i,d-1}(\epsilon)x_{d-1}+\eta_{i,d}(\epsilon),$$

with

$$\eta_{i,j}(\epsilon)=\eta_{i,j}+\epsilon^{2^{di+d-j}},$$

for $1\leq j\leq d$ and $0\leq i\leq n-1$. It is important to see that the exponents of ϵ assume all powers of 2, from 2^0 through 2^{dn-1}, and that they are therefore pairwise different. In fact, we have

$$\sum_{(i,j)\in I} 2^{di+d-j}\neq\sum_{(i,j)\in J} 2^{di+d-j}$$

unless $I=J$.

Finally, we set

$$H(\epsilon)=\{h_i(\epsilon)|0\leq i\leq n-1\}.$$

Obviously, $H(0)=H$. Notice furthermore that the amount of perturbation experienced by a hyperplane decreases double-exponentially with increasing index. Below, we prove that the arrangement $\mathcal{A}(H(\epsilon))$ is simple if ϵ is small enough but positive, no matter how degenerate $\mathcal{A}(H)$ is.

Lemma 9.3: Let $H=\{h_0,h_1,\ldots,h_{n-1}\}$ be a set of non-vertical hyperplanes in E^d. Arrangement $\mathcal{A}(H(\epsilon))$ is simple if $\epsilon>0$ is sufficiently small.

Proof: First, we argue that any d hyperplanes in $H(\epsilon)$ intersect in a common point, that is, their normal vectors are linearly independent. A representative of all normal vectors of hyperplane $h_i(\epsilon)$ is

$$(\eta_{i,1}(\epsilon),\eta_{i,2}(\epsilon),\ldots,\eta_{i,d-1}(\epsilon),-1).$$

Thus, d hyperplanes $h_{i_0}(\epsilon),h_{i_1}(\epsilon),\ldots,h_{i_{d-1}}(\epsilon)$ intersect in a common point if and only if the determinant Δ' of the matrix

$$\Gamma'=\begin{pmatrix} \eta_{i_0,1}(\epsilon) & \eta_{i_0,2}(\epsilon) & \ldots & \eta_{i_0,d-1}(\epsilon) & -1 \\ \eta_{i_1,1}(\epsilon) & \eta_{i_1,2}(\epsilon) & \ldots & \eta_{i_1,d-1}(\epsilon) & -1 \\ \ldots & \ldots & \ldots & \ldots & \ldots \\ \eta_{i_{d-1},1}(\epsilon) & \eta_{i_{d-1},2}(\epsilon) & \ldots & \eta_{i_{d-1},d-1}(\epsilon) & -1 \end{pmatrix}$$

does not vanish. Formally, the determinant of this matrix is a polynomial in ϵ. By the choice of the terms added to the coefficients, each term in Δ' is of the form

$$c\cdot\epsilon^{\ell},$$

where c is a constant and ℓ is the sum of distinct powers of 2. Furthermore, the values of the ℓ's of two distinct terms are different. It follows that the terms in Δ' cannot cancel if ϵ is small enough. The determinant is thus non-zero if at least one constant c does not vanish. Such a constant that is guaranteed to be non-zero is, for example, the coefficient of the term with greatest exponent ℓ.

Next, we prove that the intersection of any $d+1$ hyperplanes $h_{i_0}(\epsilon),h_{i_1}(\epsilon),\ldots,h_{i_d}(\epsilon)$ is empty. The hyperplanes contain a common point if and only if the determinant of the matrix

$$\Gamma=\begin{pmatrix} \eta_{i_0,1}(\epsilon) & \eta_{i_0,2}(\epsilon) & \ldots & \eta_{i_0,d}(\epsilon) & -1 \\ \eta_{i_1,1}(\epsilon) & \eta_{i_1,2}(\epsilon) & \ldots & \eta_{i_1,d}(\epsilon) & -1 \\ \ldots & \ldots & \ldots & \ldots & \ldots \\ \eta_{i_d,1}(\epsilon) & \eta_{i_l,2}(\epsilon) & \ldots & \eta_{i_d,d}(\epsilon) & -1 \end{pmatrix}$$

vanishes. By the same argument as above, this is not possible if ϵ is small enough. □

Matrices Γ' and Γ will play significant roles in the actual computation that deals with the perturbed hyperplanes. This will be described in the next two sections.

9.4.2. Simulating the Perturbation

When we simulate the construction of a skeleton in an arrangement $\mathcal{A}(H(\epsilon))$, we need to be able to perform primitive operations for the hyperplanes in the conceptually perturbed set $H(\epsilon)$. As it turns out, there is only one such non-trivial primitive operation needed; it decides on which side of a given hyperplane $h_{i_d}(\epsilon)$ some other d hyperplanes $h_{i_0}(\epsilon), h_{i_1}(\epsilon), \ldots, h_{i_{d-1}}(\epsilon)$ intersect. In this section, we sketch how such a decision can be found without calculating a specific sufficiently small value of ϵ and without even computing the powers of ϵ as defined in the previous section.

First, we derive a characterization of when d hyperplanes $h_{i_0}(\epsilon)$ through $h_{i_{d-1}}(\epsilon)$ intersect above $h_{i_d}(\epsilon)$. By definition, a point $p=(\pi_1,\pi_2,\ldots,\pi_d)$ lies above $h_{i_d}(\epsilon)$ if and only if

$$\eta_{i_d,1}(\epsilon)\pi_1+\eta_{i_d,2}(\epsilon)\pi_2+\ldots+\eta_{i_d,d-1}(\epsilon)\pi_{d-1}+\eta_{i_d,d}(\epsilon)-\pi_d<0. \qquad (1)$$

We set p equal to the common intersection of the hyperplanes $h_{i_0}(\epsilon)$ through $h_{i_{d-1}}(\epsilon)$ and we express the coordinates of p in terms of the coefficients of the hyperplanes. By Cramer's rule, we have

$$\pi_i=\frac{\Delta_i'}{\Delta'},$$

with $\Delta'=\det M'$,

$$M'=\begin{pmatrix} \eta_{i_0,1}(\epsilon) & \eta_{i_0,2}(\epsilon) & \ldots & \eta_{i_0,d-1}(\epsilon) & -1 \\ \eta_{i_1,1}(\epsilon) & \eta_{i_1,2}(\epsilon) & \ldots & \eta_{i_1,d-1}(\epsilon) & -1 \\ \ldots & \ldots & \ldots & \ldots & \ldots \\ \eta_{i_{d-1},1}(\epsilon) & \eta_{i_{d-1},2}(\epsilon) & \ldots & \eta_{i_{d-1},d-1}(\epsilon) & -1 \end{pmatrix},$$

and $\Delta_i'=\det M_i'$, where M_i' is the matrix that we obtain from M' when we replace the ith column from the left by the vector

$$\begin{pmatrix} -\eta_{i_0,d}(\epsilon) \\ -\eta_{i_1,d}(\epsilon) \\ \cdots \\ -\eta_{i_{d-1},d}(\epsilon) \end{pmatrix},$$

for $1 \leq i \leq d$. Condition (1) now becomes

$$\eta_{i_d,1}(\epsilon)\frac{\Delta_1{}'}{\Delta'}+\eta_{i_d,2}(\epsilon)\frac{\Delta_2{}'}{\Delta'}+\ldots+\eta_{i_d,d-1}(\epsilon)\frac{\Delta_{d-1}{}'}{\Delta'}+\eta_{i_d,d}(\epsilon)-\frac{\Delta_d{}'}{\Delta'}<0.$$

If Δ' is positive, this is equivalent to

$$\eta_{i_d,1}(\epsilon)\Delta_1{}'+\eta_{i_d,2}(\epsilon)\Delta_2{}'+\ldots+\eta_{i_d,d-1}(\epsilon)\Delta_{d-1}{}'+\eta_{i_d,d}(\epsilon)\Delta'-\Delta_d{}'<0. \tag{2}$$

Lemma 9.4: Condition (2) is true if and only if $\Delta > 0$, for $\Delta = \det M$ and

$$M = \begin{pmatrix} \eta_{i_0,1}(\epsilon) & \eta_{i_0,2}(\epsilon) & \cdots & \eta_{i_0,d-1}(\epsilon) & -1 & -\eta_{i_0,d}(\epsilon) \\ \eta_{i_1,1}(\epsilon) & \eta_{i_1,2}(\epsilon) & \cdots & \eta_{i_1,d-1}(\epsilon) & -1 & -\eta_{i_1,d}(\epsilon) \\ \cdots & \cdots & \cdots & \cdots & \cdots & \cdots \\ \eta_{i_d,1}(\epsilon) & \eta_{i_d,2}(\epsilon) & \cdots & \eta_{i_d,d-1}(\epsilon) & -1 & -\eta_{i_d,d}(\epsilon) \end{pmatrix}.$$

Proof: We show Lemma 9.4 by evaluating Δ using the Laplace expansion for the last row. Thus,

$$\Delta = \left(\sum_{j=1}^{d-1}(-1)^{d+1+j}\eta_{i_d,j}(\epsilon)\Delta_j\right)+(-1)^{2d+1}(-1)\Delta_d+(-1)^{2d+2}(-1)\eta_{i_d,d}(\epsilon)\Delta_{d+1},$$

where $\Delta_i = \det M_i$ and matrix M_i is obtained from M by deleting the bottommost row and the ith column from the left.

For $1 \leq i \leq d$, we get matrix M_i from matrix $M_i{}'$ when we exchange the jth column with the $(j+1)^{\text{st}}$ column, for $j = i, i+1, \ldots, d-1$. Matrix M_{d+1} is the same as M'. If we exchange two columns of a matrix, then its determinant changes sign. Notice that $d-i$ exchanges of columns are necessary to go from matrix $M_i{}'$ to matrix M_i which implies

$$\Delta_i = (-1)^{d-i}\Delta_i{}',$$

for $1 \leq i \leq d$. Furthermore, we have

$$\Delta_{d+1} = \Delta'.$$

Consequently,

$$\Delta = \left(\sum_{j=1}^{d-1}(-1)^{2d+1}\eta_{i_d,j}(\epsilon)\Delta_j{}'\right)+(-1)^{2d+2}\Delta_d{}'+(-1)^{2d+3}\eta_{i_d,d}(\epsilon)\Delta'.$$

Observe that $-\Delta$ is the same as the left side of condition (2). Thus, (2) is equivalent to the condition $\Delta > 0$. □

In order to unify the appearance of the matrix given in Lemma 9.4 we perform several straightforward operations that preserve the sign of its determinant. More specifically, we multiply each entry of the last column of M with -1 and then exchange the last two columns of M. This leads to matrix

$$\Gamma = \begin{pmatrix} \eta_{i_0,1}(\epsilon) & \eta_{i_0,2}(\epsilon) & \dots & \eta_{i_0,d-1}(\epsilon) & \eta_{i_0,d}(\epsilon) & -1 \\ \eta_{i_1,1}(\epsilon) & \eta_{i_1,2}(\epsilon) & \dots & \eta_{i_1,d-1}(\epsilon) & \eta_{i_1,d}(\epsilon) & -1 \\ \dots & \dots & \dots & \dots & \dots & \dots \\ \eta_{i_d,1}(\epsilon) & \eta_{i_d,2}(\epsilon) & \dots & \eta_{i_d,d-1}(\epsilon) & \eta_{i_d,d}(\epsilon) & -1 \end{pmatrix},$$

already defined in Section 9.4.1. Both operations change the sign of the determinant which implies $\det\Gamma = \Delta$. Furthermore, we have $M' = \Gamma'$ as defined in Section 9.4.1, which allows us to formulate the following theorem which summarizes the results of this section.

Theorem 9.5: Let $h_{i_0}(\epsilon)$ through $h_{i_d}(\epsilon)$ be $d+1$ hyperplanes with

$$h_{i_j}(\epsilon)\colon x_d = \eta_{i_j,1}(\epsilon)x_1 + \eta_{i_j,2}(\epsilon)x_2 + \dots + \eta_{i_j,d-1}(\epsilon)x_{d-1} + \eta_{i_j,d}(\epsilon),$$

for $0 \le j \le d$, and let the matrices Γ and Γ' be as defined above. Then, hyperplanes $h_{i_0}(\epsilon)$ through $h_{i_{d-1}}(\epsilon)$ intersect above (below) hyperplane $h_{i_d}(\epsilon)$ if and only if

$$\det\Gamma' \cdot \det\Gamma$$

is positive (negative).

9.4.3. Computing the Sign of a Determinant of Polynomials

This section describes how we can determine the sign of the determinant of a matrix whose entries are polynomials in ϵ as defined in Section 9.4.1.

For example, consider matrix Γ defined above. The determinant of Γ is again a polynomial in ϵ. To decide whether or not $\det\Gamma$ is positive, it is sufficient to compute the terms of this polynomial in order of decreasing significance, that is, in order of increasing exponents. Each term is of the form ϵ to some power times some coefficient. The first non-zero coefficient decides the sign of $\det\Gamma$ since ϵ to some power is always positive. In fact, we will see that only a small fraction of all coefficients has to be computed since we necessarily encounter a non-zero coefficient fairly early in our computation.

To simplify the computation of the correct order of the coefficients, we sort the rows of Γ such that

$$i_0 < i_1 < \ldots < i_d.$$

If we need an odd number of exchanges of rows to sort Γ in this way, then the determinant of Γ is -1 times the determinant of the sorted matrix; otherwise, the determinant is not affected. By definition of the perturbation, the exponent of ϵ is smallest in polynomial $\eta_{i_0,d}(\epsilon)$, second smallest in polynomial $\eta_{i_0,d-1}(\epsilon)$, and so on.

To illustrate the computation of the sign of $\det\Gamma$, we consider the two–dimensional case. Here, we have

$$\Gamma = \begin{pmatrix} \eta_{i_0,1}(\epsilon) & \eta_{i_0,2}(\epsilon) & -1 \\ \eta_{i_1,1}(\epsilon) & \eta_{i_1,2}(\epsilon) & -1 \\ \eta_{i_2,1}(\epsilon) & \eta_{i_2,2}(\epsilon) & -1 \end{pmatrix}.$$

The determinant of Γ is equal to

$$\det\begin{pmatrix} \eta_{i_0,1} & \eta_{i_0,2} & -1 \\ \eta_{i_1,1} & \eta_{i_1,2} & -1 \\ \eta_{i_2,1} & \eta_{i_2,2} & -1 \end{pmatrix} - \epsilon^{2^{2i_0}} \cdot \det\begin{pmatrix} \eta_{i_1,1} & -1 \\ \eta_{i_2,1} & -1 \end{pmatrix} + \epsilon^{2^{2i_0+1}} \cdot \det\begin{pmatrix} \eta_{i_1,2} & -1 \\ \eta_{i_2,2} & -1 \end{pmatrix}$$

$$+ \epsilon^{2^{2i_1}} \cdot \det\begin{pmatrix} \eta_{i_0,1} & -1 \\ \eta_{i_2,1} & -1 \end{pmatrix} + \epsilon^{2^{2i_0+1}} \cdot \epsilon^{2^{2i_1}} \cdot \det(-1) + \ldots .$$

Notice that already the coefficient of the fifth term is necessarily non–zero, that is, it is always equal to -1. This implies that in two dimensions the determination of the sign $\det\Gamma$ requires the evaluation of one three by three determinant and of three two by two determinants in the worst case. This is fairly good if one keeps in mind that the determinant of the three by three matrix has to be computed even if the simulation of simplicity is not used.

It is worthwhile to note here that the use of common single or double precision arithmetic will not allow us to decide when the determinant of a matrix with real or integer entries is equal to zero, unless we use additional techniques that increase the accuracy of the calculation. To avoid these difficulties, we suggest to represent a hyperplane in $\boldsymbol{E}^d$ by a d–tuple of integer numbers and to use local long integer arithmetic for evaluating determinants.

9.5. Penetration Search and Extremal Queries

This section offers static and dynamic solutions to the penetration search problem in two and three dimensions. We recall the definition of the *penetration search problem* in d dimensions, d an arbitrary positive integer.

> Let H be a finite set of non-vertical hyperplanes in E^d, each one bounding an open half-space such that the common intersection T of these half-spaces is non-empty. A *penetration query* is specified by a point p in T and a non-zero vector u, and it asks for a hyperplane in H that is hit first as p moves into the direction defined by u (see Figure 9.3(a)).

Since Algorithm 9.2 uses the simulation of simplicity to cope with degenerate cases (see Section 9.4), we can neglect these cases and assume that

(i) arrangement $\mathcal{A}(H)$ is simple,
(ii) $u \neq (0,0,\ldots,0,v_d)$, and
(iii) the line $l_{p,u} = \{p+\lambda u\}$ intersects each hyperplane of H in a unique point different from its intersection with any other hyperplane in H.

It follows from (iii) that the answer to a penetration query is a unique hyperplane.

It so happens that penetration queries are easier to understand in dual space than in the original space. Therefore, we define $Q = \mathcal{D}(H)$, the point set dual to H, and $h = \mathcal{D}(p)$ and $h' = \mathcal{D}(p+u)$, the hyperplanes dual to points p and $p+u$. When p moves along line $l_{p,u}$ in the original space then h performs a rotation about the $(d-2)$-flat $h \cap h'$ in dual space (see Figure 9.3). If we think of both actions happening simultaneously, then p runs into point $p+u$ at the same time as h becomes coincident with hyperplane h'. Furthermore, p encounters a hyperplane g in H at the same time as h meets point $\mathcal{D}(g)$ in Q. If p does not encounter any hyperplane of H before it escapes to infinity, that is, if the ray $\{p+\lambda u \mid \lambda \geq 0\}$ avoids all hyperplanes of H, then h does not meet any point of Q before it reaches the unique vertical hyperplane through $h \cap h'$.

In the dual setting, it is convenient to distinguish the points of Q above hyperplane h from those below h. We will treat each set separately and therefore define the *extremal search problem*, the dual of the penetration search problem, as follows.

> Let P be a finite set of points in E^d. An *extremal query* is specified by two non-vertical hyperplanes h and h' such that
>
> $$P \subseteq h^+ \text{ or } P \subseteq h^-.$$
>
> The query asks for a point in P that is hit first when h rotates about

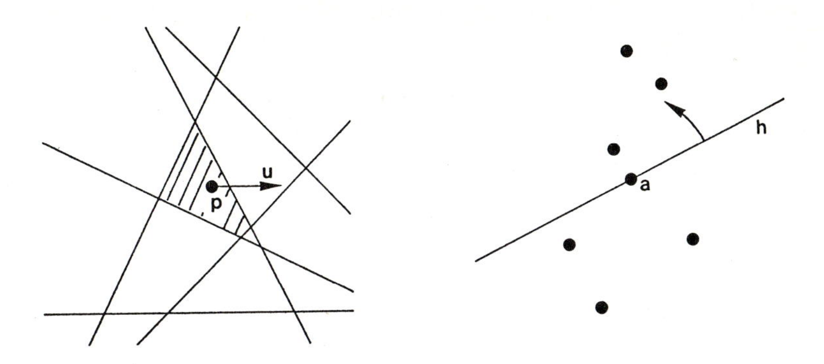

(a) Penetration query. (b) Extremal query.

Figure 9.3. A search problem in primal and in dual space.

the $(d-2)$–flat $h \cap h'$ in a direction such that it reaches h' without passing through a vertical position.

Again, only non–degenerate cases are of interest. This allows us to assume that

(i) no $d+1$ points of P lie on a common hyperplane,
(ii) h and h' are not parallel to each other, and
(iii) no two points of P lie in a hyperplane that also contains the intersection of h and h'.

In Sections 9.5.1 through 9.5.3, we present two data structures that solve the extremal search problem in two and three dimensions. The two–dimensional instance serves mainly as an introductory case which develops some ideas that appear again when we discuss the more complicated three–dimensional case. Finally, Section 9.5.4 gives a general method that can be used to modify the static data structures of Sections 9.5.1 through 9.5.3 such that they can process an intermixed sequence of extremal queries, insertions of points, and deletions of points.

9.5.1. Extremal Queries in the Plane

In two dimensions, an extremal query consists of a line l rotating in clockwise or counterclockwise direction about some anchor point a on l. All points of the finite set P lie on one side of l's initial position. The *answer* to this query is the point p in P that is hit first by line l. We denote point p by $ans_{l,a}(P)$ and we omit the two indices when they are understood. Recall that $ans(P)$ is unique since we can assume that no two points of P are collinear with point a.

Evidently, point p can be the answer to an extremal query only if it is extreme in P. This suggests that we use some representation of the convex hull of P as our data structure. Let

$$S_0=(p_0,p_1,\ldots,p_{m-1})$$

be a counterclockwise enumeration of the vertices of conv$P$. Furthermore, let S_i be the sequence obtained by deleting every other point from S_{i-1} starting with the second point, for $1 \leq i \leq k$ and k the greatest integer such that $2^k \leq m-1$. Thus, we have $k=\lfloor \log_2(m-1) \rfloor$. Note that S_i contains exactly those points of S_0 whose indices are multiples of 2^i. This implies that S_k contains exactly two points, p_0 and p_{2^k}. We call the sequence of sequences

$$\mathcal{H}(P)=(S_k,S_{k-1},\ldots,S_0)$$

the *hierarchical representation of* conv$P$ and we use it to answer extremal queries for P. Figure 9.4 indicates the hierarchical representation of a convex polygon with 15 vertices; every other polygon is shown with broken edges.

Let P_i be the set of points in sequence S_i, for $0 \leq i \leq k$. We have $P_0=\text{ext}P$ which implies that the answer to an extremal query for P is the same as the answer to the same query for P_0. To answer the extremal query for P_0 defined by line l and its anchor point a, we first answer the same extremal query for set P_1 recursively and then use $ans(P_1)$ to compute $ans(P_0)$. By the following result, which follows from straightforward convexity considerations, the step from P_1 to P_0 can be done in constant time.

Observation 9.6: Either $ans(P_{i-1})=ans(P_i)$ or $ans(P_i)$ is the successor or the predecessor of $ans(P_{i-1})$ in S_{i-1}, for $1 \leq i \leq k$.

Thus, in order to compute $ans(P_{i-1})$, we just need to answer the extremal

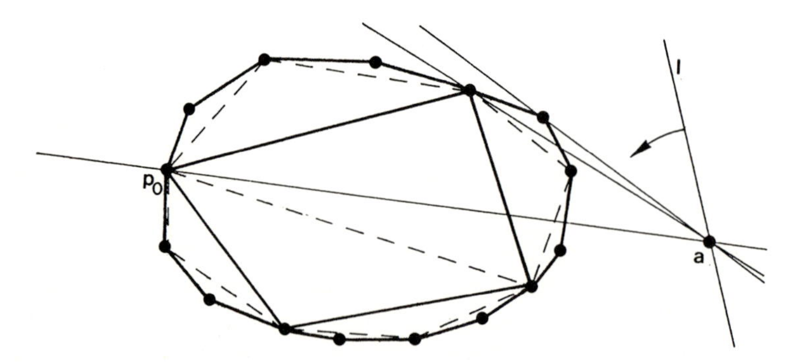

Figure 9.4. Hierarchical representation of a convex polygon.

query for the three points $p_j = ans(P_i)$, $p_{j-2^{i-1}}$, and $p_{j+2^{i-1}}$, with index calculations taken modulo m. Evidently, this can be done in constant time.

To specify the search algorithm more formally, we notice that a linear array storing sequence S_0 suffices to represent $\mathcal{H}(P)$. This is because S_i is conveniently embedded in S_0. The algorithm that we propose is recursive and it takes as input two integer numbers i and ℓ, where i keeps track of the level of recursion and $\ell = 2^i$. Initially, we have $i = 0$ and therefore $\ell = 1$.

Procedure 9.3 (Extremal search in two dimensions):

if $2\ell > m-1$, that is, $\ell = 2^k$ **then**

Answer the extremal query for the set $\{p_0, p_\ell\}$ and define j such that $p_j = ans(\{p_0, p_\ell\})$.

else

Call Procedure 9.3 recursively with parameters $i+1$ and 2ℓ. {This call computes index j such that $p_j = ans(P_{i+1})$.} Next, compute the new index j such that $p_j = ans(\{p_{j-\ell}, p_j, p_{j+\ell}\})$, doing all index calculations modulo m.

endif.

Since $\mathcal{H}(P)$ consists of at most $1+\log_2(n-1)$ sequences S_i, Procedure 9.3 takes time $O(\log n)$, n the cardinality of set P. The construction of the linear array that represents $\mathcal{H}(P)$ takes time $O(n \log n)$ if we use any one of the two algorithms described in Section 8.3. This implies the following result.

Theorem 9.7: Let P be a set of n points in the plane. There exists a data structure that requires $O(n)$ storage and takes time $O(n \log n)$ for construction such that time $O(\log n)$ suffices to answer any extremal query for P.

9.5.2. Extremal Queries in Three Dimensions: the Data Structure

An extremal query in E^3 is given by a plane h and an anchor line a contained in h about which h rotates. In order to specify the direction of the rotation, we let a be directed and we let h rotate such that the rotation is counterclockwise if we look in the direction determined by a. All points of the given set P lie on one side of the initial position of h, and the *answer* $ans_{h,a}(P)$ *to* this query is the point in P hit first by h. Without causing confusion, we will omit the two indices when we refer to $ans_{h,a}(P)$. We also recall that we can assume that no four points of P are coplanar and that no two points of P lie in a common plane with a. The latter implies that $ans(P)$ is unique.

2 (a) Map each point p in any of the sets P_i, $1 \leq i \leq k$, to the plane tangent to paraboloid U in E^3 such that p and the point where the plane touches U lie on a common vertical line. Characterize $\mathcal{V}$ in terms of a skeleton in the arrangement defined by the planes.

3 (b) Prove that the number of edges of $\mathcal{V}$ is in $O(k)$ if $\{\text{conv}P_i | 1 \leq i \leq k\}$ is a set of k pairwise non-intersecting line segments.

9.7. Bibliographic Notes

Most of the material in this chapter is taken from Edelsbrunner (1986). This includes in particular the general idea of computing a skeleton using a graph traversal algorithm. The ingredients to this general algorithm come from various sources as explained below.

A definition of eulerian tours in directed graphs can be found in Harary (1969) and other textbooks in graph theory. The depth-first-search strategy used to compute a eulerian tour of a special type of directed graph (see Procedure 9.1) is due to Tarjan (1972) and can also be found in Tarjan (1983). Algorithm 9.2 is modeled on this strategy. Each of its steps is supported by a dynamic data structure that answers extremal queries. In two dimensions, a linear array that contains the vertices of the convex hull of the points in sorted order suffices to answer extremal queries efficiently, as observed by Chazelle, Dobkin (1980). The static data structure for the three-dimensional case is taken from Dobkin, Kirkpatrick (1983). Applications of this data structure to other problems can be found in Dobkin, Kirkpatrick (1985) and in Edelsbrunner, Maurer (1985). The technique used in Section 9.5.4 to convert these static data structures to dynamic ones is taken from Maurer, Ottmann (1979). An extension of their method to a situation where no information about the executable sequence of requests is known beforehand is given in van Leeuwen, Wood (1980).

In two dimensions, there are more efficient dynamic data structures for the extremal search problem as indicated in Problem 9.6(a). Such a data structure can be found in Overmars, van Leeuwen (1981); Edelsbrunner, Welzl (1986a) applied their data structure to the construction of skeletons in two-dimensional arrangements. General sources for dynamization techniques applicable to static data structures are Bentley, Saxe (1980) and Overmars (1983). The data structure for the maintenance of the convex hull of a point set in three dimensions indicated in Problem 9.7(a) can be found in Gowda, Kirkpatrick (1980). Overmars (1981) sacrifices the $O(n \log\log n)$ bound for the amount of storage to get a more general variant of the same data structure.

The simulation of simplicity which helps us to cope with degenerate cases in an arrangement is taken from Edelsbrunner (1986). Implementation details of the method are reported in Mücke (1985). Similar ideas for overcoming the complication caused by degenerate cases can already be found in the linear programming literature (see Dantzig (1963) or Chvátal (1983)). These ideas lead to the problem of computing the sign of the determinant of a square matrix with integer entries. The algorithm for this problem indicated in Exercise 9.5(a) can be found in Aho, Hopcroft, Ullman (1974).

CHAPTER 10

LINEAR PROGRAMMING

Solving linear programs is one of the best investigated and most important problems in mathematical optimization. Formally, it is the problem of minimizing a linear objective function

$$v_1x_1+v_2x_2+\ldots+v_dx_d$$

subject to a collection of constraints, where each constraint is a linear inequality

$$\omega_{i,1}x_1+\omega_{i,2}x_2+\ldots+\omega_{i,d}x_d \leq \omega_{i,d+1},$$

for real numbers v_1 through v_d and $\omega_{i,1}$ through $\omega_{i,d+1}$. If the linear program involves d variables, $x_1,x_2,\ldots,x_d$, then each inequality can be interpreted as a closed half-space in E^d, and the collection of constraints becomes a convex polyhedron P that is the intersection of all half-spaces. The problem is then to find a point $x=(\xi_1,\xi_2,\ldots,\xi_d)$ in P, if it exists, that minimizes

$$\langle u,x\rangle = v_1\xi_1+v_2\xi_2+\ldots+v_d\xi_d,$$

where $u=(v_1,v_2,\ldots,v_d)$ is the vector which defines the objective function. Intuitively, x is a point of P which lies furthest in the direction determined by the vector $-u=(-v_1,-v_2,\ldots,-v_d)$. However, such a point does not always exist. We will return to this issue in Section 10.1, where we discuss the different types of linear programs existing and where we specify what we require from a solution to a linear program.

This chapter describes an algorithmic technique, called prune-and-search, which solves a linear program in time linear in the number of constraints. The amount of time needed is, however, double-exponential in d, the number of dimensions, which limits its practical use to low-dimensional cases. It is exactly these cases that we are interested in. In Section 10.2, we briefly examine the meaning of a linear program in dual space and mention a few applications of linear programming to other geometric problems. Section 10.3 presents the algorithm for the two-dimensional case; here, we stress the intuitive development of the global strategy. In Section 10.4, we generalize the algorithm to three and higher dimensions. The techniques used to solve two- and higher-dimensional linear programs are of independent interest since they can be used to solve a number of problems which are different from linear programming.

10.1. The Solution to a Linear Program

In this section, we specify precisely what it means to "solve" a linear program $\mathcal{L}$. For reasons of recursive applications, this specification is going to be somewhat non-standard. Let the non-zero vector

$$u=(\upsilon_1,\upsilon_2,\ldots,\upsilon_d)$$

define the objective function of $\mathcal{L}$, and let the half-space s_i of $\mathcal{L}$ be defined by the non-zero vector

$$w_i=(\omega_{i,1},\omega_{i,2},\ldots,\omega_{i,d})$$

and the real number $\omega_{i,d+1}$, such that

$$s_i=\{x\,|\langle w_i,x\rangle\leq\omega_{i,d+1}\},$$

for $1\leq i\leq n$. We partition the set of half-spaces into three sets S^+, S^0, and S^-, such that $s_i\in S^+$ if $\langle w_i,u\rangle>0$, $s_i\in S^0$ if $\langle w_i,u\rangle=0$, and $s_i\in S^-$ if $\langle w_i,u\rangle<0$. For example, if we have $u=(0,0,\ldots,0,1)$, then S^+ contains each half-space bounded from below by a hyperplane, S^- contains each half-space bounded from above by a hyperplane, and S^0 contains all half-spaces bounded by vertical hyperplanes. Now, we define

$$P(\mathcal{L})=\bigcap_{i=1}^{n}s_i,$$

the *feasible domain of* $\mathcal{L}$, and

$$\Sigma(\mathcal{L})=\bigcap_{s\in S^0}s,$$

the *range of* $\mathcal{L}$.

We distinguish four different types of linear programs. The classification discriminates between the linear programs depending on whether the feasible regions and the ranges are empty or non-empty. If $P(\mathcal{L})\neq\emptyset$ then $\mathcal{L}$ is said to be *feasible*, and it is *infeasible* if $P(\mathcal{L})=\emptyset$. Even if $\mathcal{L}$ is feasible, there is no optimal point if P is unbounded in the direction $-u$, that is, if there is no half-space g: $\langle u,x\rangle\geq\gamma$ which contains P. In this case, $\mathcal{L}$ is said to be *unbounded*, and it is *bounded* if P is contained in such a half-space g. If $\mathcal{L}$ is infeasible then it may or may not have an empty range $\Sigma(\mathcal{L})$. We will distinguish both cases in our forthcoming investigations, although we do not introduce special terms for the two types of linear programs.

A *solution to* $\mathcal{L}$ decides the type of $\mathcal{L}$, and it provides a small witness of the geometry of $\mathcal{L}$ when $\mathcal{L}$ is feasible and bounded and when $\mathcal{L}$ is infeasible with non-empty range. In the first case, this witness is a point x^* in the feasible domain $\mathcal{P}(\mathcal{L})$ that minimizes $\langle u, x^* \rangle$. We call x^* a *feasible solution to* $\mathcal{L}$. If $\mathcal{L}$ is infeasible and $\Sigma(\mathcal{L}) \neq \emptyset$, then the witness is a point x^* of the set $\mathcal{P}_\sigma$ defined as follows: for any real number α define $s_\alpha = s-(0,0,\ldots,0,\alpha)$ if $s \in S^+$ and $s_\alpha = s+(0,0,\ldots,0,\alpha)$ if $s \in S^-$, define

$$\mathcal{P}_\alpha = (\bigcap_{s \in S^0} s) \cap (\bigcap_{s \in S^+ \cup S^-} s_\alpha),$$

and let σ be the smallest real number α such that $\mathcal{P}_\alpha$ is non-empty. Such a real number σ exists since $\Sigma(\mathcal{L}) \neq \emptyset$, and σ is positive since $\mathcal{L}$ is infeasible which is equivalent to $\mathcal{P}_\alpha = \emptyset$, for $\alpha = 0$. Intuitively, σ is the smallest real number such that the linear program $\mathcal{L}$ becomes feasible when we move every half-space in S^+ vertically downward by σ units of length and every half-space in S^- vertically upward by the same amount. The point x^* is termed an *infeasible solution to* $\mathcal{L}$ and σ is called the *infeasibility of* x^*.

Notice that only a few half-spaces of $\mathcal{L}$ determine the feasible or infeasible solutions to $\mathcal{L}$, if they exist. We introduce some terminology to distinguish these half-spaces from the ones which do not have any effect on the solution to $\mathcal{L}$.

A half-space s of $\mathcal{L}$ is said to be *tight* if there exists a solution x^* such that
(i) $x^* \in \text{bd}s$, if x^* is a feasible solution, and
(ii) $x^* \in \text{bd}s$ for $s \in S^0$ and $x^* \in \text{bd}s_\sigma$ for $s \in S^+ \cup S^-$, if x^* is an infeasible solution and σ is its infeasibility.

We will not use "tightness" as a synonym for "non-redundancy" of half-spaces. In fact, we say that a half-space s of $\mathcal{L}$ is *redundant* if the feasible domain of $\mathcal{L}$ does not change as we remove s from the set of constraints. So a half-space can be tight but redundant, and it can be non-redundant but still not tight.

The requirement of computing an infeasible solution x^*, when the linear program is infeasible and has non-empty range, is rather unusual. This piece of information is needed only for recursive applications of our algorithm for linear programming. It will enable us to decide on which side of a chosen hyperplane (in a space with one more dimension) the solutions to a higher-dimensional linear program lie. To assist in the description of such recursive applications, we introduce the following terminology. Let $\mathcal{L}$ be a linear program in E^d as specified above, and let h be a hyperplane parallel to vector u which defines the objective function of $\mathcal{L}$. The *restriction of* $\mathcal{L}$ *to* h is the $(d-1)$-dimensional linear program $\mathcal{L}(h)$ defined by vector u and the set of constrains $\{s \cap h | s \text{ a half-space of } \mathcal{L}\}$. We can assume that no half-space s of $\mathcal{L}$ contains h or completely avoids h. In the first case, s turns into a void constraint of $\mathcal{L}(h)$, and in the second case, it

A *solution to* $\mathcal{L}$ decides the type of $\mathcal{L}$, and it provides a small witness of the geometry of $\mathcal{L}$ when $\mathcal{L}$ is feasible and bounded and when $\mathcal{L}$ is infeasible with non-empty range. In the first case, this witness is a point x^* in the feasible domain $\mathcal{P}(\mathcal{L})$ that minimizes $\langle u, x^* \rangle$. We call x^* a *feasible solution to* $\mathcal{L}$. If $\mathcal{L}$ is infeasible and $\Sigma(\mathcal{L}) \neq \emptyset$, then the witness is a point x^* of the set $\mathcal{P}_\sigma$ defined as follows: for any real number α define $s_\alpha = s - (0,0,...,0,\alpha)$ if $s \in S^+$ and $s_\alpha = s + (0,0,...,0,\alpha)$ if $s \in S^-$, define

$$\mathcal{P}_\alpha = (\bigcap_{s \in S^0} s) \cap (\bigcap_{s \in S^+ \cup S^-} s_\alpha),$$

and let σ be the smallest real number α such that $\mathcal{P}_\alpha$ is non-empty. Such a real number σ exists since $\Sigma(\mathcal{L}) \neq \emptyset$, and σ is positive since $\mathcal{L}$ is infeasible which is equivalent to $\mathcal{P}_\alpha = \emptyset$, for $\alpha = 0$. Intuitively, σ is the smallest real number such that the linear program $\mathcal{L}$ becomes feasible when we move every half-space in S^+ vertically downward by σ units of length and every half-space in S^- vertically upward by the same amount. The point x^* is termed an *infeasible solution to* $\mathcal{L}$ and σ is called the *infeasibility of* x^*.

Notice that only a few half-spaces of $\mathcal{L}$ determine the feasible or infeasible solutions to $\mathcal{L}$, if they exist. We introduce some terminology to distinguish these half-spaces from the ones which do not have any effect on the solution to $\mathcal{L}$.

> A half-space s of $\mathcal{L}$ is said to be *tight* if there exists a solution x^* such that
> (i) $x^* \in \mathrm{bd}s$, if x^* is a feasible solution, and
> (ii) $x^* \in \mathrm{bd}s$ for $s \in S^0$ and $x^* \in \mathrm{bd}s_\sigma$ for $s \in S^+ \cup S^-$, if x^* is an infeasible solution and σ is its infeasibility.

We will not use "tightness" as a synonym for "non-redundancy" of half-spaces. In fact, we say that a half-space s of $\mathcal{L}$ is *redundant* if the feasible domain of $\mathcal{L}$ does not change as we remove s from the set of constraints. So a half-space can be tight but redundant, and it can be non-redundant but still not tight.

The requirement of computing an infeasible solution x^*, when the linear program is infeasible and has non-empty range, is rather unusual. This piece of information is needed only for recursive applications of our algorithm for linear programming. It will enable us to decide on which side of a chosen hyperplane (in a space with one more dimension) the solutions to a higher-dimensional linear program lie. To assist in the description of such recursive applications, we introduce the following terminology. Let $\mathcal{L}$ be a linear program in $\boldsymbol{E}^d$ as specified above, and let h be a hyperplane parallel to vector u which defines the objective function of $\mathcal{L}$. The *restriction of* $\mathcal{L}$ *to* h is the $(d-1)$-dimensional linear program $\mathcal{L}(h)$ defined by vector u and the set of constrains $\{s \cap h | s \text{ a half-space of } \mathcal{L}\}$. We can assume that no half-space s of $\mathcal{L}$ contains h or completely avoids h. In the first case, s turns into a void constraint of $\mathcal{L}(h)$, and in the second case, it makes $\mathcal{L}(h)$ an infeasible linear program for trivial reasons.

10.2. Linear Programming and Duality

A linear program $\mathcal{L}$ in E^d is specified by n half-spaces, the constraints, and a vector $u=(v_1,v_2,\ldots,v_d)$, which determines the objective function. For the time being, assume that

$$u=(0,0,\ldots,0,1),$$

and assume that no hyperplane bounding a half-space of $\mathcal{L}$ is vertical, that is, parallel to the x_d-axis. Let H^+ be the set of hyperplanes h such that $\mathrm{cl}h^+$, the closed half-space above h, is a half-space of the linear program, and define H^- analogously. The linear program is feasible if

$$P(\mathcal{L})=(\bigcap_{h\in H^+}\mathrm{cl}h^+)\cap(\bigcap_{h\in H^-}\mathrm{cl}h^-)\neq\emptyset,$$

and it is bounded if it is feasible and there is a horizontal hyperplane g: $x_d=\gamma$, such that g^+ contains $P(\mathcal{L})$. If the linear program is bounded, then a feasible solution is a point $x^*=(\xi_1,\xi_2,\ldots,\xi_d)$ in $P(\mathcal{L})$ which minimizes ξ_d. Intuitively, x^* belongs to the intersection of $P(\mathcal{L})$ with the bottommost horizontal hyperplane such that the intersection is non-empty. It follows that the set of solutions is the closure of a face of $P(\mathcal{L})$.

We gain new insights into the problem of solving a linear program when we investigate its geometric interpretation under the dual transform $\mathcal{D}$. This transform maps a non-vertical hyperplane

$$h\colon x_d=\eta_1x_1+\eta_2x_2+\ldots+\eta_{d-1}x_{d-1}+\eta_d$$

to the point

$$\mathcal{D}(h)=(\frac{\eta_1}{2},\frac{\eta_2}{2},\ldots,\frac{\eta_{d-1}}{2},-\eta_d),$$

and it maps a point p to the hyperplane $\mathcal{D}(p)$ such that $p=\mathcal{D}(\mathcal{D}(p))$. Define the point sets

$$P^+=\mathcal{D}(H^+)=\{\mathcal{D}(h)|h\in H^+\} \text{ and } P^-=\mathcal{D}(H^-)=\{\mathcal{D}(h)|h\in H^-\}.$$

If the linear program is feasible and bounded, then there is a solution $x^*=(\xi_1,\xi_2,\ldots,\xi_d)$, and this point corresponds to the hyperplane

$$\mathcal{D}(x^*)\colon x_d=2\xi_1x_1+2\xi_2x_2+\ldots+2\xi_{d-1}x_{d-1}-\xi_d.$$

Since the transform is incidence- and order-preserving (see Section 1.4), we have

$$P^+\subseteq\mathrm{cl}\mathcal{D}(x^*)^+ \text{ and } P^-\subseteq\mathrm{cl}\mathcal{D}(x^*)^-;$$

thus, $\mathcal{D}(x^*)$ is a hyperplane that separates P^+ and P^-. In addition, since x^*

minimizes the dth coordinate, $\mathcal{D}(x^*)$ is a separating hyperplane which has the topmost intersection with the x_d–axis. The linear program is infeasible if and only if there is no non–vertical hyperplane h with $P^+ \subseteq \text{cl}h^+$ and $P^- \subseteq \text{cl}h^-$.

If the vector u, which defines the objective function, has coordinates $(v_1, v_2, \ldots, v_{d-1}, v_d)$ not necessarily equal to $(0,0,\ldots,0,1)$, then the dual hyperplane $\mathcal{D}(h)$ of a feasible solution x^* is extreme with respect to a vertical line not necessarily equal to the x_d–axis. If $v_d > 0$ then $\mathcal{D}(x^*)$ has the topmost intersection with the line $l(u)$ of points $p = (\pi_1, \pi_2, \ldots, \pi_d)$, where

$$\pi_i = \frac{-v_i}{2v_d}, \quad \text{for } 1 \leq i \leq d-1,$$

and π_d varies over all real numbers. If $v_d < 0$ then $\mathcal{D}(x^*)$ realizes the bottommost intersection with line $l(u)$, and if $v_d = 0$ then $\mathcal{D}(x^*)$ minimizes or maximizes the slope of the line that is the intersection of $\mathcal{D}(x^*)$ with the (two–dimensional) plane that contains the point $(\frac{v_1}{2}, \frac{v_2}{2}, \ldots, \frac{v_{d-1}}{2}, 0)$ and the x_d–axis.

By the observations in the above discussion, it is easy to see that the following two geometric problems can be solved using a dual transform and an algorithm for linear programming.

Linear separability of two point sets: Let P_1 and P_2 be two finite and non–empty sets of points in E^d. Determine whether or not P_1 and P_2 can be separated by a hyperplane. Here, we say that a hyperplane h *separates* P_1 and P_2 if P_1 and P_2 are contained in different closed half–spaces determined by h.

To solve the problem, we solve two linear programs with arbitrary objective function. For one program, we map each point p in P_1 to the half–space $\text{cl}\mathcal{D}(p)^+$ and each point q in P_2 to the half–space $\text{cl}\mathcal{D}(q)^-$. For the second program, we use the same hyperplanes but define the half–spaces on the respective opposite sides. There is no non–vertical hyperplane that separates P_1 and P_2 if and only if both linear programs are infeasible. To check the existence of a vertical separation, we project the points vertically onto the plane h_0: $x_d = 0$ and repeat the process in $d-1$ dimensions.

Topmost convex combination on a line: Let P be a set of n points in E^d. For a vertical line l, we define the *topmost convex combination* $p(l)$ *of* l as the upper endpoint of the interval $l \cap \text{conv}P$ on line l. This point $p(l)$ is the intersection of l with a hyperplane h such that $P \subseteq \text{cl}h^-$ and h has the bottommost intersection with l.

To compute this point, if it exists, we map each point p of P to the half-space cl$\mathcal{D}(p)^-$. Notice that the resulting linear program is feasible since the intersection of these half-spaces cannot be empty. If l intersects the hyperplane h_0: $x_d = 0$ in the point $(\lambda_1, \lambda_2, \ldots, \lambda_{d-1}, 0)$, then we solve the linear program with the objective function determined by the vector $(2\lambda_1, 2\lambda_2, \ldots, 2\lambda_{d-1}, -1)$. The linear program is unbounded if and only if the line l does not intersect conv$P$. If it is bounded then the dual hyperplane of a feasible solution x^* contains the topmost convex combination of l, that is, $p(l) = l \cap \mathcal{D}(x^*)$.

There are many more examples of geometric problems which can be solved by transforming them to linear programs. A few of these problems can be found in the exercise section which also sketches a two-dimensional convex hull algorithm that uses the above solution to the convex combination problem.

10.3. Linear Programming in Two Dimensions

A linear program in two dimensions is one which involves only two variables, x_1 and x_2. In this case, each constraint can be interpreted as a half-plane in E^2. This section is devoted to the description of an algorithm which solves a linear program defined by n half-planes in time $O(n)$. The presentation is meant to be tutorial and thus stresses the intuitive ideas that lead to the algorithm. Nevertheless, we give all details which are necessary for an implementation. Section 10.4 will generalize the algorithm to three and higher dimensions, and every single step of the two-dimensional algorithm will be replaced by a considerably more involved step. It is therefore important to comprehend the intuition behind the algorithm when we discuss its specialization to two dimensions.

The global structure of the algorithm follows the so-called prune-and-search paradigm, where a search step decreases the range of possible solutions, and a prune step eliminates data which is irrelevant in this range. After a search and a prune step, we simply recur with the smaller set of data until the problem becomes trivial. A prune step is typically straightforward, while a fair deal of sophistication is usually required to design an effective search step. The main goal in a search step is to decrease the range of possible solutions in a way that allows us to eliminate a proportional amount of the data. Thus, a search step consists of two steps which may be iterated a constant number of times: first, we find a suitable test, and second, we answer this test. In our case, a test comes as a vertical line, and we decide on which side of this line we are going to continue the search for a solution. Below, we give a more formal description of the prune-and-search paradigm. We write D for the set of data and Σ for the range which is known to contain all solutions. Section 10.3.4 will rewrite this algorithm with the specifics for linear programming in the plane added.

Algorithm 10.1 (Prune-and-search):

```
if the size of D is at most some constant then
    Use a trivial procedure to solve the problem.
  else
    SEARCH: Iterate the following two steps some constant number of
        times:
        FIND_TEST: Find an appropriate test t.
        BISECT: Decrease the range Σ which contains all solutions by
            answering the test t.
    PRUNE: Eliminate some subset of D which is irrelevant in Σ.
    RECUR: Repeat the computation for the new sets D and Σ.
endif.
```

There are two possible ways to handle the range Σ of possible solutions. We can keep track of the answers to all tests and thus decrease Σ at each BISECT step, as suggested by Algorithm 10.1. We will follow this strategy when we discuss linear programs in E^2. As will be seen in Section 10.4, which is concerned with linear programs in E^d, $d \geq 3$, it is also possible to ignore the accumulated information about Σ. The advantage of the first strategy which records all answers to tests is that it leads to more eliminations of constraints in a single PRUNE step because our knowledge about the location of the solutions is more precise than without the records. The advantage of the second strategy, which can be called the "forgetful" approach, is that we do not have the additional overhead of maintaining Σ. This will lead to a slightly less complicated algorithm for the d-dimensional case.

In the following three sections, we will discuss the specifics of the steps called FIND_TEST, BISECT, and PRUNE for the case of two-dimensional linear programming in reverse order. Throughout, we assume that the vector $u=(0,1)$ defines the objective function of the linear program $\mathcal{L}$ considered. Finally, we will put the pieces together and give an analysis of the overall algorithm in Section 10.3.4. It is important to keep in mind that the efficiency of the algorithm not only relies on the efficiency of every single step but also on the proper composition of their interdependence.

10.3.1. Prune: Eliminate Redundant Half-Planes

A half-plane is redundant if its bounding line does not contain an edge of the feasible domain of $\mathcal{L}$. It is certainly too expensive to detect all redundant half-planes in a single PRUNE step, and we will see that it is actually not necessary for an efficient overall performance of the algorithm. We decide on the redundancy of a half-plane based on our knowledge about the location of the feasible or infeasible solutions to the linear program $\mathcal{L}$. This knowledge is expressed in the range $\Sigma(\mathcal{L})$ of $\mathcal{L}$. We will give sufficient conditions which define when a half-plane is redundant, and we will eliminate a half-plane if and only if it meets these conditions.

The information that we derive from the preceding steps of the algorithm consists of the range $\Sigma(\mathcal{L})$ of the linear program $\mathcal{L}$ and a set of pairs of half-planes. $\Sigma(\mathcal{L})$ is determined by two vertical lines l_1: $x_1=a$ and l_2: $x_2=b$ such that $\Sigma(\mathcal{L})=\{(\xi_1,\xi_2)|a\leq\xi_1\leq b\}$ (a may be equal to $-\infty$, and b may be $+\infty$). Initially, $\Sigma(\mathcal{L})$ is defined as the intersection of all half-planes that are bounded by vertical lines. We can therefore assume that the half-planes which are paired are all bounded by non-vertical lines. Furthermore, two half-planes with bounding lines g and h are paired only if they are similarly directed, that is, if they are equal to $\mathrm{cl}g^+$ and $\mathrm{cl}h^+$ or to $\mathrm{cl}g^-$ and $\mathrm{cl}h^-$.

Now, let g and h be the bounding lines of a pair of half-planes, and assume without loss of generality that $\mathrm{cl}g^+$ and $\mathrm{cl}h^+$ are the corresponding half-planes. Let g: $x_2=\gamma_1x_1+\gamma_2$ and h: $x_2=\eta_1x_1+\eta_2$ be the equations of the two lines. We distinguish several cases depending on the slopes and intercepts of the lines g and h.

Case 1: The lines are parallel, that is, $\gamma_1=\eta_1$. In this case, one of the half-planes is redundant independent of the range $\Sigma(\mathcal{L})$: this half-plane is $\mathrm{cl}g^+$, if $\gamma_2<\eta_2$, and it is $\mathrm{cl}h^+$, if $\gamma_2>\eta_2$ (see Figure 10.1(a)).

Case 2: The lines are not parallel, that is, $\gamma_1\neq\eta_1$. Without loss of generality, assume that $\gamma_1<\eta_1$. Let $p=g\cap h$ be the point of intersection with coordinates (π_1,π_2), that is,

$$\pi_1=\frac{\eta_2-\gamma_2}{\gamma_1-\eta_1} \quad\text{and}\quad \pi_2=\frac{\gamma_1\eta_2-\gamma_2\eta_1}{\gamma_1-\eta_1}.$$

Here, we distinguish three subcases.

Case 2.1: Point p lies to the left of the interior of $\Sigma(\mathcal{L})$, that is, $\pi_1\leq a$. In this case, $\mathrm{cl}g^+$ is redundant (see Figure 10.1(b)).

Case 2.2: Point p belongs to $\mathrm{int}\Sigma(\mathcal{L})$, that is, $a<\pi_1<b$. In this case, both $\mathrm{cl}g^+$ and $\mathrm{cl}h^+$ may be non-redundant (see Figure 10.1(c)).

Case 2.3: Point p lies to the right of $\mathrm{int}\Sigma(\mathcal{L})$, that is, $b\leq\pi_1$. This is symmetric to Case 2.1, and the redundant half-plane in this case is $\mathrm{cl}h^+$ (see Figure 10.1(d)).

Notice that in each case, with the exception of Case 2.2, one half plane of the pair is found to be redundant. We will see in Section 10.3.3 how we can

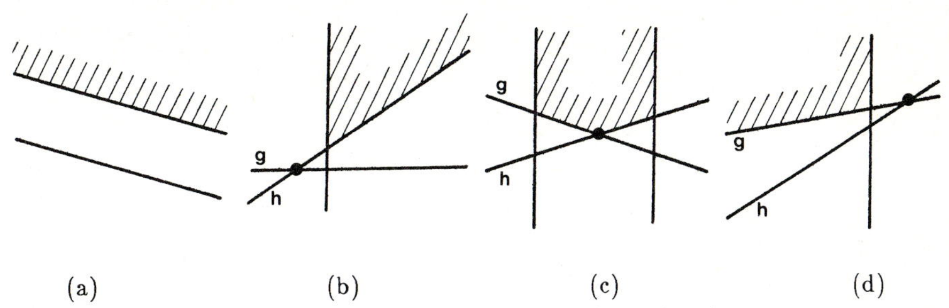

Figure 10.1. Sufficient conditions for redundancy.

guarantee that Case 2.2 applies to at most half of all pairs constructed.

10.3.2. Bisect: Decrease the Range of the Linear Program

We have seen earlier that the range $\Sigma(\mathcal{L})$ of $\mathcal{L}$ is the closed region between two vertical lines l_1: $x_1=a$ and l_2: $x_1=b$, with $a \leq b$. The non-vertical lines bounding half-planes of $\mathcal{L}$ come in two sets H^+ and H^- such that $\mathrm{cl}h^+$ is a constraining half-plane if $h \in H^+$, and $\mathrm{cl}h^-$ is one if $h \in H^-$. Given a vertical test line t: $x_1=\tau$, this section considers the problem of deciding whether the solutions to $\mathcal{L}$, if they exist, lie to the left or the right of t, or whether t itself contains a solution to $\mathcal{L}$. The set of feasible solutions is convex, since it is the closure of a face of

$$P=(\bigcap_{h \in H^+} \mathrm{cl}h^+) \cap (\bigcap_{h \in H^-} \mathrm{cl}h^-),$$

and the set of infeasible solutions is convex since it is the intersection of half-planes. Therefore, the decision is unambiguous if t contains no feasible or infeasible solution. Deciding on which side of t the solutions lie does not necessarily mean that we have to know anything about the type of the linear program.

We can assume that line t belongs to $\Sigma(\mathcal{L})$, that is, $a \leq \tau \leq b$. Otherwise, the answer will be the side of t which contains $\Sigma(\mathcal{L})$; it can be found in constant time. To solve our problem, consider the curves

$$C^+=bd(\bigcap_{h \in H^+} h^+) \text{ and } C^-=bd(\bigcap_{h \in H^-} h^-)$$

(see Figure 10.2). Keep in mind, however, that both curves are implicitly given as unordered sets of lines. Notice that the boundary of the polyhedron P is contained in the union of C^+ and C^-, and that the regions above C^+ and below C^- are both convex. In the first step, we compute the intersections

$$p^+(t)=t \cap C^+ \text{ and } p^-(t)=t \cap C^-$$

of line t with both curves. If $p^+(t)$ lies above $p^-(t)$ then $x^*(t)=\frac{p^+(t)+p^-(t)}{2}$ is

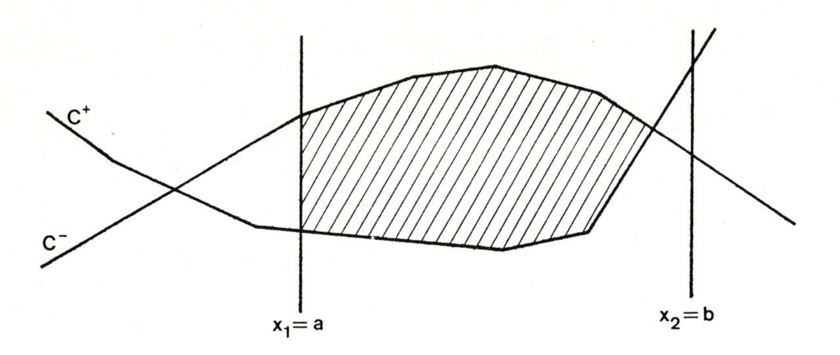

Figure 10.2. The curves C^+ and C^-.

the infeasible solution to the restriction $\mathcal{L}(t)$ of $\mathcal{L}$ to line t. If $p^+(t)=p^-(t)$ or if $p^+(t)$ lies below $p^-(t)$, then $x^*(t)=p^+(t)$ is the feasible solution to $\mathcal{L}(t)$. Next, we want to find out whether the solution "improves" when we move t towards the left, or towards the right. By a "better" solution, we mean an infeasible solution with smaller infeasibility or a smaller feasible solution. Because of convexity, the solution can improve only on one side of t. It is straightforward to prove this fact in two dimensions, and we omit the proof since Section 10.4 will show the generalization of this statement to three and higher dimensions. One way to find the direction is to solve two additional one-dimensional linear programs for two vertical lines arbitrarily close to the left and to the right of t. We get the same effect at lower cost if we solve two one-dimensional linear programs $\tilde{\mathcal{L}}(t')$ and $\tilde{\mathcal{L}}(t'')$ for the vertical lines t': $x_1=\tau-1$ and t'': $x_1=\tau+1$, whose only constraints are the intersections with the half-planes of $\mathcal{L}$ that correspond to tight constraints of $\mathcal{L}(t)$. Recall that such a half-plane contains point $p^+(t)$ in its bounding line if $p^+(t)$ is a feasible solution to $\mathcal{L}(t)$, and it contains $p^+(t)$ or $p^-(t)$ in its bounding line if $\mathcal{L}(t)$ is infeasible. We write $x^*(t')$ and $x^*(t'')$ for the solutions to $\tilde{\mathcal{L}}(t')$ and $\tilde{\mathcal{L}}(t'')$.

To solve our problem, we distinguish several cases which depend on the solutions to $\mathcal{L}(t)$, $\tilde{\mathcal{L}}(t')$, and $\tilde{\mathcal{L}}(t'')$. The various cases are illustrated in Figure 10.3.

Case 1: $\mathcal{L}(t)$ is unbounded. This implies that $\mathcal{L}$ is also unbounded (see Figure 10.3(a)).

Case 2: Point $x^*(t)$ is a feasible solution to $\mathcal{L}(t)$.

Case 2.1: If $\tilde{\mathcal{L}}(t')$ has a feasible solution $x^*(t')$ with smaller x_2-coordinate than $x^*(t)$, then the feasible solutions to $\mathcal{L}$ lie to the left of t (see Figure 10.3(b)).

Case 2.2: If $\tilde{\mathcal{L}}(t'')$ has a feasible solution $x^*(t'')$ with smaller x_2-coordinate than $x^*(t)$, then the feasible solutions to $\mathcal{L}$ lie to the right of t.

Case 2.3: If neither $x^*(t')$ nor $x^*(t'')$ have smaller x_2–coordinate than $x^*(t)$, then $x^*(t)$ is a feasible solution to $\mathcal{L}$ (see Figure 10.3(c)).

Case 3: Point $x^*(t)$ is an infeasible solution to $\mathcal{L}(t)$.

Case 3.1: If $\tilde{\mathcal{L}}(t')$ has a feasible solution or if its infeasible solution has smaller infeasibility than $x^*(t)$, then the solutions to $\mathcal{L}$ lie to the left of t.

Case 3.2: If $\tilde{\mathcal{L}}(t'')$ has a feasible solution or its infeasible solution has smaller infeasibility than $x^*(t)$, then the solutions to $\mathcal{L}$ lie to the right of t.

Case 3.3: If Cases 3.1 and 3.2 do not apply, then $x^*(t)$ is an infeasible solution to $\mathcal{L}$ (see Figure 10.3(d)).

The above case–analysis relies on the convexity of the region above curve C^+ and the region below curve C^-. It is interesting to see what computations are needed to perform the above analysis. Essentially, they consist of solving three one–dimensional linear programs: $\mathcal{L}(t)$, $\tilde{\mathcal{L}}(t')$, and $\tilde{\mathcal{L}}(t'')$. The two latter linear programs are only non–trivial in rather degenerate cases when many lines which bound half–planes contain point $p^+(t)$ or point $p^-(t)$. In any case, all actions can be performed in time linear in the number of half–planes involved in the problem.

10.3.3. Find_Test: Determine the Median

It is the responsibility of the FIND_TEST step to determine a vertical line which allows us to eliminate a proportional number of half–planes. Recall that Section 10.3.1 demonstrates that the only case where no half–plane of a given pair can be eliminated occurs if the intersection of the two bounding lines belongs to the interior of the vertical strip $\Sigma(\mathcal{L})$. The idea is to pair the half–planes first and to

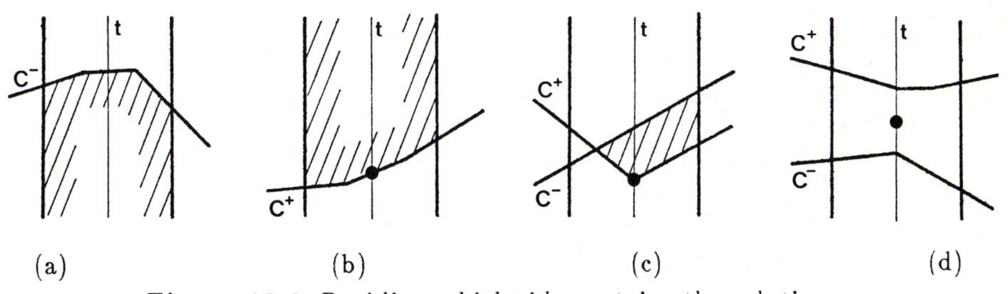

Figure 10.3. Deciding which side contains the solutions.

choose the test line t according to the locations of the intersections between the bounding lines of matched half-planes. This idea leads to the strategy described below.

In the first step, we match each but possibly one line in H^+ with another unique line in H^+. This creates a set H^+_{pair} of $\lfloor\frac{\text{card}H^+}{2}\rfloor$ pairs of lines. Now we define $P^+=\{g\cap h \mid \{g,h\}$ is a pair in $H^+_{pair}\}$. It is important to keep all copies of a multiple point in P^+ which implies that we have to think of P^+ as a multiset rather than a set. Analogously, we match the lines in H^- and thus create a multiset P^- of at most $\lfloor\frac{\text{card}H^-}{2}\rfloor$ intersection points. Next, we determine the point p in the multiset $P^+\cup P^-$ with median x_1-coordinate τ, where the *median of* a multiset of m elements is defined as the $\lfloor\frac{m}{2}\rfloor^{\text{th}}$ element in a sorted ordering of the multiset. The test line is t: $x_1=\tau$. By the construction of line t, each open half-plane determined by t contains at most half of the intersection points, which implies that at most half of the intersection points end up in the interior of the new range $\Sigma(\mathcal{L})$ which is updated in the BISECT step.

10.3.4. Assembling the Algorithm

In this section, we amalgamate the algorithmic portions developed in Sections 10.3.1 through 10.3.3 into a single algorithm. Afterwards, we will show that this algorithm works in linear time. The input to the algorithm consists of two sets of non-vertical lines H^+ and H^- and a vertical strip $\Sigma(\mathcal{L})$. The sets H^+ and H^- are such that a line h is in H^+ if and only if clh^+ is a constraining half-plane of the linear program $\mathcal{L}$, and h is in H^- if and only if clh^- is a half-plane of $\mathcal{L}$. $\Sigma(\mathcal{L})$ is a vertical strip which is known to contain all solutions to $\mathcal{L}$, and initially, it is the intersection of all constraining half-planes bounded by vertical lines. We assume that $\Sigma(\mathcal{L})\neq\emptyset$; otherwise, $\mathcal{L}$ is infeasible for trivial reasons.

Algorithm 10.2 (Linear programming in two dimensions):

if card$H^+\leq 1$ and card$H^-\leq 1$ **then**

Use a trivial procedure to solve $\mathcal{L}$ and halt. The cases that have to be distinguished are indicated in Figure 10.4.

else

FIND_TEST: Match the lines in H^+ and H^- and let t be the vertical line through the intersection point with median x_1-coordinate (see Section 10.3.3).

BISECT: If t avoids the range $\Sigma(\mathcal{L})$ then skip further computations and continue with the PRUNE step. Otherwise, solve the one–dimensional linear programs $\mathcal{L}(t)$, $\tilde{\mathcal{L}}(t')$, and $\tilde{\mathcal{L}}(t'')$, and decide whether the solution to $\mathcal{L}(t)$ is the same as that to $\mathcal{L}$, and, if not, decide on which side of t the solutions to $\mathcal{L}$ lie (see Section 10.3.2). We distinguish four cases.

Case 1: $\mathcal{L}(t)$ is unbounded. In this case, $\mathcal{L}$ is unbounded too, and we halt.

Case 2: Line t contains a feasible or infeasible solution $x^*(t)$ to $\mathcal{L}$. Report this solution and halt.

Case 3: The solutions lie to the left of line t. Restrict the range $\Sigma(\mathcal{L})$ to the left side of t, that is, set $b := \min\{b, \tau\}$.

Case 4: The solutions lie to the right of line t. Restrict the range $\Sigma(\mathcal{L})$ to the right side of t, that is, set $a := \max\{a, \tau\}$.

PRUNE: Eliminate a line from each matched pair whose intersection does not lie in the interior of $\Sigma(\mathcal{L})$ (see Section 10.3.1).

RECUR: Repeat Algorithm 10.2 for the new sets H^+, H^-, and $\Sigma(\mathcal{L})$.

endif.

Notice that each step of Algorithm 10.2, with the exception of the recursive call, takes time linear in the number of lines in sets H^+ and H^-. To prove that the entire algorithm takes linear time, we need to show that the sets H^+ and H^- are considerably reduced after each iteration of the algorithm.

Lemma 10.1: Let $n^+ = \text{card}H^+$ and $n^- = \text{card}H^-$, for sets H^+ and H^- at the beginning of some iteration of Algorithm 10.2. This iteration either solves the problem and halts, or it eliminates a total of at least $\frac{n^+ + n^-}{6}$ lines from H^+ and H^-.

Proof: Assume that the current iteration of Algorithm 10.2 does not halt, and

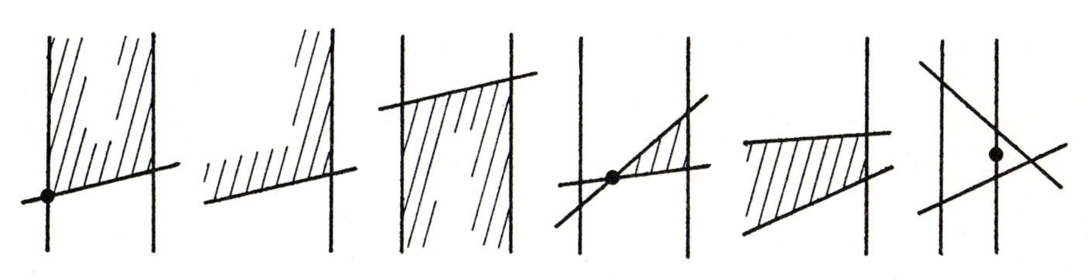

Figure 10.4. Analysis of the trivial cases.

consider the steps in sequence. The FIND_TEST step matches lines in H^+ with lines in H^+, and matches lines in H^- with lines in H^-. It therefore constructs

$$m = \lfloor \frac{\mathrm{card} H^+}{2} \rfloor + \lfloor \frac{\mathrm{card} H^-}{2} \rfloor = \lceil \frac{n^+ - 1}{2} \rceil + \lceil \frac{n^- - 1}{2} \rceil$$

pairs of lines. By the choice of the vertical test line t, at most $\frac{m}{2}$ pairs have their intersection in the interior of the range $\Sigma(\mathcal{L})$ after it is updated in the BISECT step. Thus, at least $\frac{m}{2}$ pairs of lines either intersect outside $\mathrm{int}\Sigma(\mathcal{L})$ or they are parallel. The PRUNE step eliminates one line of each such pair which shows that at least

$$\lceil \frac{m}{2} \rceil = \lceil \frac{\lceil \frac{n^+ - 1}{2} \rceil + \lceil \frac{n^- - 1}{2} \rceil}{2} \rceil$$

lines are eliminated. The assertion follows since

$$\lceil \frac{\lceil \frac{n^+ - 1}{2} \rceil + \lceil \frac{n^- - 1}{2} \rceil}{2} \rceil \geq \frac{n^+ + n^-}{6},$$

unless $n^+ \leq 1$ and $n^- \leq 1$ (see Exercise 10.5). □

Lemma 10.1 shows that the total size of H^+ and H^- decreases unless $\mathrm{card} H^+ \leq 1$ and $\mathrm{card} H^- \leq 1$; it follows that Algorithm 10.2 terminates. Let n be the number of half-planes of the linear program $\mathcal{L}$, and let n_i, for $1 \leq i$, be the number of lines in H^+ and H^- at the beginning of the ith iteration of Algorithm 10.2. We have $n_1 \leq n$ ($n_1 = n$ if and only if no half-plane is bounded by a vertical line), and

$$n_i \leq \frac{5 n_{i-1}}{6}, \quad \text{for } i \geq 2,$$

by Lemma 10.1. Now, let $T(n)$ denote the amount of time needed by Algorithm 10.2 to solve a two-dimensional linear program for n half-planes. Then we have

$$T(n) = T(\frac{5n}{6}) + \mathrm{O}(n),$$

which implies that $T(n) = \mathrm{O}(n)$ since the argument of function T decreases geometrically. We therefore conclude the main result of this section.

Theorem 10.2: Let H be a set of n half-planes in E^2. Algorithm 10.2 solves a linear program defined by H and a non-zero vector in time $\mathrm{O}(n)$.

For the generalization of Algorithm 10.2 to three and higher dimensions, it is important to see that it is not necessary to update the range of the linear program whenever we decide on which side of a vertical test line the solutions lie. Certainly, the PRUNE step eliminates enough half-planes if it just uses the information on which side of the last test line the solutions lie and thus disregards the earlier tests. We will adapt this "forgetful" strategy in handling the answers to tests in three and higher dimensions. The main reason for this approach is to avoid an increase in the number of constraints, even if it is only by a constant number. Remember that in two dimensions, the range $\Sigma(\mathcal{L})$ can always be represented by two constraints, while in three dimensions the number can be arbitrarily large.

10.4. Linear Programming in Three and Higher Dimensions

The generalization of Algorithm 10.2 to dimensions $d>2$ is by no means straightforward and requires a number of new ideas. In particular the SEARCH step is going to be rather involved since it needs to make sure that the PRUNE step eliminates a proportional number of half-spaces. This will be achieved by 2^{d-2} iterations of finding a vertical test hyperplane t and deciding which side of t can be excluded from further considerations. We will describe this iteration in Sections 10.4.3 for $d=3$ and in Section 10.4.4 for $d>3$. The strategy for deciding a single test, which comes as a vertical hyperplane, is the same as in two dimensions. Sections 10.4.2 offers a brief description of this procedure and proves its correctness. To eliminate half-spaces in the PRUNE step, we consider pairs of half-spaces, exactly as in the planar case, and we eliminate a half-space of a pair only if the intersection of the two bounding hyperplanes does not intersect the interior of the cell determined in the preceding SEARCH step.

Throughout this section, we assume that the objective function of the linear program $\mathcal{L}$ is defined by the vector $u=(0,0,...,0,1)$. There are two justifications for this assumption. First, we can apply a suitable transformation that takes vector u to $(0,0,...,0,1)$. The disadvantage of this approach is the introduction of numerical errors and instability, especially, if a sequence of transformations is performed. Alternatively, all computations can be done without transformation by suitably changing the meaning of notions like "vertical", "bounded from above", "bounded from below", etc. In addition, we assume that H^+ (H^-) contains a hyperplane h if and only if $\mathrm{cl}h^+$ ($\mathrm{cl}h^-$) is a half-space of $\mathcal{L}$. The half-spaces of $\mathcal{L}$ which are bounded by vertical hyperplanes are stored in a set S^0. The range $\Sigma(\mathcal{L})$ is defined throughout as the intersection of the half-spaces in S^0, and it will increase when we eliminate half-spaces from S^0.

10.4.1. The Geometry of Pruning

The PRUNE step of the algorithm detects some of the redundant half-spaces of the linear program $\mathcal{L}$ and eliminates them. The elimination is based on the information provided by the SEARCH step (see Algorithm 10.1 for the global strategy). This information consists of a set of pairs of half-spaces, none of which is bounded by a vertical hyperplane, of a set of half-spaces all bounded by vertical hyperplanes, and of a set S^* of 2^{d-2} new constraining half-spaces bounded by vertical hyperplanes. The 2^{d-2} new half-spaces result from testing 2^{d-2} vertical hyperplanes in the SEARCH step. We will base the elimination of half-spaces on

$$\Sigma^* = \bigcap_{s \in S^*} s$$

rather than on $\Sigma(\mathcal{L})$ as in Section 10.3 because the description of Σ^* is independent of n, the number of half-spaces of $\mathcal{L}$. Except in two dimensions, this is not true for $\Sigma(\mathcal{L})$.

Let g and h be two bounding non-vertical hyperplanes of a pair of half-spaces, and, without loss of generality, let these half-spaces be $\text{cl}g^+$ and $\text{cl}h^+$. Below, we describe sufficient conditions for when $\text{cl}g^+$ or $\text{cl}h^+$ are redundant. We eliminate $\text{cl}g^+$ or $\text{cl}h^+$ from the set of constraints if and only if they meet one of these sufficient conditions. The case-analysis involves solving linear programs of constant size, where the constant depends on the number of dimensions, d.

Case 1: The hyperplanes g and h are parallel. In this case, one of the associated half-spaces is redundant independent of Σ^*: this half-space is $\text{cl}g^+$ if $h \subseteq g^+$, and it is $\text{cl}h^+$ if $g \subseteq h^+$.

Case 2: The hyperplanes are not parallel and therefore intersect in a $(d-2)$-flat. Denote by f the vertical hyperplane through this $(d-2)$-flat, let $a = (\alpha_1, \alpha_2, \ldots, \alpha_{d-1}, 0)$ be a normal vector of f, and let α_{d+1} be the real number such that f contains all points x with $\langle x, a \rangle = \alpha_{d+1}$. We now define

$$f^{pos}: \langle x, a \rangle > \alpha_{d+1} \text{ and } f^{neg}: \langle x, a \rangle < \alpha_{d+1},$$

and we assume without loss of generality that every vertical line l in f^{pos} intersects hyperplane g below hyperplane h (see Figure 10.5). We solve the linear programs $\mathcal{L}^{pos}$ and $\mathcal{L}^{neg}$: both are defined for the half-spaces in S^*, vector a defines the objective function of $\mathcal{L}^{pos}$, and $-a$ defines the objective function of $\mathcal{L}^{neg}$. Both linear programs are $(d-1)$-dimensional and of constant size. By the construction of S^*, which is described in Sections 10.4.3 and 10.4.4, we can assume that both linear programs are feasible.

Case 2.1: $\mathcal{L}^{neg}$ has a feasible solution x^{neg} which is contained in

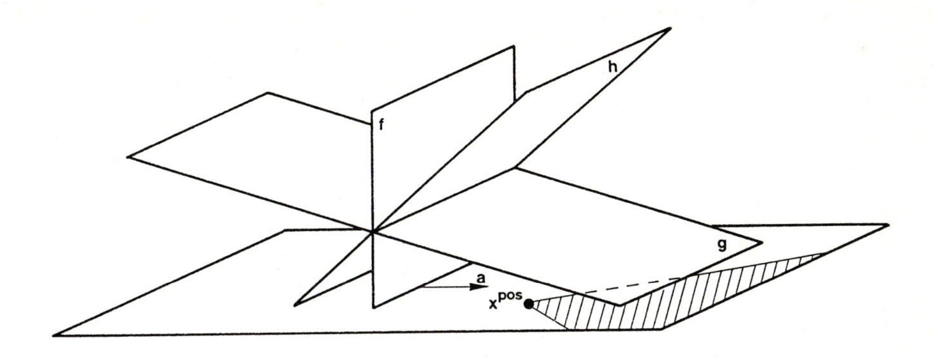

Figure 10.5. Detecting redundant half-spaces.

half-space f^{neg}. It follows that f^{neg} contains Σ^* and that every vertical line in Σ^* intersects hyperplane h below hyperplane g; thus, $\mathrm{cl}h^+$ is redundant and can be eliminated.

Case 2.2: $\mathcal{L}^{neg}$ is unbounded or its feasible solution x^{neg} belongs to f^{pos}, and $\mathcal{L}^{pos}$ is unbounded or its feasible solution x^{pos} belongs to f^{neg}. Thus, hyperplane f intersects Σ^* and both $\mathrm{cl}g^+$ and $\mathrm{cl}h^+$ may be non-redundant.

Case 2.3: $\mathcal{L}^{pos}$ has a feasible solution x^{pos} which is contained in half-space f^{pos}. It follows that f^{pos} contains Σ^* and that every vertical line in Σ^* intersects hyperplane h above hyperplane g; thus, $\mathrm{cl}g^+$ is redundant and can be eliminated (see Figure 10.5).

A half-space bounded by a vertical hyperplane can be tested in the same way as the half-spaces f^{pos} and f^{neg} above. In fact, such a half-space needs less effort than a pair of half-spaces since we can omit the construction of the vertical hyperplane f and take the bounding hyperplane of the half-space as a substitute.

As in the two-dimensional case, we find that one half-space is redundant, unless the vertical hyperplane f intersects $\mathrm{int}\Sigma^*$, see Case 2.2. The SEARCH step suggested in Sections 10.4.3 and 10.4.4 will guarantee that Σ^* is chosen such that Case 2.2 does not apply to at least a proportional number of all cases, where the factor of proportionality is $1/2^{2^{d-2}}$.

10.4.2. The Geometry of Bisecting

The problem that we solve in this section is as follows: given a linear program $\mathcal{L}$, defined by a vector $u=(0,0,\ldots,0,1)$ and by sets S^+, S^0, and S^- of half spaces, and given a vertical hyperplane t, decide whether t contains a solution to $\mathcal{L}$ and, if not, decide which side of t can be excluded from further considerations. As usual, we define the sets S^+, S^0, and S^- such that each half-space in S^+ (S^-) is bounded from below (above) by a hyperplane and each half-space in S^0 is bounded by a vertical hyperplane.

If t does not intersect the interior of the range $\Sigma(\mathcal{L}) = \bigcap_{s \in S^0} s$ of $\mathcal{L}$, then the half-space defined by t that avoids $\Sigma(\mathcal{L})$ can be trivially excluded. To check this case, we solve two $(d-1)$-dimensional linear programs with objective functions defined by vectors a and $-a$, where a is a normal vector of t, and with the constraints equal to the intersections of the half-spaces in S^0 with the hyperplane h_0: $x_d = 0$. Notice the similarity of this step with the actions taken in Case 2 of the PRUNE step discussed in Section 10.4.1.

If t intersects the interior of $\Sigma(\mathcal{L})$ then we solve up to three $(d-1)$-dimensional linear programs. The first is the restriction $\mathcal{L}(t)$ of $\mathcal{L}$ to t. Depending on the solution to $\mathcal{L}(t)$, we may solve two additional linear programs $\tilde{\mathcal{L}}(t')$ and $\tilde{\mathcal{L}}(t'')$, where t' and t'' are two hyperplanes parallel to t and on different sides of t. The constraints of $\tilde{\mathcal{L}}(t')$ and of $\tilde{\mathcal{L}}(t'')$ are the intersections of t' and t'' with the half-spaces of $\mathcal{L}$ that correspond to tight constraints in $\mathcal{L}(t)$. Having solved these linear programs, we perform exactly the same case-analysis as in two dimensions (see Section 10.3.2). Below, we prove that this case-analysis is correct. First, we show that all solutions to $\mathcal{L}$ lie on one side of a hyperplane t if t does not contain any solution to $\mathcal{L}$. Second, we argue that the case-analysis picks the correct side of t.

Lemma 10.3: The set of solutions to a linear program $\mathcal{L}$, whether feasible or infeasible, is a convex set.

Proof: If $\mathcal{L}$ is unbounded or infeasible with empty range then the set of solutions is empty and therefore trivially convex. If $\mathcal{L}$ is feasible, then the set of solutions is the intersection of $\mathcal{P}(\mathcal{L})$, the feasible domain of $\mathcal{L}$, and a hyperplane h normal to the vector which defines the objective function of $\mathcal{L}$. In fact, h must be tangent to $\mathcal{P}(\mathcal{L})$, for otherwise, no point of h minimizes the objective function. Thus, the set of feasible solutions to $\mathcal{L}$ is the closure of a face of $\mathcal{P}(\mathcal{L})$ which is more than we had to prove. Finally, if $\mathcal{L}$ is infeasible and has non-empty range $\Sigma(\mathcal{L})$ then the set of infeasible solutions is the intersection of suitable translations of the half-spaces of $\mathcal{L}$ and is therefore convex. □

Lemma 10.3 implies that the set of solutions is connected and that the answer to a test, which comes as a vertical hyperplane, is therefore unique.

Lemma 10.4: Let $\mathcal{L}$ be a linear program in E^d defined by a non-zero vector u and by sets of half-spaces S^+, S^0, and S^-. Let t be a hyperplane parallel to u and let $t' \neq t$ be a hyperplane parallel and sufficiently close to t.

(i) If $\mathcal{L}(t)$, the restriction of $\mathcal{L}$ to t, is unbounded then $\mathcal{L}$ is unbounded.

(ii) Assume that $\mathcal{L}(t)$ has a feasible solution $x^*(t)$. Then $\mathcal{L}(t')$ has a feasible solution $x^*(t')$ with $\langle u, x^*(t')\rangle < \langle u, x^*(t)\rangle$ if and only if all solutions to $\mathcal{L}$ lie on the same side of t as t'.

(iii) Assume that $\mathcal{L}(t)$ has an infeasible solution $x^*(t)$ with infeasibility $\sigma(t)$. $\mathcal{L}(t')$ has an infeasible solution $x^*(t')$ with infeasibility $\sigma(t') < \sigma(t)$ if and only if the solutions to $\mathcal{L}$ lie on the same side of t as t'.

Proof: Part (i) is trivial. We prove (ii) using the convexity of $\mathcal{P}(\mathcal{L})$, which is the feasible domain of $\mathcal{L}$. If there is a point x^* on the same side of t as t' with

$$\langle u, x^*\rangle < \langle u, x^*(t)\rangle,$$

then every convex combination $y^*(\lambda) = \lambda x^* + (1-\lambda)x^*(t)$, $0 < \lambda \leq 1$, belongs to the feasible domain $\mathcal{P}(\mathcal{L})$, and we have

$$\langle u, y^*(\lambda)\rangle < \langle u, x^*(t)\rangle.$$

Since hyperplane t' is sufficiently close to t, it contains such a point. To prove the other direction, notice that the existence of a point x^* on the side of t other than t' with

$$\langle u, x^*\rangle < \langle u, x^*(t)\rangle$$

contradicts

$$\langle u, x^*(t')\rangle < \langle u, x^*(t)\rangle,$$

for $x^*(t')$ a solution to $\mathcal{L}(t')$: every convex combination of $x^*(t')$ and of x^* belongs to $\mathcal{P}(\mathcal{L})$, and the convex combination in hyperplane t improves upon $x^*(t)$.

To prove part (iii), we consider the polyhedra

$$\Sigma(\mathcal{L}) = \bigcap_{s \in S^0} s, \quad \mathcal{P}^+ = \bigcap_{s \in S^+} s, \quad \text{and} \quad \mathcal{P}^- = \bigcap_{s \in S^-} s.$$

For any point y in the hyperplane h_0: $x_d = 0$, we define

$$\delta(y) = \frac{\pi^+(y) - \pi^-(y)}{2},$$

where $\pi^+(y)$ is the vertical projection of y onto the boundary of $\mathcal{P}^+$, and $\pi^-(y)$ is the vertical projection of y onto the boundary of $\mathcal{P}^-$. Since $\mathcal{P}^+$ and $\mathcal{P}^-$ are both convex, the set $\Delta = \{x = (\xi_1, \xi_2, \ldots, \xi_d) \mid y = (\xi_1, \xi_2, \ldots, \xi_{d-1}, 0) \text{ and } \xi_d \geq \delta(y)\}$ is convex. The relation between function δ and the infeasible solutions to $\mathcal{L}$ is as follows: if $x^*(t)$ is an infeasible solution to $\mathcal{L}(t)$ then $\delta(y^*(t))$ is the infeasibility of $x^*(t)$, with $y^*(t)$ the vertical projection of $x^*(t)$ onto hyperplane h_0. Since Δ is convex, the same reasoning as used above to prove (ii) now implies (iii). □

To see that the case–analysis which decides on which side of a vertical hyperplane t the solutions lie is correct, we just need to observe that all constraints of $\mathcal{L}$ which are not tight in t cannot be tight in hyperplanes t' and t'' if they are sufficiently close to t. Finally, if we remove all non–tight constraints then how far t' and t'' are from t is immaterial.

10.4.3. Searching Lines in the Plane

This section considers the SEARCH step for a three–dimensional linear program $\mathcal{L}$. We use this discussion to introduce the ideas that will lead to the SEARCH step in arbitrary dimensions. As usual, we assume $u=(0,0,1)$ and we partition the set of half–spaces of $\mathcal{L}$ into sets S^+, S^0, and S^-. Our goal is to find and test two vertical planes such that the PRUNE step is guaranteed to eliminate at least about one sixteenth of the half–spaces.

In the first step, we construct a multiset L of lines in the plane h_0: $x_3=0$. For each half–space s in S^0, we add the line $h_0 \cap \text{bd}s$ to L. Notice that this line is well defined since $\text{bd}s$ is a vertical plane. Furthermore, we pair the bounding planes of half–spaces in S^+ and, separately, in S^-. If two such paired planes g and h intersect, then they intersect in a line, and we add the vertical projection of this line onto h_0 to L. Note that two parallel non–vertical planes do not contribute a line to L which is constructed to determine redundant half–spaces. This case, however, is to our advantage since one of the half–spaces bounded by two parallel planes is eliminated independently of what the further computations are.

For an arbitrary line l in h_0, we define its *extension* $t(l)$ as the vertical plane that intersects h_0 in l. Given a line l in h_0, we can find a closed half–space bounded by plane $t(l)$ which contains all solutions to $\mathcal{L}$, provided $t(l)$ contains no solution to $\mathcal{L}$ (see Section 10.4.2). We denote this open half–space by $t(l)^{pos}$, and we write $t(l)^{neg}$ for the other open half–space defined by $t(l)$. Our next goal is to find two lines $\hat{l}$ and $\bar{l}$ in h_0 such that at least one eighth of the lines in L avoid the open quadrant

$$q=h_0 \cap t(\hat{l})^{pos} \cap t(\bar{l})^{pos}.$$

Figure 10.6 illustrates this problem: it indicates a set L of seventeen lines and two lines $\hat{l}$ and $\bar{l}$ such that four lines of L avoid the shaded quadrant q. The idea of the construction of $\hat{l}$ and $\bar{l}$ is to pair the lines in L, to project the intersection points along two directions, and to solve two one-dimensional median finding

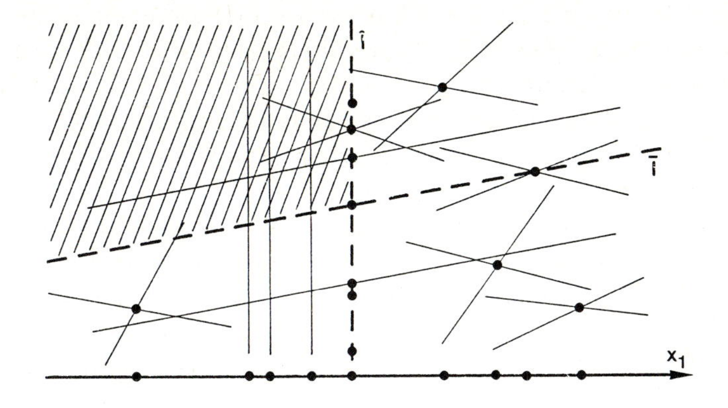

Figure 10.6. Finding suitable lines.

problems. Recall that the median of a multiset of m elements is the $\lfloor \frac{m}{2} \rfloor^{\text{th}}$ element in a sorted ordering of the multiset. Line $\hat{l}$ will be chosen parallel to the first direction such that it contains the median determined perpendicular to this direction. Line $\bar{l}$ is analogously defined parallel to the second direction. We will not pair lines which are parallel to any of the two chosen directions. To prove the effectiveness of our choice of lines, we will argue that at least one line of each pair that intersects in the closure of the quadrant $\bar{q}$ diagonally opposite to quadrant q necessarily avoids q. This is not true in general, but it is true if we pair the lines of L in a specific manner which matches lines which are steeper than $\bar{l}$ with lines which are less steep than $\bar{l}$. We give a more formal description of this process below.

Procedure 10.3 (Searching lines in the plane):

Step 1: Determine the median line l_{med} in an ordering of the non-vertical lines in L by slope. Partition L into sets L_1 of vertical lines, L_3 of lines parallel to l_{med}, L_2 of non-vertical lines that are less steep than l_{med}, and set L_4 of non-vertical lines that are steeper than l_{med}.

Step 2: Match each (non-vertical) line in L_2 with a unique line in L_4, if available. This leaves $|\text{card}L_2 - \text{card}L_4|$ lines of either L_2 or L_4 unmatched. Let L_{pair} be the set of pairs of matched lines.

Step 3: Construct a multiset $\hat{P}$ of points on the x_1-axis as follows: for each pair of lines in L_{pair}, add the vertical projection of the intersection point to $\hat{P}$, and for each (vertical) line in L_1, add the intersection with the x_1-axis to $\hat{P}$ (see Figure 10.6).

Step 4: Determine the median point in $\hat{P}$ and let $\hat{l}$ be the vertical line through this point (see Figure 10.6). Perform a BISECT step for the extension $t(\hat{l})$ of $\hat{l}$ which either finishes the overall algorithm or it determines which half-space bounded by $t(\hat{l})$ is $t(\hat{l})^{pos}$ and which one is $t(\hat{l})^{neg}$.

Step 5: Construct a multiset $\overline{P}$ of points on line $\hat{l}$ as follows: for each pair of lines in L_{pair} whose intersection point belongs to $t(\hat{l})^{neg}$, add the projection of this point onto $\hat{l}$ to $\overline{P}$, where the projection is done along the direction determined by l_{med}, and for each line in L_3 (which is parallel to l_{med}), add the intersection with $\hat{l}$ to $\overline{P}$.

Step 6: Determine the median point in $\overline{P}$ and let $\overline{l}$ be the line parallel to l_{med} that contains this point (see Figure 10.6). Perform a BISECT step for the extension $t(\overline{l})$ of $\overline{l}$; this either finishes the overall algorithm or it determines which half-space bounded by $t(\overline{l})$ is $t(\overline{l})^{pos}$ and which one is $t(\overline{l})^{neg}$.

The half-spaces $t(\hat{l})^{pos}$ and $t(\overline{l})^{pos}$ computed in Procedure 10.3 define the set S^* of constraints needed in the PRUNE step. In Section 10.4.4, we will prove that such a PRUNE step eliminates at least

$$\frac{\text{card}S^0 + \lfloor \frac{\text{card}S^+}{2} \rfloor + \lfloor \frac{\text{card}S^-}{2} \rfloor}{8}$$

half-spaces of $\mathcal{L}$. Notice that this guarantees that at least one constraining half-space is eliminated, unless S^0 is empty and both S^+ and S^- contain at most one half-space each.

10.4.4. The Geometry of Searching

The generalization of the three-dimensional SEARCH step, discussed in the previous section, to dimensions $d > 3$ is relatively straightforward. In the first step, we pair bounding hyperplanes, unless they are vertical. Next, we construct a multiset G of $(d-2)$-flats in the hyperplane h_0: $x_d = 0$ by vertically projecting the intersections of pairs of hyperplanes and the vertical hyperplanes. Then, we perform what we call a "search of G" as follows:

1. we choose two directions in h_0,
2. we pair $(d-2)$-flats which are not parallel to either direction in a specific manner,
3. we project the intersections of paired $(d-2)$-flats and we project the $(d-2)$-flats which are parallel to the first direction along the first direction,
4. we "search" the projections recursively,
5. we project some of the intersections between paired $(d-2)$-flats

and we project all $(d-2)$-flats which are parallel to the second direction along the second direction, and

6. we "search" these projections recursively.

This section gives the details of these steps and an analysis of the construction.

Let $\mathcal{L}$ be a linear program in E^d determined by vector $u=(0,0,\ldots,0,1)$ and by sets S^+, S^0, and S^- of half-spaces. For each hyperplane h bounding a half-space in S^0 (h is therefore vertical), we add the intersection $h \cap h_0$ to the multiset G. Furthermore, we pair each hyperplane which bounds a half-space in S^+ with another hyperplane bounding a half-space in the same set and add the vertical projection onto h_0 of their intersection to G, provided this intersection is a $(d-2)$-flat. We do the same for the hyperplanes bounding half-spaces of S^-. This leaves at most one hyperplane bounding a half-space in S^+ and one hyperplane bounding a half-space in S^- unmatched. We will use G to determine a set S^* of 2^{d-2} half-spaces with vertical bounding hyperplanes such that a constant fraction of the half-spaces of $\mathcal{L}$ avoid the interior of their intersection. It follows that the PRUNE step will eliminate a constant fraction of the constraining half-spaces of $\mathcal{L}$.

For a k-flat f in the $(k+1)$-flat spanned by the first $k+1$ coordinate-axes, we define its extension $t(f)$ to be the hyperplane which contains f and which is parallel to the last $d-k-1$ coordinate-axes. To find the half-spaces in S^* it suffices to find a set G^* of $(d-2)$-flats in h_0 such that a proportional number of $(d-2)$-flats of G avoid a specified cell defined by the $(d-2)$-flats of G^*. This cell is determined by performing a BISECT step for each extension $t(g)$ with g in G^*. In the construction of G^*, we will alternate between determining a $(d-2)$-flat and testing its extension. The remainder of the strategy uses recursive calls to lower dimensions, as in E^3 where the two-dimensional subproblem was solved by recurring to two one-dimensional problems.

For convenience, we identify hyperplane h_0 with the $(d-1)$-dimensional Euclidean space E^{d-1}; so the $(d-1)^{\text{st}}$ coordinate-axis defines the vertical direction in h_0. Without causing confusion, we will refer to the elements of G as hyperplanes in E^{d-1}. Every non-vertical hyperplane g in E^{d-1} can be specified by an equation of the form

$$x_{d-1}=\gamma_1 x_1+\gamma_2 x_2+\ldots+\gamma_{d-2}x_{d-2}+\gamma_{d-1}.$$

We call γ_{d-2} the *slope of* g. We give a formal description of the determination of G^* below.

Procedure 10.4 (Searching hyperplanes in $d-1$ dimensions):

if $d-1=1$ **then**

Compute the median g of G, set $G^*:=\{g\}$, compute the extension $t(g)$ of g, which is a hyperplane in E^d, and set $S^*:=\{t(g)^{pos}\}$ after performing a BISECT step for $t(g)$.

else

Step 1: Determine the hyperplane g_{med} with median slope among the non–vertical hyperplanes of G. Partition G into multisets G_i, for $1 \leq i \leq 4$, such that G_1 contains all vertical hyperplanes of G, G_3 contains all hyperplanes of G which have the same slope as g_{med}, G_2 contains all non–vertical hyperplanes with slope less than the slope of g_{med}, and G_4 contains all non–vertical hyperplanes of G whose slope is greater than the slope of g_{med}.

Step 2: Match each hyperplane of G_2 with a unique hyperplane of G_4, if available. Let G_{pair} be the set of pairs of matched hyperplanes.

Step 3: Construct a multiset $\hat{F}$ of $(d-3)$–flats in the hyperplane $\hat{g}$: $x_{d-1}=0$ as follows: for each pair of hyperplanes in G_{pair}, add the vertical projection of their intersection to $\hat{F}$, and for each hyperplane in G_1, add the intersection with $\hat{g}$ to $\hat{F}$.

Step 4: Recur for $\hat{F}$ replacing G and let $\hat{F}^*$ be the set of determined $(d-3)$–flats in $\hat{g}$, and let $\hat{S}$ be the associated set of half–spaces.

Step 5: Construct a multiset $\overline{F}$ of $(d-3)$–flats in the hyperplane $\overline{g}$: $x_{d-2}=0$ as follows: for each pair of hyperplanes in G_{pair} whose intersection avoids the interior of the intersection of the half–spaces in $\hat{S}$, add the projection of this intersection onto $\overline{g}$ to $\overline{F}$, where the projection is done along the line l_{med} which is the intersection of g_{med} with the (two–dimensional) plane spanned by the x_{d-1}– and the x_{d-2}–axes. Additionally, for each hyperplane in G_3, add the intersection with $\overline{g}$ to $\overline{F}$.

Step 6: Recur for $\overline{F}$ replacing G, let $\overline{F}^*$ be the $(d-3)$–flats computed for $\overline{F}$, and let $\overline{S}$ be the associated set of half–spaces. Set $S^* := \hat{S} \cup \overline{S}$.

endif.

The set S^* of half–spaces will be used to eliminate constraining half–spaces of $\mathcal{L}$. The remainder of this section demonstrates that S^* is computed such that a PRUNE step eliminates at least

$$\frac{\text{card}S^0 + \lfloor \frac{\text{card}S^+}{2} \rfloor + \lfloor \frac{\text{card}S^-}{2} \rfloor}{2^{2^{d-1}-1}}$$

half-spaces of $\mathcal{L}$. We will express this result in terms of hyperplanes in E^{d-1} which avoid a particular cell specified by 2^{d-2} hyperplanes constructed in

Procedure 10.4. We write m for $\text{card}S^0+\lfloor\frac{\text{card}S^+}{2}\rfloor+\lfloor\frac{\text{card}S^-}{2}\rfloor$, and we define $c(G)=\Sigma^*\cap h_0$, the cell defined by the 2^{d-2} hyperplanes.

Lemma 10.5: Let G be a multiset of m hyperplanes in E^{d-1}, $d-1\geq 1$, and let $c(G)$ be the cell as constructed in Procedure 10.4. At least $m/2^{2^{d-1}-1}$ of the hyperplanes in G avoid $c(G)$.

Proof: This proof uses induction on the number of dimensions. The assertion is trivial for $d-1=1$, since Procedure 10.4 constructs $c(G)$ as an open interval determined by the median of G which is a multiset of points on a line.

To prove Lemma 10.5 in arbitrary $d-1$ dimensions, we define m_i to be the cardinality of the multiset G_i, for $1\leq i\leq 4$, and we assume without loss of generality that $m_2\leq m_4$. So there are at least as many non-vertical hyperplanes with greater slope than hyperplane g_{med} as there are non-vertical hyperplanes with smaller slope than g_{med}. This implies that m_4-m_2 hyperplanes of G_4 remain unmatched. Thus,

$$\text{card}\,G_{pair}=m_2 \text{ and } \text{card}\hat{F}=m_1+m_2.$$

Since g_{med} is the median hyperplane in a sorted order of the non-vertical hyperplanes of G, we also have $m_2+m_3\geq m_4$. A $(d-3)$-flat of $\hat{F}$ that avoids the $(d-2)$-face $c(\hat{F})$, recursively constructed for $\hat{F}$, either corresponds to a vertical hyperplane of G or to a pair of non-vertical hyperplanes of G. In the first case, the corresponding vertical hyperplane does not intersect the largest subset of E^{d-1} that projects vertically onto $c(\hat{F})$, and, therefore, it does not intersect $c(G)$ which is a subset of this set. In the second case, we will be able to show that at most one hyperplane of the corresponding pair avoids $c(G)$. Thus, in a worst-case scenario, each $(d-3)$-flat of $\hat{F}$ that avoids $c(\hat{F})$ is the projection of the intersection of two hyperplanes in G. By the induction hypothesis, there are at least $(m_1+m_2)/2^{2^{d-2}-1}$ such $(d-3)$-flats in $\hat{F}$ which implies

$$\text{card}\overline{F}\geq\frac{m_1+m_2}{2^{2^{d-2}-1}}+m_3.$$

Once again using the induction hypothesis, at least

$$\frac{m_1+m_2}{2^{2^{d-1}-2}}+\frac{m_3}{2^{2^{d-2}-1}}$$

$(d-3)$-flats of $\overline{F}$ miss the $(d-2)$-face $c(\overline{F})$ constructed recursively in the hyperplane $\overline{g}$: $x_{d-2}=0$. For each such $(d-3)$-flat, we have an associated hyperplane in G which avoids both $c(\hat{F})$ and $c(\overline{F})$ and, therefore, also their intersection $c(G)$. Consequently, at least

$$\frac{m_1+m_2+(2^{2^{d-2}}-1)m_3}{2^{2^{d-1}}-2} \geq \frac{m_1+m_3+m_4}{2^{2^{d-1}}-2} \geq \frac{m_1+m_2+m_3+m_4}{2^{2^{d-1}}-1} \geq \frac{m}{2^{2^{d-1}}-1}$$

hyperplanes of G avoid cell $c(G)$. □

Lemma 10.5 implies that each iteration of the main algorithm eliminates a constant fraction of the containing half-spaces. In addition, it guarantees that at least one half-space is eliminated, unless the set S^0 is empty and both S^+ and S^- contain at most one half-space each. A nice side-effect of this feature is that we can iterate until the linear program is solvable by a trivial case-analysis. In most cases, however, the algorithm will halt earlier in a lower-dimensional sub-computation.

10.4.5. The Overall Algorithm

The computations described in Sections 10.4.1, 10.4.2, and 10.4.4 need to be combined in an appropriate manner so as to yield an efficient algorithm for linear programming. This section offers such a careful composition, following the prune-and-search paradigm introduced in Section 10.3. The algorithm that we are going to develop is highly recursive. In essence, it reduces a linear program $\mathcal{L}$ in d dimensions to a complicated schedule of one-dimensional tasks, where each such task involves the determination of a minimum, maximum, or median of a multiset. The input to the algorithm consists of d, the number of dimensions, and of sets S^+, S^0, and S^- of half-spaces, as usual. The vector u which defines the objective function of $\mathcal{L}$ is assumed to be equal to $(0,0,\ldots,0,1)$.

Algorithm 10.5 (Linear programming in d dimensions):

if card$S^0=0$, card$S^+\leq 1$, and card $S^-\leq 1$ **then**

Solve the linear program by a straightforward case-analysis.

elseif $d=1$, that is, each half-space is an interval **then**

Solve the one-dimensional linear program $\mathcal{L}$ by determining a largest endpoint of an interval in S^+ and a smallest endpoint of an interval in S^-.

elseif $d\geq 2$ **then**

SEARCH: Match the half-spaces in S^+ and, separately, in S^-. Iterate the construction of a vertical hyperplane t and the determination of the side of t which does not contain any solution to $\mathcal{L}$ as described in Section 10.4.4. This yields a set S^* of 2^{d-2} half-spaces, unless one of the tested hyperplanes t contains a solution to $\mathcal{L}$ or the restriction $\mathcal{L}(t)$ of $\mathcal{L}$ to t is unbounded; in both of these cases, the algorithm halts. The set S^* defines a range $\Sigma^* = \bigcap_{s\in S^*} s$.

PRUNE: Eliminate a half-space in S^0 if it contains Σ^*, and eliminate one half-space of a pair of half-spaces in S^+ or S^- if the intersection of their boundaries avoids Σ^*.

RECUR: Repeat Algorithm 10.5 for the new sets S^+, S^0, and S^-.

endif.

To analyze the time-complexity of Algorithm 10.5, it is important to understand its recursive structure. There is a recursive call hidden in the PRUNE step which we can ignore since it solves only a linear program of constant size. The SEARCH step itself is recursive and produces 2^{d-2} vertical hyperplanes to be tested. Each such hyperplane gives rise to up to five $(d-1)$-dimensional linear programs in a BISECT step. This implies that one iteration of Algorithm 10.5 solves up to $5 \cdot 2^{d-2}$ $(d-1)$-dimensional linear programs which reduces to up to

$$5^{d-1} 2^{(d-1)(d-2)/2}$$

one-dimensional linear programs. We assume inductively that a $(d-1)$-dimensional linear program with n constraining half-spaces can be solved in time $O(n)$, where the constant of proportionality depends on d. This is true if $d-1=1$. Under this assumption, it is easy to see that one iteration of Algorithm 10.5 including all recursive calls that involve lower-dimensional linear programs takes time $O(n)$.

Let n_i be the sum of the cardinalities of the sets S^+, S^0, and S^- at the beginning of the ith iteration of Algorithm 10.5. The above discussion implies that the ith iteration of Algorithm 10.5 takes time $O(n_i)$. By definition of n and n_1, we have $n_1 = n$, and by Lemma 10.5, we have

$$n_i \leq \frac{(2^{2^{d-1}})n_{i-1}}{2^{2^{d-1}}+1}, \quad \text{for } i \geq 2,$$

unless $n_{i-1} < 2^{2^{d-1}+1}+2$ (see Exercise 10.6). Thus, the sequence of the n_i is geometrically decreasing with a factor of $1 - \frac{1}{2^{2^{d-1}}+1}$, until the number of half-spaces is smaller than some constant. Lemma 10.5 also implies that $n_i < n_{i-1}$ unless $\text{card}S^0 = 0$, $\text{card}S^+ \leq 1$, and $\text{card}S^- \leq 1$. Thus, Algorithm 10.5 computes until at most two half-spaces are left, or it halts in some lower-dimensional subcomputation. We therefore conclude with the main result of this chapter.

Theorem 10.6: Let S be a set of n half-spaces in E^d, for some constant d. Algorithm 10.5 solves a linear program defined by S and a non-zero vector in time $O(n)$.

If one uses Algorithm 10.5 in dimensions $d \geq 4$, then the double-exponential growth in d of the constant hidden in the big-Oh notation of its time-complexity is a real drawback unless n is astronomically large. There is some hope to decrease the constant to a single–exponential growth in d if one is willing to cope with additional complications in the implementation (see also Exercise 10.10). This aspiration is based on the imbalance of the constants contributed by recursive calls of Algorithm 10.5 to $(d-1)$–dimensional linear programs versus those contributed by calls to d–dimensional linear programs: the first factor is single–exponential in d^2 while the second is double–exponential in d.

10.5. Exercises and Research Problems

Exercise 10.1: Let P be a set of n points in E^d, and let $\delta(h,P)$ be a distance function defined for hyperplane h. We call h a *regression hyperplane of* P *(with respect to* δ*)* if it minimizes $\delta(g,P)$, for all hyperplanes g.

3 (a) For a point $p=(\pi_1,\pi_2,...,\pi_d)$ and a non-vertical hyperplane h given by the equation $x_d = \sum_{i=1}^{d-1} \eta_i x_i + \eta_d$, we call the absolute value of $\pi_d - \sum_{i=1}^{d-1} \eta_i \pi_i - \eta_d$ the *vertical distance between* p and h. Define $\delta_v(h,P)$ as the maximal vertical distance between h and any point in set P. Show that Algorithm 10.5 can be used to determine a regression hyperplane of P with respect to δ_v in time $O(n)$. *(Hint: this problem resists a formulation as a linear program with n constraints and d variables, but it is possible to formulate two constraints for each point in P such that a regression hyperplane is an infeasible solution to the linear program.)*

2 (b) Define $\delta_o(h,P)$ as the maximal normal distance of any point of P from hyperplane h. Prove that a regression hyperplane with respect to the distance function δ_v is not necessarily a regression hyperplane with respect to δ_o.

3 (c) Give an algorithm which takes time $O(n\log n)$ to find a regression line with respect to δ_o for a set P of n points in E^2. *(Hint: use the fact that the length of the projection of the convex hull of P is minimized along the direction determined by the regression line.)*

3 **Exercise 10.2:** Let P be a set of n points in E^2 and let h be the number of extreme points in P. Prove that the following algorithm reports the sequence of vertices on the upper part of the convex hull of P in time $O(n\log h)$.

Algorithm (Upper boundary of conv$P$):
  if card$P=1$ **then**
    Report the only point in set P.
  else
    DIVIDE: Find a vertical line which partitions P into two equal–sized subsets.
    MERGE: Find the upper edge e of the boundary of the convex hull of P that intersects the vertical line.
    RECUR: Let P_l be the subset of P to the left of the vertical line through the left endpoint v_l of edge e, and let P_r be the subset of P to the right of the vertical line through the right endpoint v_r of e. First, recur for set $P_l \cup \{v_l\}$ and then recur for set $P_r \cup \{v_r\}$.
  endif.

Problem 10.3: Let P be a set of n points in E^d. We define the *depth of* a point p in P equal to one if p is extreme. Otherwise, the depth of p in P is one more than the depth of point p in set P reduced by all extreme points.

2 (a) Show that the set of extreme points of P can be determined in time $O(n^2)$. *(Hint: the set of points of P can be mapped to a set of half-spaces such that a point is extreme if and only if the corresponding half-plane is not redundant.)*

2 (b) Show that time $O(n^3)$ suffices to compute the depth of each point in P.

5 (c) Is there an algorithm that computes the depth of every point of P in time $o(n^3)$, if $d \geq 4$?

2 (d) Suppose all points of P lie on the unit-sphere in E^d. Show that time $O(n)$ suffices to decide whether or not there is a half-space that contains P and whose bounding hyperplane contains the origin.

Exercise 10.4: Show that the function $T(n)$ is in $O(n)$ if $T(n)$ obeys the following recurrence relations:

2 (a) $T(n) = T(\alpha n) + O(n)$, with $0 < \alpha < 1$ and $T(1) = O(1)$.

2 (b) $T(n) = \sum_{i=1}^{k} T(\alpha_i n) + O(n)$, with $\alpha_i \geq 0$, for $1 \leq i \leq k$, $\sum_{i=1}^{k} \alpha_i < 1$, and $T(1) = O(1)$.

2 **Exercise 10.5:** Prove

$$\left\lceil \frac{\lceil \frac{i-1}{2} \rceil + \lceil \frac{j-1}{2} \rceil}{2} \right\rceil \geq \frac{i+j}{6},$$

for all non-negative integers i and j, unless $i \leq 1$ and $j \leq 1$.

Exercise 10.6: Define a sequence of real numbers as follows: $a_0 = 3$, and

$$a_j = \frac{c-1}{c} a_{j+1} + \frac{2}{c},$$

for integers $j \geq 0$ and for $c > 1$ a real number.

2 (a) Prove $a_j < a_{j+1}$, for all $j \geq 0$.

2 (b) Prove $a_j \leq \frac{c}{c+1} a_{j+1}$, unless $a_{j+1} < 2(c+1)$.

3 **Exercise 10.7:** Implement an algorithm that computes the median of a multiset of n points on a line in time $O(n)$.

4 **Exercise 10.8:** Implement Algorithm 10.5 for linear programming in d dimensions such that it runs in time $O(n)$ if the linear program is defined by n half-spaces.

4 **Exercise 10.9:** Design an algorithm that solves the following problems in time $O(n)$: given a set P of n points in E^2, determine the smallest disc that covers P. *(Hint: use the prune-and-search paradigm described in Section 10.3.)*

4 **Exercise 10.10:** The constant hidden in the big-Oh notation of the time-complexity of Algorithm 10.5 is on the order of 2^{2^d}, with d the number of dimensions. Modify the SEARCH strategy of Algorithm 10.5 such that the constant is "only" single-exponential in d^2.

5 **Problem 10.11:** Can Algorithm 10.5 be modified such that the SEARCH step works on d-tuples of hyperplanes instead of on 2^{d-1}-tuples, as it does in the description of Section 10.4.4?

Problem 10.12: Define $h_k^{(d)}$ as the infimum over all real numbers $0 \leq \alpha \leq 1$ such that for any set H of n hyperplanes in E^d there are k hyperplanes with the property that each cell defined by the k hyperplanes intersects at most αn of the hyperplanes in H.

2 (a) Prove $h_2^{(2)} \geq \frac{3}{4}$. *(Hint: if no two lines of H are parallel and no three are concurrent then there are at most four lines in H that intersect less than three regions defined by any two lines.)*

3 (b) Prove $h_2^{(2)} \leq \frac{3}{4}$. *(Hint: let the slope of the first line l be equal to the median slope in the set and choose its intercept such that l contains a point p with the property that at most half of the steeper (less steep) lines intersect l on either side of p. The second line is the vertical line through p.)*

5 (c) Is $h_3^{(3)}$ smaller than 1?

2 (d) Prove $h_4^{(3)} \leq \frac{7}{8}$. *(Hint: dualize the fact that n* points in E^3 *can be partitioned by three planes such that each octant contains at most n/8 of the points, see Theorem 4.12.)*

3 (e) Let $k(d)$ be the smallest integer such that $h_{k(d)}^{(d)} < 1$. Prove that $k(d)$ exists and prove $k(d) \geq d$.

5 (f) Determine $k(d)$ for $d \geq 3$.

10.6. Bibliographic Notes

The problem of linear programming has created a rich literature over the last forty years, see Dantzig (1963), Papadimitriou, Steiglitz (1982), Chvátal (1983), and others. However, this literature is almost exclusively concerned with the case when the number of variables is of approximately the same order of magnitude as the number of constraints. The most widely used method for solving linear programs is undoubtedly the simplex algorithm introduced by Dantzig in 1947. Klee, Minty (1972) showed that the time-complexity of the simplex algorithm in the worst case is exponential in the size of the input, where the size of the input is defined by d, the number of variables, and n, the number of constraints. Nevertheless, if is efficient for "almost every" input. The longstanding open problem as to whether or not there is a worst-case polynomial time algorithm for linear programming was partially answered in the affirmative by Khachiyan (1979): he gives an algorithm that runs in worst-case time which is polynomial in d, n, and L, where L is the largest number of bits used to specify any single coefficient. Karmarkar (1984) offers a variation of Khachiyan's algorithm which is expected to be efficient even in a "practical sense". Neither algorithm answers the question as to whether or not there is an algorithm that solves a linear program in a number of arithmetic operations which is polynomial in n and d.

The techniques used to solve linear programs in low dimensions efficiently are fundamentally different from the techniques mentioned above. Dyer (1984) and, independently, Megiddo (1983b) developed algorithms that run in time $O(n)$ if $d=2$ or $d=3$. Later, Megiddo (1984) generalized this result to any fixed d; the time-complexity of his method grows double-exponentially with d. Most of the material of this chapter is taken from these three publications. Dyer (1986) and, independently, Clarkson (1986) improved Megiddo's linear time algorithm such that the time-complexity grows exponentially with d^2; this solves Exercise 10.10.

The algorithmic techniques used by Dyer and Megiddo follow a novel design scheme called prune-and-search, a term used for the first time by Lee, Preparata (1984b). This scheme can be viewed as a generalization of the, by now classic, method for finding a median of a multiset of points on a line in worst-case time linear in the cardinality of the multiset, see

Aho, Hopcroft, Ullman (1974) and other textbooks on algorithms. This method for finding a median originates in Blum, Floyd, Pratt, Rivest, Tarjan (1972) and was improved by Schönhage, Paterson, Pippenger (1976). Floyd, Rivest (1975) discuss simple algorithms that determine a median in fast expected time. These algorithms are useful in practice since the worst-case optimal methods all suffer from poor average behavior. From a different perspective, the prune-and-search method can be seen to be related to the general search techniques described in Megiddo (1983a). Here, we also mention Cole (1986b) who describes an improvement of Megiddo's search techniques. His improvement is, however, fundamentally different from the idea of prune-and-search which combines searching in the data with eliminating parts of the data.

Algorithms that are efficient for low-dimensional linear programs have a large potential to lead to efficient algorithms for common geometric problems. The relationship between linear programming and other geometric problems is addressed from a general point of view in Dobkin, Reiss (1980). Most of the geometric problems discussed in this paper relate to linear programming by duality. One such example is the problem of determining whether or not a given point of a finite set is extreme. This problem is equivalent to the one of determining whether or not some given half-space of a linear program is redundant. The reduction of the problem of finding a regression hyperplane which minimizes the largest vertical distance to a point (see Exercise 10.1(a)) is mentioned in Megiddo (1984). It is interesting to note that the computation of regression hyperplanes that minimize other distance functions can be very different (see Exercise 10.1(b) and Edelsbrunner (1985) for a solution to Exercise 10.1(c)). Kirkpatrick, Seidel (1986) exploit the possibility of solving a two-dimensional linear program in time $O(n)$ for the design of an algorithm that constructs the convex hull of a set of n points in the plane in time $O(n \log h)$, where h is the number of extreme points in the set (see Exercise 10.2). An application of the line-searching procedure discussed in Section 10.4.3 to constructing a ham-sandwich line for two separated sets with a total of n points in E^2 can be found in Megiddo (1985) (see also Section 14.1). Megiddo (1983b) showed that the problem of finding the smallest disc that covers a given set of n points in E^2 can be solved in time $O(n)$ using the prune-and-search paradigm; this solves Exercise 10.9.

CHAPTER 11

PLANAR POINT LOCATION SEARCH

One of the more fundamental problems in computational geometry is that of locating a specific point in a given two–dimensional subdivision. It is thus necessary to define a subdivision appropriately. An *embedding* ϵ of a graph $\mathcal{G}$ maps each node v of $\mathcal{G}$ to a point $\epsilon(v)$ in E^2 and each arc $a=\{v,w\}$ to a simple connected curve $\epsilon(a)$ with endpoints $\epsilon(v)$ and $\epsilon(w)$. The embedding ϵ is *plane* if $\epsilon(v)\neq\epsilon(w)$, for any two nodes $v\neq w$ of $\mathcal{G}$, and if $\epsilon(a)\cap\epsilon(b)=\emptyset$, for any two arcs $a\neq b$ of $\mathcal{G}$; so we assume that a curve $\epsilon(a)$ does not contain its endpoints. Graph $\mathcal{G}$ is *planar* if it admits a plane embedding of itself. The embedding of a node of $\mathcal{G}$ is called a *vertex*, the embedding of an arc is called an *edge*, and a connected component of E^2 reduced by all vertices and edges is called a *region*. For the sake of generality, we admit one node of $\mathcal{G}$ to be embedded at infinity; thus, all incident arcs correspond to unbounded edges of the subdivision, and all unbounded edges of the subdivision correspond to arcs incident upon this node. The embedding of a node at infinity is called an *improper vertex*. In formal terms, the *point location search problem* can now be defined as follows:

> let S be a subdivision of E^2; for a given query point q, determine which of the regions, edges, and vertices of S contains point q.

We assume that the subdivision S is used for a large number of such queries; so, we can reduce the total time required for this task by preprocessing S into a data structure suitable for the search. There are many applications of such a data structure; some can be found in Exercise 11.5 and others in Chapters 12 and 13. The performance of a proposed solution is measured in terms of three quantities: the amount of storage, the time needed for a search, and the time required for constructing the data structure. These quantities are analyzed as functions which are dependent on the number of vertices of the subdivision at hand.

We will show that the number of vertices is actually a good measure of the complexity of a subdivision in Section 11.1. Section 11.2 presents the basic geometric properties of so–called monotone subdivisions. These properties are used in Section 11.3 which presents an efficient (although not optimal) solution to the point location search problem for monotone subdivisions. This solution is improved to optimal in Section 11.6. In Sections 11.4 and 11.5 we discuss how a subdivision can be represented and how the data structure of Section 11.3 can be constructed from such a representation. In Section 11.7, we show how this structure can be modified to the optimal data structure described in Section 11.6. Finally, Section 11.8 addresses the problem of refining general subdivisions to monotone ones.

11.1. Euler's Relation for Planar Graphs

In this section, we substantiate the claim that the number n of vertices is a good measure for the complexity of a subdivision. This is done by showing that the number of edges and regions of a subdivision is at most proportional to n. For a subdivision S in E^2, we use $f_0(S)$ to represent the number of proper vertices, $f_1(S)$ for the number of edges, and $f_2(S)$ for the number of regions of S. Similar to Euler's famous theorem on convex polytopes (compare Theorem 6.8 for $d=3$ and Theorem 6.11), we have the following result.

Theorem 11.1: Let S be a subdivision of E^2 which is the plane embedding of a connected graph, where no node is embedded at infinity. Then
(i) $f_0(S)-f_1(S)+f_2(S)=2$,
(ii) $f_1(S)\leq 3f_0(S)-6$, if $f_0(S)\geq 3$, and
(iii) $f_2(S)\leq 2f_0(S)-4$, if $f_0(S)\geq 3$.

Proof: First, we prove equation (i) inductively and then derive the inequalities (ii) and (iii) by straightforward counting arguments. Define $n=f_0(S)$ and $m=f_1(S)$, and let $(e_1,e_2,\ldots,e_m)$ be an enumeration of the edges of S. For each integer $1\leq i\leq m$, define $E_i=\{e_1,e_2,\ldots,e_i\}$, and let $\mathcal{G}_i=(N_i,A_i)$ be the graph with N_i the set of vertices of S incident upon at least one edge of E_i, and with arc $\{v,w\}\in A_i$ if v and w are the endpoints of a common edge in E_i. We can choose the enumeration of the edges of S such that $\mathcal{G}_i$ is connected, for all $1\leq i\leq m$, and such that $\mathcal{G}_i$ is a tree if $i\leq n-1$. Let S_i be a plane embedding of $\mathcal{G}_i$. Obviously,

$$f_0(S_i)=f_1(S_i)+1 \text{ and } f_2(S_i)=1,$$

if $i\leq n-1$, and therefore

$$f_0(S_i)-f_1(S_i)+f_2(S_i)=2,$$

if $i\leq n-1$. By the choice of graph $\mathcal{G}_{n-1}$, every arc of A_m-A_{n-1} has both endpoints in N_{n-1}. It follows that edge e_j, $j>n-1$, cuts one region of S_{j-1} into two connected portions. Thus, $f_1(S_j)=f_1(S_{j-1})+1$ and $f_2(S_j)=f_2(S_{j-1})+1$, for $n \leq j \leq m$. Since the number of vertices of S_j is the same as the number of vertices of S_{j-1}, we have

$$f_0(S_j)-f_1(S_j)+f_2(S_j)=2,$$

for $1 \leq j \leq m$, and in particular, for $j=m$.

If we keep the number of vertices of S fixed and add as many edges as possible without violating the property of the embedding being plane, then we obtain a new subdivision S' with the same number of vertices. If $f_0(S) \geq 3$ then each region of S' is bounded by exactly three edges, and each edge of S' bounds exactly two regions. This implies

$$3f_2(S')=2f_1(S').$$

Substituting this equality into (i) results in

$$f_1(S')=3f_0(S')-6 \text{ and } f_2(S')=2f_0(S')-4.$$

The inequalities (ii) and (iii) then follow, since $f_0(S)=f_0(S')$, $f_1(S) \leq f_1(S')$, and $f_2(S) \leq f_2(S')$. □

If the underlying graph of a subdivision S consists of $c>1$ connected components, then we have

$$f_0(S)+f_1(S)-f_2(S)=c+1$$

instead of equation (i) in Theorem 11.1 (see also Exercise 11.1). The inequalities (ii) and (iii) still hold, for when we add a new arc the number of connected components does not increase. Theorem 11.1 can be easily extended to subdivisions which are obtained by embedding one node at infinity. The number of edges and regions is the same as for an embedding without an improper vertex and there is one proper vertex fewer.

Theorem 11.1 guarantees that the amount of storage required for a subdivision with n vertices is in $O(n)$. Notice, however, that there are "natural" cell complexes in E^2 which are not subdivisions according to the above definition. One such example consists of a single finite vertex from which $m>1$ rays emanate. This is not a subdivision since the underlying structure is a multigraph: it contains two nodes (one embedded at infinity) and $m>1$ arcs incident upon this pair of nodes. It is, however, easy to refine this subdivision by adding an artificial vertex that cuts each ray into a segment and a ray. The subdivision would then contain $m+1$ proper vertices, one improper vertex, $2m$ edges, and m regions. Similar refinements can be used to modify other subdivisions which do not quite meet our requirements.

11.2. The Geometry of Monotone Subdivisions

A subset of E^2 is said to be *monotone* (*with respect to* x_1) if its intersection with any vertical line is connected. A subdivision of E^2 is *monotone* (*with respect to* x_1) if all of its regions are monotone. The solution to the point location problem to be presented in this chapter requires that the subdivision be monotone. For technical reasons, we also require that all edges are straight and that no edge belongs to a vertical line. These conditions are imposed to simplify the presentation and can be removed with some care. Figure 11.1 shows a monotone subdivision.

With minor caveats and modifications, monotone subdivisions include triangulations, and subdivisions into convex regions among others. Section 11.8 treats the problem of refining an arbitrary subdivision into one which is monotone, and uses methods that are based on the following observation.

Observation 11.2: A subdivision S with straight edges, none of which belongs to a vertical line, is monotone if and only if every vertex v is incident upon at least two edges which lie on different sides of the vertical line through v.

Notice that a monotone subdivision must have unbounded edges and thus has an improper vertex. Furthermore, this subdivision cannot be the embedding of a graph with more than one connected component.

A monotone subdivision allows its regions to be vertically ordered, and this

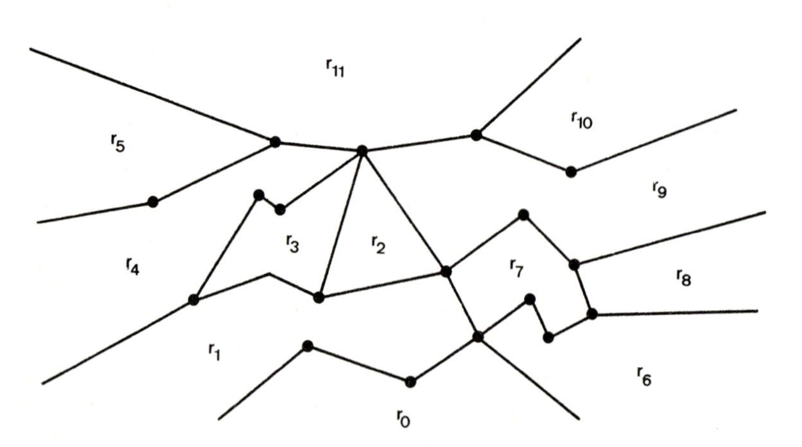

Figure 11.1. A monotone subdivision.

will be crucial for our solution to the point location problem. More generally, we say that a subset r of E^2 is *below* another subset s, and we write $r \ll s$, if $r \neq s$, and for any two points $p = (\pi_1, \pi_2)$ in r and $q = (\psi_1, \psi_2)$ in s we have $\pi_2 \leq \psi_2$ if $\pi_1 = \psi_1$. If r is below s, we also say that s is *above* r, and we write $s \gg r$. For instance, we have $r_0 \ll r_1$ and $r_0 \ll r_{11}$ in Figure 11.1. If we restrict our attention to regions of a monotone subdivision then this relation has several interesting properties.

Observation 11.3: Any two regions r_i and r_j of a monotone subdivision of E^2 satisfy exactly one of $r_i = r_j$, $r_i \ll r_j$, $r_j \ll r_i$, or there is no vertical line that intersects both r_i and r_j.

If three regions r_i, r_j, and r_k intersect a common vertical line then $r_i \ll r_j$ and $r_j \ll r_k$ implies $r_i \ll r_k$. Transitivity may not hold without this restriction, but we have the following result.

Lemma 11.4: The relation "$\ll$" is acyclic if applied to the regions of a monotone subdivision of E^2.

Proof: Suppose the relation is not acyclic. Let $r_0 \ll r_1 \ll \ldots \ll r_k = r_0$ be a cycle with minimum length. From Observation 11.3 it follows immediately that $k > 2$. Regions r_i and r_{i+1} intersect a common vertical line, but r_{i-1} and r_{i+1} do not (with indices taken modulo k); otherwise, there is a shorter cycle without r_i. So for each i, there is a vertical line that intersects region r_i but which does not intersect r_{i-1} or r_{i+1}. Let r_j be the region such that the rightmost point in its boundary is the leftmost. Then r_{j-1} and r_{j+1} must both intersect the rightmost vertical line that meets the boundary of r_j, a contradiction. □

Since "$\ll$" is acyclic, there must be regions r in every monotone subdivision such that there is no region r' below r. This region r is necessarily unbounded, and since we forbid vertical edges, the boundary of r must intersect every vertical line in exactly one point. It follows that r is the unique region with no region below it.

The relation "$\ll$" is computationally unattractive since it can contain a quadratic number of related pairs which is too expensive to compute. We therefore introduce a sparser relation which can be used as a substitute for "$\ll$" when it comes to algorithms. We say that a region r_i is *immediately below* another region r_j, and we write $r_i < r_j$, if $r_i \ll r_j$ and both regions share a common edge. In the example of Figure 11.1, we have $r_0 < r_1$ but not $r_0 < r_{11}$. In general, the relation "$<$" is stronger than "$\ll$", but the following is easily seen to be true.

Observation 11.5: The transitive closures of "$\ll$" and "$<$" are the same if they are applied to the regions of a monotone subdivision.

Roughly, the idea for point location is, first, to compute a linear ordering of the regions which is compatible with the relation "$\ll$" or "$<$", and second, to do some kind of binary search in this linear ordering. We therefore need some means by which we can efficiently answer questions like "does point q belong to the closure of a region in the first half of the ordering?". This leads to the following definitions.

A *separator for* a monotone subdivision S of E^2 is the connected union of edges and vertices of S with the property that it meets every vertical line in exactly one point. Let s be a separator for S and let r be a region of S. Then r is either below s $(r \ll s)$ or above s $(r \gg s)$. Figure 11.2 shows a separator for the subdivision depicted in Figure 11.1. A *complete family of separators for* a monotone subdivision S with ℓ regions is a sequence of $\ell-1$ distinct separators $s_1 \ll s_2 \ll \ldots \ll s_{\ell-1}$. There must be at least one region between any consecutive pair of separators, and also one below s_1 and one above $s_{\ell-1}$. We conclude that, if S admits a complete family of separators, its regions can be enumerated as $r_0, r_1, \ldots, r_{\ell-1}$ such that

$$r_0 \ll s_1 \ll r_1 \ll s_2 \ll \ldots \ll s_{\ell-1} \ll r_{\ell-1}.$$

One edge of S is contained in at least one separator and in at most all separators of a complete family. To be able to be more specific about which separators contain a given edge e, we define integer numbers $b(e) < a(e)$ such that e is incident upon regions $r_{b(e)}$ and $r_{a(e)}$. Given a complete family of separators and an

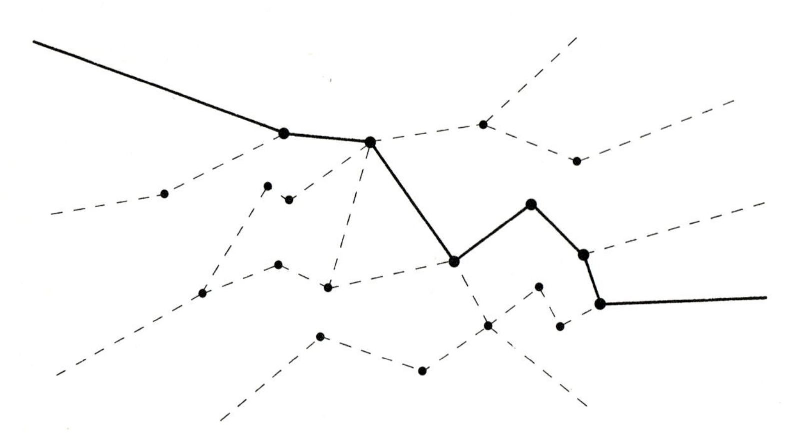

Figure 11.2. A separator.

enumeration of the regions as above, we have the following result.

Observation 11.6: Let e be an edge of a monotone subdivision S of E^2 with a complete family of $\ell-1$ separators s_1 through $s_{\ell-1}$ and a corresponding enumeration of the regions. Then the separators containing e are $s_{b(e)+1}, s_{b(e)+2}, \ldots, s_{a(e)}$.

We are now ready to prove the main result of this section by means of a construction.

Lemma 11.7: Every monotone subdivision S of E^2 admits a complete family of separators.

Proof: Let $r_0, r_1, \ldots, r_{\ell-1}$ be a linear ordering of the regions of S that is compatible with the relation "$\ll$", that is, $r_i \ll r_j$ only if $i<j$. For $1 \leq i \leq \ell-1$, define

$$s_i = \bigcup_{j<i\leq k} \mathrm{bd}r_j \cap \mathrm{bd}r_k .$$

For example, Figure 11.2 shows s_8. We show below that $s_i \ll s_{i+1}$, for $1 \leq i \leq \ell-2$, and that s_i is a separator, that is, it intersects every vertical line in exactly one point and it is connected.

Consider any vertical line l, and let $r_{i_1}, r_{i_2}, \ldots, r_{i_t}$ be the regions it meets, from bottom to top. Since l meets regions r_0 and $r_{\ell-1}$, and since $r_{i_1} \ll r_{i_2} \ll \ldots \ll r_{i_t}$, we have $0 = i_1 < i_2 < \ldots < i_t = \ell-1$. Therefore, for any $1 \leq i \leq \ell-1$, there is exactly one point p on l such that there are indices $j < i \leq k$ with point p in the common boundary of regions r_j and r_k, that is, p belongs to s_i. Furthermore, s_i intersects l in the same point or vertically above s_{i-1}, and if line l meets region r_{i-1} then the two intersections are distinct. We thus have $s_1 \ll s_2 \ll \ldots \ll s_{\ell-1}$. All that now remains is to show that s_i is connected. Suppose it is not. Then there is a vertical line that intersects $\mathrm{cl}s_i$ in at least two points p and q (see Figure 11.3). But then we have $p \ll r \ll q$, for some region r, which contradicts the construction of s_i. □

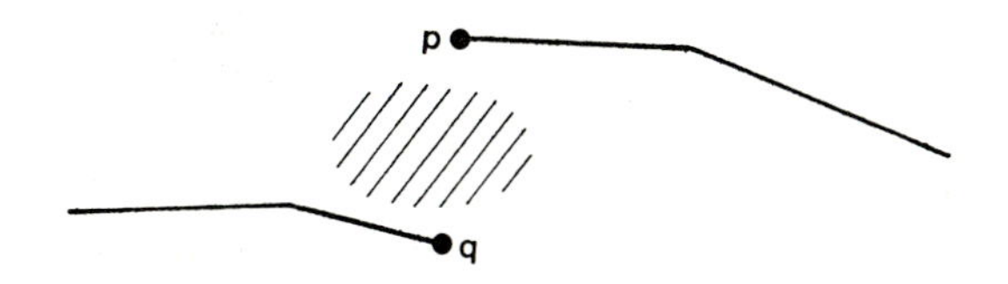

Figure 11.3. The constructed curves are connected.

11.3. A Tree of Separators for Point Location

Section 11.5 will tackle the problem of efficiently computing a useful representation of a complete family of separators for a monotone subdivision S of E^2. So we assume for now that we have such a family $s_1, s_2, \ldots, s_{\ell-1}$, with a corresponding enumeration of the regions $r_0, r_1, \ldots, r_{\ell-1}$ such that

$$r_0 \ll s_1 \ll r_1 \ll s_2 \ll r_2 \ll \ldots \ll s_{\ell-1} \ll r_{\ell-1}.$$

This section demonstrates that the family of separators can be used to efficiently determine the unique region, edge, or vertex of S that contains a query point q.

The algorithm uses two levels of binary search. The inner loop takes a separator s_i (as a linear array of edges and vertices, sorted from left to right), and determines the edge or vertex e of s_i which intersects the vertical line through q. By testing q against e, we can decide whether q is above or below s_i, or on e itself (in which case the search terminates). The outer loop performs a binary search in $s_1, s_2, \ldots, s_{\ell-1}$ in order to locate q between two consecutive separators s_i and s_{i+1}, that is, in region r_i. Apart from the existence of the separators, we assume that each region stores its index, and similarly, that the adjacent regions $r_{b(e)}$ and $r_{a(e)}$ of edge e can be obtained in constant time from e. We will see that the construction of this information is part of the process of constructing the family of separators. The search algorithm uses this information to substantially reduce the storage requirements.

For $\ell = 2^k$ and k an integer number, let T_ℓ be the *perfect binary tree* with ℓ leaves, that is, each internal node has exactly two sons and each leaf has depth k. If ℓ is not a power of two, then define k such that $2^{k-1} < \ell < 2^k$ and let T_ℓ be the tree which consists of the ℓ leftmost leaves of T_{2^k} and their ancestors (see Figure 11.4). We denote the leaves by $0, 1, \ldots, \ell-1$ from left to right, and we denote the internal nodes in inorder by $1, 2, \ldots, \ell-1$. The tree T_ℓ is used as a flowchart for the outer loop of the search algorithm, with the convention that each internal node i represents a test of the query point q against the separator s_i (if it exists), and each leaf j represents the output "q belongs to region r_j". We also use T_ℓ as the first level of a data structure which supports the search. The second level of this structure consists of representations of separators, one for each internal node of T_ℓ. The left and the right sons of an internal node i are denoted by $l(i)$ and $r(i)$, respectively. Furthermore, we let $lca(i,j)$ be the *lowest common ancestor of* the leaves i and j, that is, the root of the smallest subtree of T_ℓ that contains both leaves i and j.

When testing point q against a separator, we adopt the convention that each edge contains its left endpoint but not its right endpoint, if it exists. This is unambiguous since there are no vertical edges. If the algorithm detects that q

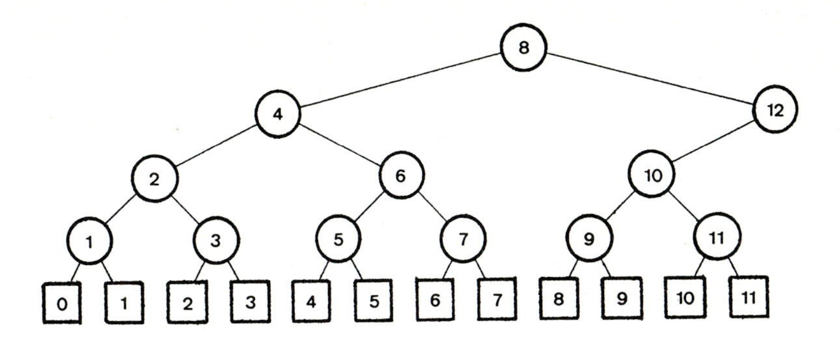

Figure 11.4. Tree $\mathcal{T}_{12}$.

lies on some edge e during a discrimination against a separator, it terminates the search and, by comparing q with the left endpoint of e, determines if q is a vertex of the subdivision. The algorithm maintains two integer numbers i and j such that at each point in time, point q is known to belong to one of the regions $r_i, r_{i+1}, \ldots, r_j$, or to some edge or vertex in the common boundary of two of them. By definition of the separators, this is equivalent to saying that q lies above the separator s_i and below s_{j+1}, whenever these separators exist. A formal description of the algorithm follows.

Algorithm 11.1 (Two–level binary search):

Initial step: Set $i:=0$, $j:=\ell-1$, and $k:=lca(0,\ell-1)$.
while $i<j$ **do**
  if $i<k\leq j$ **then**
    Using binary search in the edges of separator s_k, determine the edge e in s_k such that e intersects the vertical line through q.
    Case 1: q belongs to edge e. In this case, report e or its left endpoint, whichever contains point q, and halt.
    Case 2: $q \ll e$. In this case, set $j:=b(e)$.
    Case 3: $q \gg e$. Adjust i, that is, set $i:=a(e)$.
  else
    if $k\leq i$ **then** set $k:=r(k)$ **else** $\{k>j\}$ set $k:=l(k)$ **endif**
  endif
endwhile;
Report region r_i as the one which contains point q.

Note that by keeping track of the variables i and j, we are able to skip the binary search in some separators. This optimization does not affect the worst–case time bound which is $O(\log\ell\cdot\log m)$, since we perform binary search on up to $\log_2\ell$ separators, each in time $O(\log m)$.

If we independently represent each separator by a linear array of its edges, then we store $\ell-1$ separators, each one consisting of $\Omega(m)$ edges in the worst case. Figure 11.5 shows such a worst-case subdivision: $m-2\ell+2$ edges are common to all $\ell-1$ separators and any two separators differ only by two edges. This amounts to $O(\ell m)$ storage in the worst case. However, after point q has been tested against an edge e, the variables i and j are updated in such a way that we will never look at the edges of any other separator that contains e again. Therefore, an edge needs only be stored in the first separator containing it that can be encountered in a search down the tree. Specifically, if the edge e is incident upon the regions r_i and r_j, $i<j$, then, by Observation 11.6, e belongs to the separators $s_{i+1}, s_{i+2}, \ldots, s_j$, and so it suffices to store e in s_k, where k is the lowest common ancestor of i and j. This is the highest node in $\mathcal{T}_\ell$ whose separator contains edge e.

Note that only those edges assigned to s_k according to the above rule are actually stored in the linear array for s_k. In general, these form a subset of all the original edges of s_k, so between successively stored edges of s_k there may be *gaps* (see Figure 11.6). The ordered list of x_1-coordinates of all vertices incident upon edges stored in the linear array for s_k is called the *chain* c_k, for $1 \leq i \leq \ell-1$. More formally, c_k contains a real number α if and only if there is a vertex $u=(\alpha, v_2)$ incident upon an edge e such that $k=lca(b(e),a(e))$. If there are m_k such edges then c_k contains at most $2m_k$ real numbers which define at most $2m_k+1$ intervals; by convention, an interval contains its left endpoint but not its right endpoint. If an interval of c_k is the vertical projection of an edge e with $k=lca(b(e),a(e))$, then the interval is said to be *covered by* e. Thus, exactly m_k of the at most $2m_k+1$ intervals of c_k are covered.

The total storage needed to represent the chains c_1 through $c_{\ell-1}$ is in $O(m)$

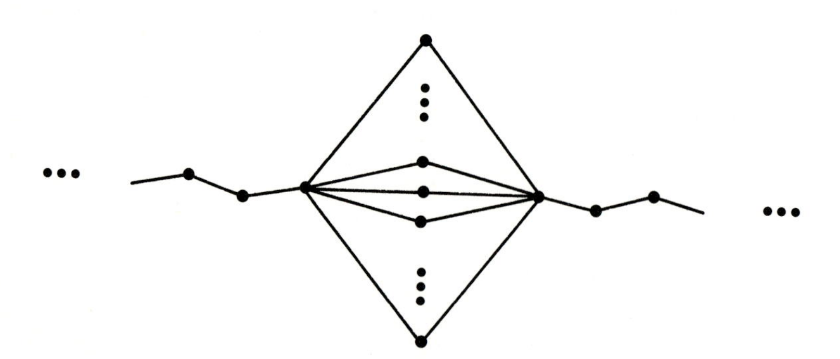

Figure 11.5. Any two separators share $m-2\ell+2$ edges.

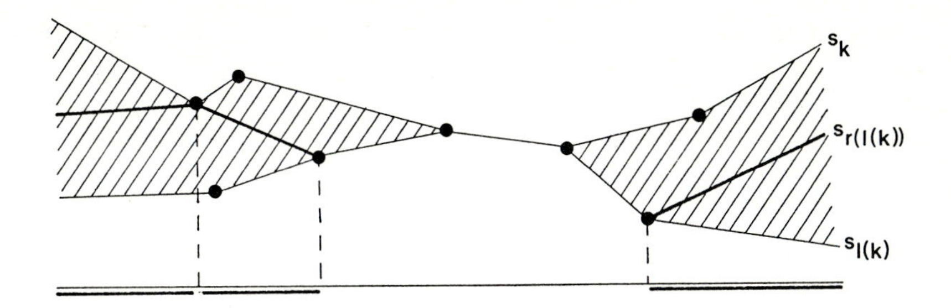

Figure 11.6. A chain consists of covered and uncovered intervals.

since each edge covers an interval of exactly one chain. We summarize the results of this section.

Lemma 11.8: There is a data structure which solves the point location search problem for a subdivision with n vertices in $O(n)$ storage and $O(\log^2 n)$ query time.

We will modify the above data structure such that it allows us to answer a point location query in time $O(\log n)$ in Section 11.6. Before, however, we discuss how a subdivision can be represented and how the above data structure can be constructed from such a subdivision.

11.4. Representing a Plane Subdivision

As for arrangements of hyperplanes and for convex polytopes, there are several possible data structures that can be used to represent a subdivision of E^2. We will choose one which seems rather fundamental in that most of the other common representations can be thought of as modifications of our structure. Nevertheless, it is rich enough so that the point location structure presented in Section 11.3 can be computed efficiently from any monotone subdivision of E^2 that is represented by this data structure.

Let S be a monotone subdivision of E^2. Our representation of S is its *incidence graph* $I(S)$, which is an undirected graph. This graph contains a node for each vertex, edge, and region of S, a node for $\emptyset$, the (-1)–face of S, and a node for S, the so–called 3–face of S. Two nodes are adjacent if the two corresponding faces are incident. Here, we say that each vertex is incident upon $\emptyset$, and that each region is incident upon S. Figure 11.7 shows the incidence graph of a subdivision with four bounded and two unbounded edges.

Let us further discuss our representation of the incidence graph by considering a collection of records with various pointers between them. For convenience,

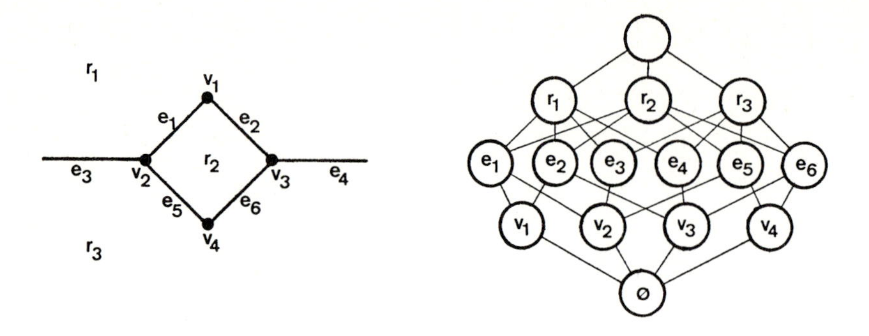

Figure 11.7. The incidence graph of a subdivision.

we will draw no distinction between a node and the record that represents it. Notice that each edge is incident upon exactly two regions and upon at most two vertices. To aid several upcoming computations, we distinguish the incidences of an edge of which there are at most four: every record which stores an edge e has fields

> *left*, which is a pointer to the left endpoint of e,
> *right*, a pointer to the right endpoint of e,
> *above*, a pointer to the region above e, and
> *below*, a pointer to the region below e.

To encode the geometry of S, we equip each record that represents a vertex v with the coordinates of v. Each record that holds an unbounded edge e contains a vector $\vec{v}$ such that

$$e = \{x \in E^2 | x = u + \lambda \vec{v},\ \lambda > 0\},$$

where u is the finite endpoint of edge e. In addition, each record that stores a region contains an integer field *index* to be used later, and each record that stores a vertex or an edge has a bit field which allows us to mark that vertex or edge.

11.5. Constructing a Family of Separators

Let S be a monotone subdivision of E^2 with n vertices, m edges, and ℓ regions represented by its incidence graph $I(S)$. This section considers the problem of constructing the data structure for the point location search presented in Section 11.3. We will refer to this data structure as the *chain tree* $T(S)$ *of* S.

The construction of $T(S)$ proceeds in two phases as follows:

Phase 1: Compute an ordering of the regions of S which is compatible with the relation "$\ll$", that is. for each region r compute an integer $index(r)$ in $\{0,1,...,\ell-1\}$ such that $index(r) < index(r')$ if $r \ll r'$.

Phase 2: Traverse the edges of S from left to right, that is, visit an edge only if its left endpoint is visited, and visit a vertex only if all edges for which the vertex is the right endpoint are visited. When an edge e is visited, then determine $b(e)$ and $a(e)$, the indices of its incident regions, and append e to the chain c_k, where $k = lca(b(e), a(e))$.

We implement Phase 1 as a topological sort of the directed graph $\mathcal{R}$ defined as follows:

the nodes of $\mathcal{R}$ are the regions of S and (r,r') is an arc of $\mathcal{R}$ if $r < r'$, that is, $r \ll r'$ and regions r and r' share a common edge.

It is not necessary to construct graph $\mathcal{R}$ since it is encoded in the incidence graph $\mathcal{I}(S)$: $\mathcal{R}$'s nodes are the nodes of $\mathcal{I}(S)$ which represent regions of S, and the arcs of $\mathcal{R}$ are the nodes of $\mathcal{I}(S)$ which represent the edges of S. As a matter of fact, this encoding represents each arc of $\mathcal{R}$ possibly more than once since two regions can share more than one common edge. This will not be inconvenient for our algorithm. The graph $\mathcal{R}$ has exactly one source, since there is exactly one region in S which is below all other regions. This allows us to implement Phase 1 as a compact recursive procedure. It takes as input the incidence graph $\mathcal{I}(S)$, with $index(r)$ undefined for all regions $r \in S$, and a region r which is initially the region below all other regions of S. Furthermore, the procedure accesses a globally available integer number i which is equal to $\ell-1$ whenever we call the procedure from the outside.

Procedure 11.2 (Phase 1):

if $index(r)$ is undefined **then**
  for each region r' with (r,r') an arc of $\mathcal{R}$ **do**
    Recur for region r'.
  endfor;
  Set $index(r) := i$, and set $i := i-1$.
endif.

Figure 11.1 shows a monotone subdivision with 12 regions. The indices of the regions can be thought of as being assigned by Procedure 11.2 if the regions immediately above any region are processed from right to left.

A similar graph traversal algorithm is used to visit the edges of S in the desired order from left to right. Such an ordering is one which is compatible with the relation defined as follows:

an edge e is *left of* another edge e', for short $e \preceq e'$, if there is a vertex of S which is the right endpoint of e and the left endpoint of e'.

We represent the inverse of this relation by the directed graph $\mathcal{E}(S)$ which contains the edges of S as its nodes, and which contains an arc (e,e') if $e' \preceq e$. Unfortunately, $\mathcal{E}(S)$ contains $k_1 \cdot k_2$ arcs for each vertex of S which is the right endpoint of k_1 edges and the left endpoint of k_2 edges. This leads us to use the following directed graph $\overline{\mathcal{E}}(S)$ which decreases the number of arcs to $O(n)$ at a negligible increase in the number of nodes:

the nodes of $\overline{\mathcal{E}}(S)$ are the edges and the vertices of S, and (a,b) is an arc of $\overline{\mathcal{E}}(S)$ if b is the left endpoint of edge a or if a is the right endpoint of edge b.

Like $\mathcal{R}(S)$, we need not construct $\overline{\mathcal{E}}(S)$, for it is encoded in the incidence graph $\mathcal{I}(S)$: the nodes of $\mathcal{I}(S)$ that store vertices or edges of S are the nodes of $\overline{\mathcal{E}}(S)$, and an arc $\{v,e\}$ of $\mathcal{I}(S)$, between a vertex v and an edge e, appears as directed arc (v,e) if v is the right endpoint of e. This arc appears as (e,v) if v is the left endpoint of e. The following recursive procedure can be used to traverse the nodes of $\overline{\mathcal{E}}(S)$ in a sequence which is the reverse of a topological order. The order of the edges in this sequence is compatible with the relation "$\preceq$". Each edge of S visited during this traversal is added to a chain of the chain tree of S. The procedure assumes that we add an artificial new vertex v_∞ to $\overline{\mathcal{E}}(S)$ which has no incoming arc and an outgoing arc to each source of $\overline{\mathcal{E}}(S)$; this new vertex thus becomes the only source of $\overline{\mathcal{E}}(S)$. The input to the procedure consists of the incidence graph $\mathcal{I}(S)$, with all edges and vertices unmarked, along with an edge or vertex f. Initially, $f = v_\infty$.

Procedure 11.3 (Phase 2):
- **if** f is unmarked **then**
  - **for** each f' with (f,f') an arc of $\overline{\mathcal{E}}(S)$ **do**
    - Recur with f'.
  - **endfor;**
  - Mark f;
  - **if** f is an edge of S **then**
    - Set $b(f) := index(below(f))$, $a(f) := index(above(f))$, and compute $k := lca(b(f), a(f))$. Append the x_1-coordinate of $left(f)$ to the chain c_k of the chain tree, unless $left(f) =$ **nil** or c_k already contains the x_1-coordinate of $left(f)$.
    - **if** $right(f) \neq$ **nil then**
      - Append the x_1-coordinate of $right(f)$ to chain c_k and establish a pointer from the penultimate interval of c_k to edge f.
    - **else**

Establish a pointer from the last interval of c_k to edge f.
endif
endif
endif.

The chains $c_1, c_2, \ldots, c_\ell$ of the chain tree must be organized as linked lists such that it is possible to add a new value at the end of a chain in constant time. When one wants to use the chain tree for point location search, these linked lists have to be replaced by linear arrays or other kinds of search structures. Alternatively, one could connect the various chains (as will be described in Section 11.6); this also improves the search time from $O(\log^2 n)$ to $O(\log n)$.

If the construction of the chain tree as described above is to run in time linear in n, the number of vertices, it is essential that the total time required to compute $lca(index(below(e)), index(above(e)))$, for all edges e, be in $O(n)$. This rules out the straightforward algorithm that starts at the leaves i and j, and moves up one level of $\mathcal{T}_\ell$ at a time, until the two paths join at a common node. This naive algorithm has running time in $\Omega(\log|i-j|)$, and it is possible to have subdivisions such that

$$index(above(e)) - index(below(e)) = \Omega(\sqrt{n}),$$

for $\Omega(n)$ edges e, thus giving an overall running time of $\Omega(n\log n)$.

Due to the regular structure of the tree $\mathcal{T}_\ell$, the value of $lca(i,j)$ has a simple interpretation in terms of the binary representations of i and j. Let $\kappa = \lfloor \log_2 \ell \rfloor$ be the number of bits needed to represent any number from 0 through $\ell-1$, and let

$$b(i) = (a_{\kappa-1}\, a_{\kappa-2} \cdots a_{k+2}\, 0\, i_k \cdots i_1\, i_0)_2 \text{ and}$$

$$b(j) = (a_{\kappa-1}\, a_{\kappa-2} \cdots a_{k+2}\, 1\, j_k \cdots j_1\, j_0)_2$$

be the binary representations of i and j. Then

$$lca(i,j) = (a_{\kappa-1}\, a_{\kappa-2} \cdots a_{k+2}\, 1\, 0 \ldots 0\, 0)_2.$$

Loosely speaking, $lca(i,j)$ is the longest common prefix of $b(i)$ and $b(j)$, followed by 1 and padded with zeros.

An efficient formula for computing $lca(i,j)$ is based on the function

$$msb(k) = lca(0,k) = 2^{\lfloor \log_2 k \rfloor},$$

the *most significant bit of* k, for $k \geq 1$. Its values for $k = 1, 2, \ldots, \ell-1$ are $1,2,2,4,4,4,4,8,\ldots,2^{\kappa-1}$. These numbers can be easily precomputed in time $O(\ell)$ and stored in a linear array with $\ell-1$ entries, so we can compute $msb(k)$ in constant time. We can then express the lowest common ancestor function as

$$lca(i,j) = j \wedge \neg(msb(i \oplus j) - 1),$$

where $\oplus$, $\wedge$, and $\neg$ are the boolean operations of bitwise "exclusive or", "and", and "complement". We assume that these boolean operations can be computed in constant time, as in the cases of addition and subtraction. This seems to be justified since their theoretical complexity is no greater than that of addition, and most present day computers have such instructions.

Subsequently, it is not difficult to see that the above procedures run in time $O(n)$. We thus summarize the results of this section.

Lemma 11.9: Given a monotone subdivision S with n vertices represented by its incidence graph $I(S)$, there is an algorithm which constructs the chain tree of S in time $O(n)$.

11.6. Optimal Search by Connecting Separators

Our hope of obtaining a faster algorithm for point location comes from the fact that there is some obvious loss of information in the method described in Section 11.3. More specifically, when we discriminate a point $q = (\psi_1, \psi_2)$ against a chain c_k, we localize ψ_1 in the x_1-coordinate to within an edge or a gap of c_k. Yet, when we continue the search in the chain $c_{l(k)}$ or in $c_{r(k)}$, we start this localization process all over again. This section presents a technique which refines the chains so that the localization of ψ_1 in chain c_k allows us to do the same localization in $c_{l(k)}$ or $c_{r(k)}$ with only a constant extra effort. Below, we describe a refinement that, for each chain c_k, produces a refined chain $\bar{c}_k$; remember that c_k is a list of x_1-values and intervals, and so is $\bar{c}_k$. By virtue of being a refinement, each interval of $\bar{c}_k$ is contained in exactly one interval of c_k. In addition, it overlaps at most two intervals of the refined chains $\bar{c}_{l(k)}$ and $\bar{c}_{r(k)}$. As we will see, this last condition is compatible with keeping the overall storage in $O(n)$.

The refined chains $\bar{c}_k$ and their interconnections can be conveniently represented by a linked data structure that we call the *layered DAG of* the subdivision. This is a directed acyclic graph whose nodes correspond to tests of three kinds: *x_1-tests*, *edge tests*, and *gap tests*. We call the corresponding nodes *x_1-nodes*, *edge nodes*, and *gap nodes*, respectively. A chain $\bar{c}_k$ is represented in the layered DAG by a collection of such nodes: each x_1-value of $\bar{c}_k$ gives rise to an x_1-node, and each interval to an edge or a gap node (see Figure 11.8).

A node ν which corresponds to an x_1-test of $\bar{c}_k$ contains an x_1-value, denoted by *value*(ν), and two pointers *behind*(ν) and *in_front*(ν) to the adjacent edge or gap nodes which represent the intervals to the left and to the right of *value*(ν). If ν is an edge or a gap node, then ν contains two pointers *down*(ν) and

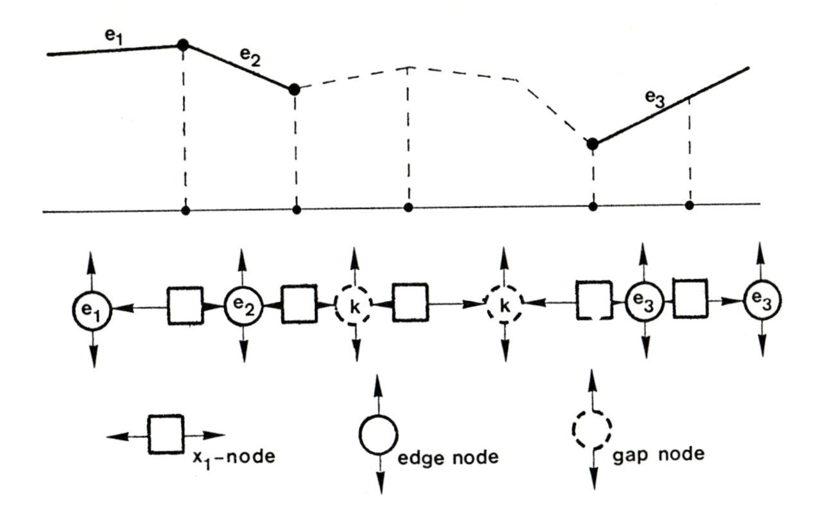

Figure 11.8. The nodes of the layered DAG for chain $\bar{c}_k$.

$up(\nu)$ to appropriate nodes of $\bar{c}_{l(k)}$ and $\bar{c}_{r(k)}$. In addition, an edge node ν contains a reference $edge(\nu)$ to the edge of separator s_k whose vertical projection covers the interval represented by ν. Rather than a pointer, a gap node contains the index $chain(\nu)$ of the chain to which it belongs.

We can now be more precise about the meaning of the pointers $down(\nu)$ and $up(\nu)$. The properties of the refined chains ensure that the interval of $\bar{c}_k$ corresponding to an edge or gap node ν overlaps either one interval ι of $\bar{c}_{l(k)}$, or two such intervals ι_1 and ι_2 separated by some x_1–value of $\bar{c}_{l(k)}$. In the first case, $down(\nu)$ points to the edge or gap node of $\bar{c}_{l(k)}$ corresponding to the interval ι; in the second case, $down(\nu)$ points to the x_1–node corresponding to the separating x_1–value of ι_1 and ι_2. Similarly, the pointer $up(\nu)$ points to a node of $\bar{c}_{r(k)}$ defined in an analogous manner. In the special case when $l(k)$ and $r(k)$ are leaves of $\mathcal{T}_\ell$, we let $down(\nu) = up(\nu) = \textbf{nil}$.

The layered DAG then consists of the nodes for all refined chains $\bar{c}_1, \bar{c}_2, \ldots, \bar{c}_{\ell-1}$, linked together in this fashion. We can use this DAG to simulate Algorithm 11.1; the main difference is that each time we move from a node of $\mathcal{T}_\ell$ to one of its children, the *down*– or *up*–pointers (possibly together with an x_1–test) allow us to locate ψ_1, the x_1–coordinate of point q, in the new chain, in constant time. As before, the variables i and j keep track of the separators which are known to lie above or below q, namely point q is above s_i and below s_{j+1}, whenever they exist. These numbers are updated after each edge test exactly as in Algorithm 11.1, and they are used during a gap test. By the time the search algorithm gets to a **nil** pointer, we will have $i = j$ which tells us that q belongs to region r_i.

Below, we give a formal description of the point location algorithm. It starts at a distinguished node *root* of the layered DAG and proceeds towards the leaves. This node is the root of a balanced tree of x_1–nodes whose leaves are the edge nodes of $\bar{c}_{lca(0,\ell-1)}$, the chain corresponding to the root $lca(0,\ell-1)$ of $\mathcal{T}_\ell$.

Algorithm 11.4 (Optimal point location search):

Initial step: Set $i:=0$, $j:=\ell-1$, and $\nu:=root$.
while $i<j$ **do**
  Case 1: ν is an x_1–node.
    if $\psi_1 < value(\nu)$ **then** set $\nu:=behind(\nu)$
              else set $\nu:=in\_front(\nu)$
    endif;
  Case 2: ν is an edge node.
    Case 2.1: $q\in edge(\nu)$. In this case, report $edge(\nu)$ or its left endpoint, whichever contains q, and halt.
    Case 2.2: $q \ll edge(\nu)$. Then set $j:=index(below(edge(\nu)))$ and set $\nu:=down(\nu)$.
    Case 2.3: $q \gg edge(\nu)$. Then set $i:=index(above(edge(\nu)))$ and set $\nu:=up(\nu)$.
  Case 3: ν is a gap node.
    if $chain(\nu)\le i$ **then** set $\nu:=up(\nu)$
              else $\{chain(\nu)>j\}$ set $\nu:=down(\nu)$
    endif
endwhile;
Report region r_i as the one which contains point q.

11.7. Constructing the Layered DAG

Now that we understand how the layered DAG is to be used, we will describe how it can be constructed. Our starting point will be the chain tree which consists of the tree $\mathcal{T}_\ell$ and the chains c_k, for $1\le k\le \ell-1$, defined in Section 11.5. Recall that c_k is defined by those edges of the separator s_k that do not belong to any separator whose index is an ancestor of k in $\mathcal{T}_\ell$.

Our construction of the layered DAG proceeds bottom up and happens simultaneously with the refinement of the chains. We first describe how the x_1–values of $\bar{c}_k$ are obtained. Note that we already have at our disposal three sorted lists of x_1–values: $\bar{c}_{l(k)}$, $\bar{c}_{r(k)}$, and c_k. The x_1–values in $\bar{c}_k$ are obtained by merging the values in c_k and every other value in $\bar{c}_{l(k)}$ and $\bar{c}_{r(k)}$. Here, the emphasis is on "every other value"; if we took every value of $\bar{c}_{l(k)}$ and $\bar{c}_{r(k)}$ then we would end up with $\Omega(n\log n)$ storage in the worst case. By convention, if k is a leaf of $\mathcal{T}_\ell$ or if $k\ge\ell$, then $\bar{c}_k$ is empty. We now imagine that this is done in a bottom–

up fashion for every node k of $\mathcal{T}_\ell$. The propagation of every other value from the children to the parent constitutes the chain refinement that was mentioned in Section 11.6. This refinement has two desirable properties.

Observation 11.10: An interval of $\bar{c}_k$ overlaps at most two intervals of $\bar{c}_{l(k)}$ and two intervals of $\bar{c}_{r(k)}$.

This implies that the refined chains $\bar{c}_k$ can actually be connected as described in Section 11.6. The next property concerns the number of nodes of the layered DAG.

Lemma 11.11: The total number of x_1-values in the refined chains $\bar{c}_k$, for $1 \leq k \leq \ell - 1$, is at most $4m$, four times the number of edges.

Proof: If n_k denotes the number of x_1-values in chain c_k, then

$$\sum_{1 \leq k \leq \ell-1} n_k \leq 2m,$$

since each edge of the subdivision contributes the x_1-coordinates of its endpoints to exactly one chain. Let $\bar{n}_k$ denote the number of x_1-values in the refined chain $\bar{c}_k$, and let N_k ($\bar{N}_k$) be the sum of the n_i ($\bar{n}_i$) over all descendents i of k in $\mathcal{T}_\ell$, including k itself. To prove the lemma, it suffices to show that

$$\bar{N}_r \leq 2N_r,$$

for $r = lca(0, \ell-1)$, the root of $\mathcal{T}_\ell$. As an inductive hypothesis assume that

$$\bar{N}_i + \bar{n}_i \leq 2N_i,$$

for $i = l(k)$ and for $i = r(k)$. This hypothesis is trivially true for the leaves of $\mathcal{T}_\ell$. Observe now that

$$\bar{N}_k = \bar{N}_{l(k)} + \bar{N}_{r(k)} + \bar{n}_k,$$

by the definition of $\bar{N}_k$, and that

$$\bar{n}_k \leq n_k + \frac{\bar{n}_{l(k)} + \bar{n}_{r(k)}}{2},$$

by the definition of chain $\bar{c}_k$. Applying the inductive hypothesis yields

$$\bar{N}_k + \bar{n}_k \leq 2N_k,$$

which proves the same inequality for $k = r$. □

Intuitively, half of the x_1-values of chain c_k propagate to the chain of the parent of k, a quarter to the chain of the grandparent of k, and so on. Although this argument is not strictly correct, it illustrates why we can expect the size of the layered DAG to remain linear in m.

The construction of the x_1–nodes, edge nodes, and gap nodes representing the refined chain $\bar{c}_k$ in the layered DAG is now straightforward, as well as the setting of the *behind*– and the *in_front*–pointers of the x_1–nodes. The *down*–pointers of $\bar{c}_k$ can be set according to the rules of Section 11.6, by simultaneously traversing the two chains $\bar{c}_k$ and $\bar{c}_{l(k)}$ from left to right. The *up*–pointers are analogous. In fact, it is possible to build $\bar{c}_k$ and link it to the part below it in a single simultaneous traversal of c_k, $\bar{c}_{l(k)}$, and $\bar{c}_{r(k)}$. This bottom–up process terminates with the construction of the chain $\bar{c}_r$, where $r = lca(0, \ell-1)$. As a final step, we organize the x_1–nodes of $\bar{c}_r$ in a sorted minimal height binary tree whose leaves are made to point to the appropriate edge nodes of $\bar{c}_r$. All nodes of the layered DAG can be reached from the root of that tree.

By Lemma 11.11, the above construction of the layered DAG from the chain tree takes only time $O(m)$. The longest path in this DAG consists of at most $\log_2 m + 2\log_2 \ell$ nodes: at most $\log_2 m$ x_1–nodes of the initial tree that represents $\bar{c}_r$, and at most two nodes of each level of T_ℓ. This implies that Algorithm 11.4 takes time $O(\log n)$ to answer a point location query. Together with Lemma 11.9 and Theorem 11.1, this implies the main result of this chapter.

Theorem 11.12: Let S be a monotone subdivision of E^2 represented by its incidence graph, and let n be the number of vertices of S. There is a data structure which takes $O(n)$ storage and time for construction such that a point location query can be answered in time $O(\log n)$.

11.8. Refining Non–Monotone Subdivisions

The layered DAG can be used to answer point location queries even for non–monotone subdivisions, provided they can be refined to monotone subdivisions. In this section, we demonstrate how a straight subdivision with n vertices can be made monotone by adding at most n new edges but no new vertices. For convenience, we assume that no edge is contained in a vertical line. This assumption can be removed with some care.

Recall that a subdivision S is not monotone if and only if there is a vertex u in S such that all incident edges of u lie on one side of the vertical line l_u through u. We say that u is *left–irregular* if all incident edges lie to the right of l_u, and we say that u is *right–irregular* if those edges all lie to the left of l_u. To eliminate a right–irregular vertex u, we add an edge to the right of l_u which connects u with another vertex of S or it extends to infinity (see Figure 11.9). Left–irregular vertices are treated symmetrically. We will manage to add at most one new edge per vertex of S, and thus we do not increase the complexity of S asymptotically.

The algorithm is based on the idea of a vertical line l sweeping from left to right. We call an edge of S *active* (*with respect to* l) if it intersects l, and *inactive*, otherwise. The set of active edges changes only when line l passes through a vertex of S, at which time some edges may become inactive and others may become active. Except at those moments, the active edges have an obvious vertical ordering. Let $e_1, e_2, \ldots, e_k$ be the active edges for some position of line l when it contains no vertex of S, and assume that the indices are chosen such that

$$e_1 \ll e_2 \ll \ldots \ll e_k.$$

The k edges cut l into $k+1$ intervals $\iota_0, \iota_1, \ldots, \iota_k$, where ι_i is delimited by edges e_i and e_{i+1}. To make this definition meaningful even for $i=0$ and $i=k$, we introduce dummy edges e_0 below all active edges and e_{k+1} above all active edges. To each interval ι_i we assign a *generator* $\gamma(e_i, e_{i+1})$ which is the last vertex passed by l that lies on or between the lines that support edges e_i and e_{i+1}. This vertex may be a left endpoint of e_i or e_{i+1}, or it may be a right–irregular vertex. If both e_i and e_{i+1} have no left endpoint, then $\gamma(e_i, e_{i+1})$ may not even exist and we take the vertex at infinity in the direction of e_i or e_{i+1} as a substitute. In any case, we have the following straightforward observation.

Observation 11.13: The relatively open segment $\mathrm{ri\,conv}\{\gamma(e_i, e_{i+1}), p\}$, for p any point of ι_i, does not intersect any edge of S.

The algorithm simulates this line sweep. It utilizes two data structures: a list $\mathcal{L}$ which contains the vertices of S in lexicographically increasing order, and a balanced binary tree $\mathcal{T}$ which contains the active edges sorted by intersection with line l in its interior nodes and the generators in the appropriate leaves. To simplify the discussion, we pretend that $\mathcal{T}$ stores the dummy edges e_{k+1} and e_0 above and below all active edges. We assume that $\mathcal{T}$ is a dictionary, that is, it allows us to search, insert, and delete in time logarithmic in the number of active edges. Below, we give a more formal description of the algorithm.

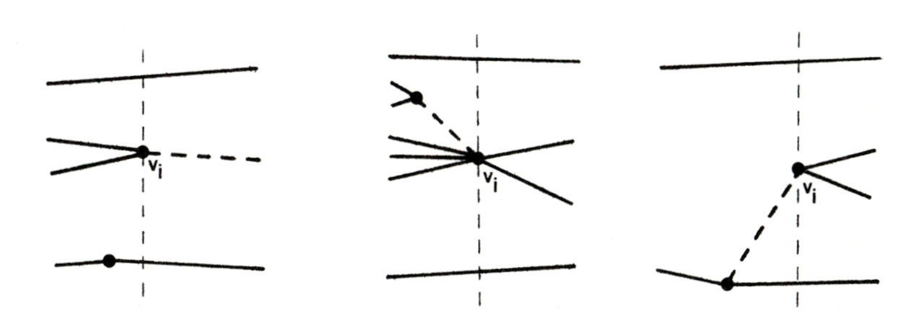

Figure 11.9. Adding new edges to get rid of irregular vertices.

Algorithm 11.5 (Plane–sweep for monotonicity):

Initial step: Sort the n vertices of S into lexicographical order and store them in list $\mathcal{L}$. Initialize $\mathcal{T}$ to the empty tree and insert into $\mathcal{T}$ all edges of S which are unbounded to the left.

Sweep:

for $i:=1$ **to** n **do**

Let v_i be the ith vertex in list $\mathcal{L}$, let $e_1,e_2,\ldots,e_{l_i}$ be the edges with right endpoint v_i, and let $f_1,f_2,\ldots,f_{r_i}$ be the edges with left endpoint v_i. We assume that both sequences of edges are sorted in increasing order by the slopes of the edges. Furthermore, let e_b be the active edge vertically below v_i, and let e_a be the active edge vertically above v_i.

unless $l_i=0$ **do**

if the generator $\gamma(e_b,e_{l_i})$ is right–irregular **then**

Add riconv$\{\gamma(e_b,e_{l_i}),v_i\}$ as a new edge to S.

endif;

Do the same for the generator of the interval between edges e_1 and e_a (see Figure 11.9).

Delete edges e_1 through e_{l_i} from $\mathcal{T}$.

else $\{l_i=0$, that is, v_i is left–irregular$\}$

Connect vertex v_i with the generator $\gamma(e_b,e_a)$ of the interval between e_b and e_a.

endunless;

Insert edges f_1 through f_{r_i} into $\mathcal{T}$ and set $\gamma(e,f):=v_1$, for all pairs $\{e,f\}$ in $\{\{e_b,f_1\},\{f_1,f_2\},\{f_2,f_3\},\ldots,\{f_{r_i},e_a\}\}$.

endfor;

Final step: If $\mathcal{T}$ contains an irregular generator, then create a ray emanating from it.

Figure 11.9 illustrates three different cases that can occur when adding a new edge to the subdivision. The time required by Algorithm 11.5 is in $O(n\log n)$, since the vertices can be sorted in time $O(n\log n)$, and each operation in the dictionary $\mathcal{T}$ takes time $O(\log n)$.

11.9. Exercises and Research Problems

2 **Exercise 11.1:** Let $\mathcal{G}$ be a graph with n nodes, m arcs, and c connected components. Prove that every plane embedding of $\mathcal{G}$ satisfies

$$n-m+\ell=c+1,$$

for ℓ the number of regions of the embedding.

3 **Exercise 11.2:** Let S be a non–straight subdivision with m edges in E^2. What are the computational properties of the edges that are necessary such that the point location algorithm of Section 11.6 can be adapted to S without any loss in efficiency?

4 **Exercise 11.3:** Let S be a plane embedding of a connected graph with m arcs such that each edge of S is straight. Design an algorithm that takes time $O(m\log\log m)$ to refine S to a monotone subdivision.

3 **Exercise 11.4:** Let S be a straight, monotone subdivision with m edges in E^2. Give an algorithm which takes time $O(m)$ to triangulate S, that is, it refines S in such a way that each region is bounded by exactly three edges. *(Hint: triangulate each region in time proportional to the number of its edges.)*

Exercise 11.5: Use the point location algorithm of Section 11.6 to solve the following problems with $O(\log n)$ search time and $O(n)$ storage.

2 (a) Given a convex polytope P with n vertices in E^3, preprocess P such that we can decide for a query point q whether $q \in P$ or $q \notin P$. *(Hint: do a central projection of P onto two planes, one above and one below P.)*

3 (b) Preprocess P such that, given a non–zero vector v, we can find the two planes with normal vector v which are tangent to P. *(Hint: dualize the problem and use the solution to (a) above.)*

3 (c) Let P be a set of n points in E^3. We say that a point $q=(\psi_1,\psi_2,\psi_3)$ not in P is *maximal* if there is no point $p=(\pi_1,\pi_2,\pi_3)$ in P such that

$$\psi_1 \leq \pi_1,\ \psi_2 \leq \pi_2, \text{ and } \psi_3 \leq \pi_3.$$

Preprocess P such that we can decide whether or not a given query point q is maximal. *(Hint: construct the set of points in E^3 that are maximal, project its boundary onto a plane, and build a layered DAG for the obtained subdivision.)*

4 **Exercise 11.6:** Let S be a straight subdivision with m edges in E^2. Prove the existence of a refinement of S to a monotone subdivision, such that every line intersects at most $O(\log m)$ of the "new" edges which decompose one of the "old" regions of S.

Exercise 11.7: Design a data structure which solves the following dynamic variant of the point location problem in E^2: in addition to point location queries, we have requests for deleting an edge and for inserting a "new" edge which connects two "old" vertices and intersects no "old" edge. Let n denote the number of vertices of the subdivision.

4 (a) Give a solution which takes time $O(n\log^2 n)$ to process a sequence of n point location queries and insertions of edges. *(Hint: modify the chain tree given in Section 11.3.)*

4 (b) Give a solution which takes time $O(\log^2 n)$ for a point location query and time $O(\log^4 n)$ for an insertion or deletion of an edge. *(Hint: use the lemma formulated as Exercise 11.6 and the chain tree described in Section 11.3.)*

Exercise 11.8: Let S be a straight subdivision with $n \geq 4$ vertices and m edges in E^2 such that there is only one unbounded region and each region of S is incident upon exactly three edges.

3 (a) Prove that there are positive constants c_1 and c_2 and a set $I(S)$ of at least $c_1 n$ vertices of S, all different from the vertices of the unbounded region, such that each vertex in $I(S)$ is incident upon at most c_2 edges and no two vertices in $I(S)$ are incident upon a common edge. Call $I(S)$ a *bounded–degree independent set of* S. *(Hint: see Section 9.5.2, Theorem 9.8.)*

3 (b) Define $S_0 = S$, and for $1 \leq i \leq k$ define S_i as a subdivision obtained from S_{i-1} by the following operation:

remove the vertices of a bounded-degree independent set of S_{i-1} together with the incident edges, and triangulate the regions of the subdivision thus obtained without adding new vertices,

where S_k is a subdivision which consists of only three vertices. Show that $k=O(\log n)$. *(Hint: see Theorem 9.8.)*

3 (c) Show that the sequence $S_0, S_1, \ldots, S_k$ of progressively coarser subdivisions takes $O(n)$ storage and allows a point location query to be answered in time $O(\log n)$. *(Hint: compare this data structure with the hierarchy of convex polytopes described in Section 9.5.2.)*

Problem 11.9: For a planar graph $\mathcal{G}=(N,A)$, call a set $N' \subseteq N$ an *independent set of* $\mathcal{G}$ if no two nodes of N' are incident upon a common arc of A. For every positive integer number k, define $\boldsymbol{i}_k(n)$ as the largest integer j such that every planar graph with n nodes has an independent set of cardinality j and no node in this independent set is adjacent to more than k nodes of $\mathcal{G}$. Notice that $\boldsymbol{i}_k(n)=\boldsymbol{i}_{k+1}(n)$, for all $k \geq n-1$. Thus, we define $\boldsymbol{i}(n)=\boldsymbol{i}_{n-1}(n)$.

5 (a) Prove $\boldsymbol{i}(n) \geq \frac{n}{4}$ without using the four-color theorem, which states that the set of nodes of every planar graph can be partitioned into four independent sets.

3 (b) Prove $\boldsymbol{i}(n) \leq \lceil \frac{n}{4} \rceil$. *(Hint: prove first that there is a largest independent set which contains all nodes of $\mathcal{G}$ which have degree three (and possibly additional nodes) if $cardN \geq 5$ and $cardA = 3cardN-6$, and then prove the assertion for the embedding of a graph which is created by repeatedly inserting a plane embedding of the complete graph of four nodes in a region of the old graph.).*

3 (c) Prove that, for infinitely many integer numbers n, there is a planar graph with n nodes and $3n-6$ arcs such that there is an independent set of cardinality at least $\frac{2n-4}{3}$. *(Hint: prove that the assertion is true for the members of the following family of iteratively constructed planar graphs: let the complete graph with three nodes be the first graph, and construct the ith graph by inserting the embedding of a new node into every region of a plane embedding of the $(i-1)^{\text{st}}$ graph; note that the nodes added are independent.)*

2 (d) Prove $\boldsymbol{i}_5(n)>0$. *(Hint: use Euler's relation.)*

3 (e) Prove $\boldsymbol{i}_{11}(n) \geq \frac{n}{8}$. *(Hint: use the four-color theorem.)*

5 (f) Determine $\boldsymbol{i}_k(n)$, for $5 \leq i \leq n-1$. *(Comment: compare (d), (e), and (f) with Problem 9.9.)*

Exercise 11.10: Let S be a monotone cell complex in $\boldsymbol{E}^3$, that is, the intersection of any cell of S with any vertical line is connected, and define $c_1 \ll c_2$, if $c_1 \neq c_2$ and there are points $p=(\pi_1,\pi_2,\pi_3)$ in cell c_1 and $q=(\psi_1,\psi_2,\psi_3)$ in cell c_2 such that

$$\pi_1=\psi_1,\ \pi_2=\psi_2, \text{ and } \pi_3 \leq \psi_3.$$

2 (a) Show that $c_1 \ll c_2$ contradicts $c_2 \ll c_1$.

3 (b) Show by example that "$\ll$" is not necessarily acyclic.

3 (c) Show that "$\ll$" is not necessarily acyclic even if all cells of S are convex.

3 (d) Solve the point location problem for S in $O(\log^2 n)$ query time and in $O(n)$ storage if "$\ll$" is acyclic, where n denotes the number of faces of S.

Exercise 11.11: Let H be a set of n non-vertical lines in $\boldsymbol{E}^2$ and let $S=\mathcal{A}(H)$ be the defined subdivision.

1 (a) Show that a level of S (as defined in Chapter 3) is a separator of S.

2 (b) Modify the search algorithm of Section 11.6 such that it uses the n levels of S instead of a complete family of separators and still takes only time $O(\log n)$ to answer a point location query.

Exercise 11.12: Let H be a set of n non-vertical planes in E^3 and let $C = A(H)$ be the defined cell complex.

2 (a) Prove that the relation "$\ll$" applied to the cells of C is acyclic (see also Exercise 11.10).

3 (b) Use the n levels of C to construct a search structure which takes $O(n^3)$ storage and $O(\log^2 n)$ time for a point location query. *(Hint: for the search perform a binary search in the sequence of levels, and decide whether a query point q is above, on, or below a level using an instance of the point location structure in Section 11.6.)*

Problem 11.13: Let H be a set of n non-vertical hyperplanes in E^d, $d \geq 3$, and define $C = A(H)$ as the defined cell complex.

4 (a) Design a data structure which takes $O(n^{d+\epsilon})$ storage and $O(\log n)$ time for a point location query.

5 (b) Is there a data structure which takes $O(n^d)$ storage and $O(\log n)$ time for a point location query?

11.10. Bibliographic Notes

The problem of locating a point in a subdivision of the plane or in a cell complex in a higher-dimensional space is one of the oldest and best understood problems in computational geometry. The special case where the subdivision consists of a simply connected polygon and its complement in E^2 is known as the point-in-polygon test, and solutions were proposed as early as 1962 (see Shimrat (1962) and Hacker (1962)).

Dobkin, Lipton (1976) considered the special case where the cell complex is an arrangement of n hyperplanes in E^d. Their solution answers a point location query in time $O(\log n)$, at the expense of a large amount of storage. Using the technique of random sampling, Clarkson (1987) recently proved the existence of a data structure with query time $O(\log n)$ and only $O(n^{d+\epsilon})$ storage, for any positive real number ϵ; this solves Exercise 11.13(a). In E^3, the methods of Chazelle (1985b) and of Edelsbrunner, O'Rourke, Seidel (1986) solve the point location problem for n planes in query time $O(\log^2 n)$ and $O(n^3)$ storage; this solves Exercise 11.12(b). Currently, the best solution for the general point location problem in arbitrary dimensions is due to Chazelle (1985c); it obtains logarithmic query time at the price of a vast amount of storage. For the special three-dimensional case indicated in Exercise 11.10(d), Chazelle (1985b) gives an algorithm that solves the exercise.

The method of Dobkin, Lipton (1976) can be generalized to general straight subdivisions of the plane, where it takes query time $O(\log n)$ and $O(n^2)$ storage, if n is the number of vertices of the subdivision. Preparata (1981) improves the amount of storage to $O(n \log n)$ and also gives a solution to Exercise 11.5(c). In a different development, Lee, Preparata (1977) propose a solution to the point location problem in E^2 which takes query time $O(\log^2 n)$ and $O(n)$ storage. Historically, the first optimal solution to the two-dimensional point location problem, that is, a solution which takes query time $O(\log n)$ and $O(n)$ storage, was described in Lipton, Tarjan (1977). Due to the unreasonably high constants in both the query time and the amount of storage, this solution is only of theoretical interest. Shortly thereafter, Kirkpatrick (1983) developed another optimal solution whose constants made it interesting even for practical purposes. This method is sketched in Exercise 11.8. The problem of fine–tuning the method such that it realizes the best performance in practice is still open. A non–trivial method for obtaining constants which improve the ones given in the original publication is indicated in Exercise 11.9(e); it relies on the four–color theorem for planar graphs proved by Appel, Haken (1977). Improvements upon this constant can be found in the exercise section of Chapter 9.

The layered DAG, which is the solution for the point location problem in E^2 presented in this chapter, is taken from Edelsbrunner, Guibas, Stolfi (1986). It is a refinement of the structure of Lee, Preparata (1977) and it improves the query time and the time for construction of this structure by a logarithmic factor each. The query time is improved to $O(\log n)$ by extensions of techniques originally described in Willard (1985) and in Chazelle (1986a). The latter technique, called "fractional cascading", was developed to its full generality in Chazelle, Guibas (1986a, 1986b). The time for construction from a monotone subdivision is improved to $O(n)$ by an application of methods for topological sorting (see Knuth (1968), Aho, Hopcroft, Ullman (1974), Mehlhorn (1984b), or other textbooks) and for evaluating the lowest common ancestor function in binary trees (see Harel (1980) for a more general treatment of this problem). The representation of the subdivision chosen in this chapter differs from the one in the original publication which borrows from Lee, Preparata (1977) and Guibas, Stolfi (1985). In an independent development, Cole (1986a) gives a refinement of the method of Dobkin and Lipton which leads to essentially the same structure as the layered DAG described in Section 11.6. In another independent development, Sarnak, Tarjan (1986) construct essentially the same data structure using persistent trees that keep track of changes during a sweep of the subdivision by means of a vertical line.

The algorithm mentioned above for constructing the layered DAG assumes that the given subdivision is monotone. If this is not the case, then it has to be refined by adding new edges. This takes time $\Omega(n\log n)$, in the general case. The $O(n\log n)$ time algorithm of Section 11.8 is a simplification of the algorithm given in Lee, Preparata (1977). If we need the subdivision to be further refined such that each region is bounded by exactly three edges (see Exercise 11.4), then we can use the triangulation algorithm of Garey, Johnson, Preparata, Tarjan (1978) to perform this task in an additional time $O(n)$. If the original subdivision is not monotone but the underlying graph is connected, then the recent method of Tarjan, van Wyk (1986) finds additional edges such that the resulting subdivision is monotone in time $O(n\log\log n)$; this solves Exercise 11.3.

The dynamic version of the point location problem allows the subdivision to change over time, that is, we allow the insertion and/or the deletion of edges. This version of the problem is considered in Mehlhorn (1984c), where he gives a solution to Exercise 11.7(a). A solution to Exercise 11.7(b) can be obtained by an application of the lemma formulated as Exercise 11.6 which is taken from Fries (1985).

PART III

GEOMETRIC AND ALGORITHMIC APPLICATIONS

Results are fundamental only if they have numerous applications of various kinds to related problems. Such applications of the results found in Parts I and II are presented in Part III. The applications are algorithmic as well as combinatorial.

CHAPTER 12

PROBLEMS FOR CONFIGURATIONS AND ARRANGEMENTS

A more appropriate but longer title for this chapter would be "Problems formulated for configurations and solved for arrangements". In fact, many problems formulated for configurations, whether combinatorial or algorithmic, are easier to approach in dual space where an arrangement of hyperplanes represents the configuration. It is safe to say that the translation of the problem into dual space makes it easier to see some aspects of the problem which are otherwise hidden; these aspects may or may not be important for a solution.

This chapter discusses several problems where the transformation reduces the difficulty of the task. In all cases, angles play an important role in the primary setting; the transformation then turns angles into faces which are closer to our intuitive understanding of the world. From the many interesting problems which qualify for inclusion in this chapter, we choose a representative sample of six. The choice is made with two goals in mind. First, each problem should have the potential to singlehandedly pique the reader's interest by itself. Second, the solution to each problem should be interestingly different from those of the other problems so as to allow for a reasonably broad exhibition of methods that are useful in dealing with problems of the type indicated above. Some other problems, which can be solved by techniques similar or identical to the methods described, appear in the exercise section of this chapter.

The organization of this chapter is as follows. Sections 12.1 and 12.2 discuss two-dimensional problems with solutions that do not seem to generalize to higher dimensions. The techniques employed to solve them are geometric transformations, a type of dynamic programming, and topological sorting. The problems treated in Sections 12.3 through 12.5 are phrased independently of their dimension and solutions that work in arbitrary dimensions are developed. In doing so, a new structure, the so-called geometric graph of an arrangement, is introduced in Section 12.3, as a tool to investigate degeneracies in configurations. Section

12.4 applies a combinatorial theorem developed in Chapter 5 to the analysis of an algorithm that finds minimum measure simplices, and Section 12.5 studies two generalizations of sorting to higher dimensions. Depth-first- and breadth-first-search are repeatedly used to modify arrangements according to the actual needs. Finally, Section 12.6 makes use of the correspondence between zonotopes in E^d and arrangements in E^{d-1} to solve a vector-sum problem.

12.1. Largest Convex Subsets

Let P be a set of n points in E^2 with no three points collinear and no two points on a common vertical line. We call Q a *convex subset of* P if $Q \subseteq P$ and $Q = \text{ext}\, Q$. Note that Q need not be convex with respect to the standard definition of the term; rather, it is the set of vertices of a convex polygon. According to Section 2.5, P contains a convex subset of size m if $n \geq \binom{2m-4}{m-2}+1$, and it is conjectured that $n \geq 2^{m-2}+1$ suffices to guarantee that a convex subset of size m exists. We hope that the algorithmic treatment offered in this section contributes some insight which might help to solve the combinatorial question of how many points are needed to guarantee the existence of a convex subset of size m.

The geometric transformation $\mathcal{D}$ introduced in Chapter 1 and known as an instance of the more general concept of duality plays an important role in our attempt to compute a largest convex subset of P. (The reader might find it helpful to recall that $\mathcal{D}$ sends a point $p = (\pi_1, \pi_2)$ to the line of points (x_1, x_2) which satisfy $x_2 = 2\pi_1 x_2 - \pi_2$. Straightforward but important properties of the transform can be found in Section 1.4.) Let Q be some convex subset containing at least two points, define $H = \mathcal{D}(P)$ and $G = \mathcal{D}(Q)$ (H and G are the dual sets of lines), and let t and b denote the topmost and the bottommost regions in arrangement $\mathcal{A}(G)$ (see Figure 12.1). Since each point of Q is extreme with respect to Q, each line of G contributes an edge to the boundary of at least one of b and t. As a matter of fact, the leftmost point of Q (and equivalently the rightmost point) corresponds to a line which contributes two unbounded edges, one to b and one to t; every other line in G contributes exactly one bounded edge. We call the boundaries $\text{bd}\, b$ and $\text{bd}\, t$ the *characteristic curves* of Q. Figure 12.1 shows the dual arrangement of a set of six points in E^2. The characteristic curves of a convex subset G of size 5 are indicated by solid edges; they also define regions b and t and the arrangement $\mathcal{A}(G)$. It is important to realize that the problem of finding largest convex subsets thus translates to finding pairs of characteristic curves with the largest number of edges.

As it turns out, the problem of computing largest convex subsets is relatively easy if we restrict our attention to subsets which have a fixed point as their leftmost point. This suggests that we perform the same procedure for each point p

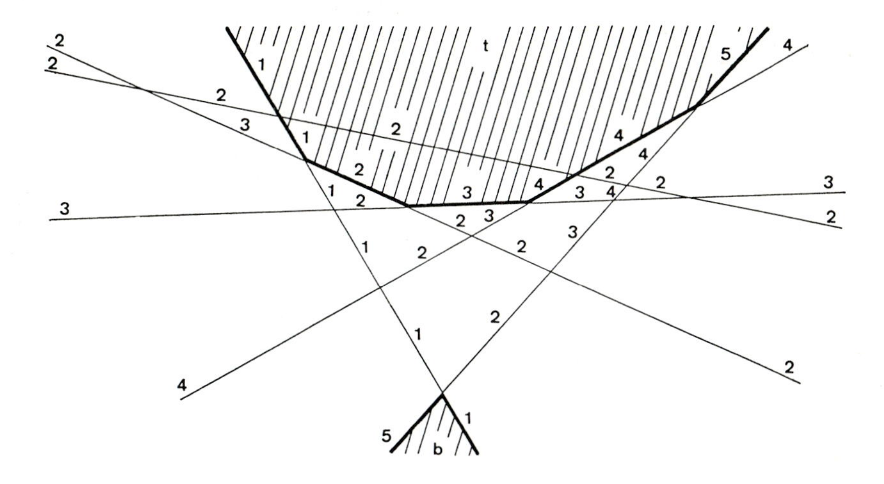

Figure 12.1. The dual image of a convex subset.

of P in turn, each time computing a largest convex subset with leftmost point p. To develop this approach, we impose a linear order on the points of any convex subset Q of P; the leftmost point of Q is smallest in this order which agrees with the counterclockwise order of the points around the convex hull of Q. Now, let p be some fixed point in P. To compute a largest convex subset with leftmost (or smallest) point p, we label each edge e of $\mathcal{A}(H)$ with the greatest integer $i = L_p(e)$ such that the dual point of its supporting line is the ith smallest point of some convex subset with leftmost point p, and e is an edge of its characteristic curves. (Figure 12.1 shows the labels for p chosen as the dual point of the least steep line.) The largest label in the arrangement equals the size of the largest convex subset with leftmost point p. To compute the labels $L_p(e)$, for all edges e of $\mathcal{A}(H)$, we use some straightforward dependencies among the labels of edges with common endpoints. These dependencies become obvious when we observe that an edge e can be part of the upper (lower) characteristic curve only if it lies above (below) line $h = \mathcal{D}(p)$. To simplify the presentation, call an endpoint of an edge its left endpoint if the edge resides to the right of the vertical line through the endpoint. Otherwise, call it the right endpoint of the edge.

Observation 12.1: If e is unbounded to the left, then let e' be the other unbounded edge on the same line. Otherwise, define edges e' and e'' such that the left endpoint of e is the right endpoint of e' and e'', and e' shares the supporting line with e (that is, $\text{aff}\, e' = \text{aff}\, e$).

(i) $L_p(e) = 0$ if $\text{aff}\, e$ is less steep than $h = \mathcal{D}(p)$,

(ii) $L_p(e) = 1$ if e belongs to h,

(iii) $L_p(e) = L_p(e')$ if e is unbounded to the left, if $e \in h^+$ and $\text{aff}\, e''$ is steeper than $\text{aff}\, e$, or if $e \in h^-$ and $\text{aff}\, e''$ is less steep than $\text{aff}\, e$, and

(iv) $L_p(e) = \max\{L_p(e'), L_p(e'')+1\}$, otherwise, unless aff$e'' = h$, in which case $L_p(e) = 2$.

Figure 12.2 illustrates cases (ii) through (iv) under the assumption that e does not belong to h^- and that labels x and y are non-zero.

Algorithm 12.1 (Largest convex subsets):

Step 1: Construct $\mathcal{A}(H)$.

Step 2: Compute the label of each edge with respect to each point:

for all points p of P **do**

Step 2.1: Compute label $L_p(e)$ for each edge e in $\mathcal{A}(H)$.

Step 2.2: Determine the greatest label m in $\mathcal{A}(H)$.

if m is greater than the current maximum **then**

Define m as the current maximum. Determine the convex subset of size m by following the labels backwards.

endif

endfor.

In order to guarantee that the label for each edge can be computed in constant time using Observation 12.1, it is sufficient to make sure that the label of an edge e does not have to be computed unless e belongs to h, the left endpoint of e belongs to h, or the labels of e' and e'' (if it exists) are already known. One way to get a sequence of edges that can safely be processed from left to right is to do a topological sort of the edges according to the following partial order "$\preceq$":

> $f \preceq f'$ if, first, the right endpoint of edge f coincides with the left endpoint of edge f' and either both f and f' are below h or neither of them is, or if, second, f and f' belong to the same line different from h such that f is unbounded to the right and f' is unbounded to the left.

Note, however, that there are sequences of edges that contradict "$\preceq$" and work well since less than the labels of both predecessors (in the sense defined by "$\preceq$") are needed in many cases. Performing the topological sort and assigning all labels for one point/line fixed takes time $O(n^2)$ which implies the following

Figure 12.2. Rules for computing labels.

result.

Theorem 12.2: A largest convex subset of a set of n points in E^2 (no three collinear and no two on a common vertical line) can be computed in $O(n^3)$ time and $O(n^2)$ storage.

We hope that the reader does not object to the restriction of the above discussion and of Theorem 12.2 to sets of points without collinear triplets and without pairs on vertical lines. In fact, it is not difficult (although tedious) to add details so that degenerate point sets can also be processed. Note, however, that there are at least two different reasonable possibilities for defining convex subsets of potentially degenerate configurations (see also Exercise 12.1).

12.2. The Visibility Graph for Line Segments

The structure studied in this section is motivated by the problem of finding shortest paths in the presence of obstacles. If the obstacles are line segments, then the shortest path between any two endpoints which does not cross any segments will consist of linear pieces either coinciding with segments or connecting two endpoints of different segments. One way to find a shortest path is thus to find all possible straight connections between endpoints and treating the problem as one defined for graphs. We start with some formal definitions.

A *(line) segment* is the convex hull of two points which are called the *endpoints of* the segment. It follows that line segments are closed and bounded. Let S be a set of n pairwise non-intersecting line segments in E^2, and denote the set of $2n$ endpoints by P. To simplify the discussion, we assume that no three points in P are collinear and that no two lie on a common vertical line. An undirected graph $\mathcal{V}=(N,A)$ is called the *visibility graph of* S if

(i) $N=P$, and
(ii) arc $\{p,q\}$ belongs to A if $\text{riint}\{p,q\}$ avoids all segments in S.

$\mathcal{V}$ can be used for finding shortest paths between two endpoints as follows: add all arcs between endpoints of the same segment, define the weight of an arc $\{p,q\}$ as the Euclidean distance between points p and q, and finally apply a standard shortest path algorithm for weighted graphs. Figure 12.3 shows the visibility graph for five line segments; the segments are indicated by broken lines.

Let p be a point in P, and for every angle α define r_α as the ray emanating from the origin such that α is the angle between the positive x_1-axis and r_α. By convention, we define r_α to be relatively open, that is, it does not contain the origin. We define the *vision* $vis(p,\alpha)$ *of* p *in direction* α as follows:

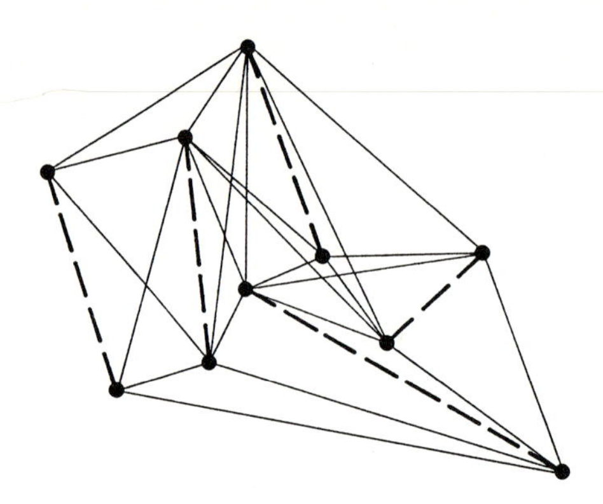

Figure 12.3. Visibility graph for five segments.

(i) $vis(p,\alpha)=\emptyset$, if $p+r_\alpha$ avoids S or if $\text{cl}(p+r_\alpha)$ contains the segment with endpoint p (denoted by $seg(p)$), and

(ii) $vis(p,\alpha)=s$, if $s \neq seg(p)$ is the segment in S which intersects $p+r_\alpha$ closest to p.

The principal idea for constructing $\mathcal{V}$ is to compute $vis(p,\alpha)$, for each point p in P and for all angles α with $-\frac{\pi}{2}<\alpha<\frac{\pi}{2}$. If $p+r_\alpha$ contains an endpoint q of segment $vis(p,\alpha)$ then $\{p,q\}$ is added to A. Intuitively, we sweep a ray anchored at p in the counterclockwise direction. At each angle α, p stores $vis(p,\alpha)$. When the ray sweeps through another endpoint q then $vis(p,\alpha)$ possibly changes. Define angles $-\frac{\pi}{2}<\alpha_1<\alpha_2<\alpha_3<\frac{\pi}{2}$ and denote the corresponding rays as r_1, r_2, and r_3 such that $p+r_2$ contains point q and no ray $p+r_\alpha$, with $\alpha_1 \leq \alpha \leq \alpha_3$, contains any point from $P-\{q\}$. In updating the vision of v, we distinguish four cases illustrated in Figure 12.4:

Case 1: $vis(p,\alpha_1) \neq seg(q)$ and p is closer to point q than to the intersection between $p+r_2$ and $vis(p,\alpha_1)$. Then $vis(p,\alpha_3)=seg(q)$, and arc $\{p,q\}$ is added to A.

Case 2: $vis(p,\alpha_1)=seg(q)$. Then $vis(p,\alpha_3)=vis(q,\alpha_2)$, and arc $\{p,q\}$ is added to A.

Case 3: $seg(p)=seg(q)$. No changes occur.

Case 4: $vis(p,\alpha_1) \neq seg(q)$ and p is closer to $(p+r_2)\cap vis(p,\alpha_1)$ than to q. No changes occur.

The only non-trivial computation occurs in Case 2 when the endpoint q of the currently visible segment is reached. The vision of p can be updated in constant time if q holds $vis(q,\alpha_2)$ at the right time. This follows automatically,

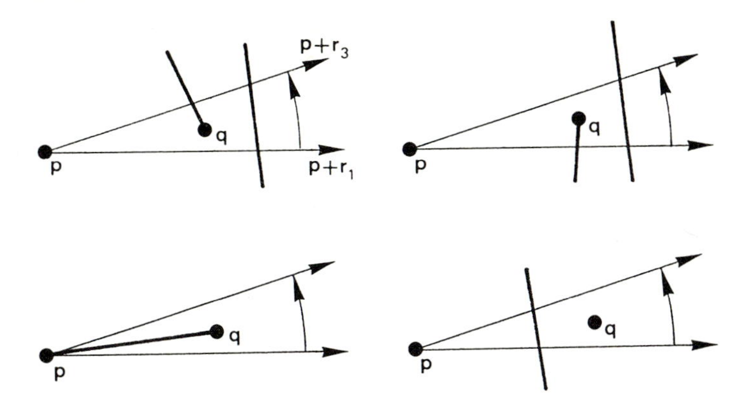

Figure 12.4. Updating the vision of p.

however, if the sequence of pairs of points processed is *legal*, that is, if for each point p the pair (p,q) is processed before (p,r) if the slope of the line through p and q is smaller than the slope of the line through p and r. ("Processing (p,q)" in this context means that the vision of p is updated if there is an angle α, with $-\frac{\pi}{2}<\alpha<\frac{\pi}{2}$, such that $p+r_\alpha$ contains q, and that the vision of q is updated, otherwise.) Such a sequence of pairs is derived most easily from the dual arrangement $\mathcal{A}(H)$, with $H=\mathcal{D}(P)$: a legal sequence of pairs of points from P corresponds to a sequence of vertices in $\mathcal{A}(H)$ such that vertex v precedes vertex w if v and w belong to a common line and v has smaller x_1-coordinate than w. We outline the algorithm below.

Algorithm 12.2 (Visibility graph):

Step 1: Compute $vis(p,-\frac{\pi}{2})$ for each point p in P.

Step 2: Construct $\mathcal{A}(H)$, with $H=\mathcal{D}(P)$.

Step 3: Do a topological sort of the vertices of $\mathcal{A}(H)$ according to the partial order "$\preceq$" defined as follows: $v\preceq w$ if v is the left and w is the right endpoint of a common edge. This yields a sequence $(v_0,v_1,\ldots,v_{m-1})$ of vertices, with $m=\binom{2n}{2}$.

Step 4: Construct $\mathcal{V}$ as follows:

for $i:=0$ **to** $m-1$ **do**
 Process the pair of points (p,q) such that $v_i=\mathcal{D}(p)\cap\mathcal{D}(q)$ following the rules given above.
endfor.

Since each step of Algorithm 12.2 takes time $O(n^2)$, we conclude with the following result.

Theorem 12.3: The visibility graph for n non-intersecting line segments in the plane, of which no three endpoints are collinear and no two endpoints lie on a common vertical line, can be constructed in time and storage $O(n^2)$.

Again we note that the restriction to non-degenerate point sets is not inherent to the method and can be removed without contradicting the bounds (see also Exercise 12.4).

12.3. Degeneracies in Configurations

Finding or reporting degenerate configurations that are subsets of a given set of points is an important issue. For example, the occurrence of such configurations could result in a contradiction to some of our programs. Apart from this pragmatic concern, degenerate configurations are of fundamental significance in computer vision since points on surfaces tend to be coplanar or cospherical (see also Exercise 12.8).

Let P be a finite set of points in E^d. For $1 \leq i \leq d-1$, a subset Q of at least $i+2$ points is termed *i-degenerate* if $\dim \operatorname{aff} Q = i$. If we report degenerate subsets of P, it is an obvious nuisance to report Q if it evolves from an $(i-1)$-degenerate subset by adding one additional point; any additional point not contained in the affine hull of the $(i-1)$-degenerate subset would produce an i-degenerate subset. Similarly, any i-degenerate subset of more than $i+2$ points contains many i-degenerate subsets which provide no additional information if they are reported. We therefore define Q to be *proper* if $\dim \operatorname{aff}(Q-\{p\}) = i-1$ for no point p, and we define Q to be *maximal* if no i-degenerate subset of P properly contains Q. This section develops an algorithm which reports all proper and maximal i-degenerate subsets, for all $1 \leq i \leq d-1$. The algorithm is best described in dual space, so we define $H = \mathcal{D}(P)$ and briefly recall how i-degenerate subsets dualize.

Observation 12.4: Let Q be an i-degenerate subset of some point set P in E^d and define $G = \mathcal{D}(Q)$.

(i) Q belongs to a non-vertical i-flat if and only if the hyperplanes in G have a $(d-i-1)$-flat in common.

(ii) Q belongs to a vertical i-flat if and only if the common intersection of all hyperplanes in G is empty.

To report i-degenerate subsets of P, we first construct the *geometric graph $\mathcal{G}(H)$ of H* defined as follows: $\mathcal{G}(H)$ is an undirected graph with set N of nodes and set A of arcs, such that N contains the common intersection of every subset of H, and

$$A = \{\{e,f\} | \dim f = \dim e + 1 \text{ and } f \text{ contains } e\}.$$

(We define $\emptyset$ as the only (–1)–flat which is therefore connected to all points, that is, 0–flats, in N.) Figure 12.5 shows the geometric graph for four lines in E^2.

It is easy to construct $\mathcal{G}(H)$ in time $O(n^d)$, $n = \text{card}H$, from the incidence graph of the arrangement $\mathcal{A}(H)$ studied in Chapters 5 and 7. Call two j–faces of $\mathcal{A}(H)$ *equivalent* if they belong to the same j–flat. Each node of $\mathcal{G}(H)$ represents an equivalence class of faces, and two nodes define an arc in A if the corresponding classes contain incident faces. So $\mathcal{G}(H)$ can be obtained by merging equivalent faces of $\mathcal{A}(H)$.

In order to use $\mathcal{G}(H)$ for our purposes, we attach a set of hyperplanes to each node of $\mathcal{G}(H)$. Notice that a node in N is also a flat in E^d, so the following definitions make sense. For two adjacent nodes e and f, we call e a *successor of* f (and f a *predecessor of* e) if $\dim f = \dim e + 1$. For each node e in $\mathcal{G}(H)$, define $H_e = \{h \in H | h \text{ contains } e\}$. Obviously, H_e is the union of sets H_f, for all predecessors f of e, unless $\dim e = d-1$, in which case $H_e = \{e\}$. Furthermore, H_e contains all hyperplanes that contain e. Consequently, $\mathcal{D}(H_e)$ is a maximal i–degenerate subset if it is i–degenerate at all. To determine which sets H_e correspond to i–degenerate subsets of P which are also proper, we need the following result.

Observation 12.5: Let e be a node of $\mathcal{G}(H)$, with $d-i-1 = \dim e$.

(i) $\mathcal{D}(H_e)$ is a maximal i–degenerate subset of P if and only if $\text{card}H_e \geq i+2$.

(ii) $\mathcal{D}(H_e)$ is a proper i–degenerate subset of P if and only if

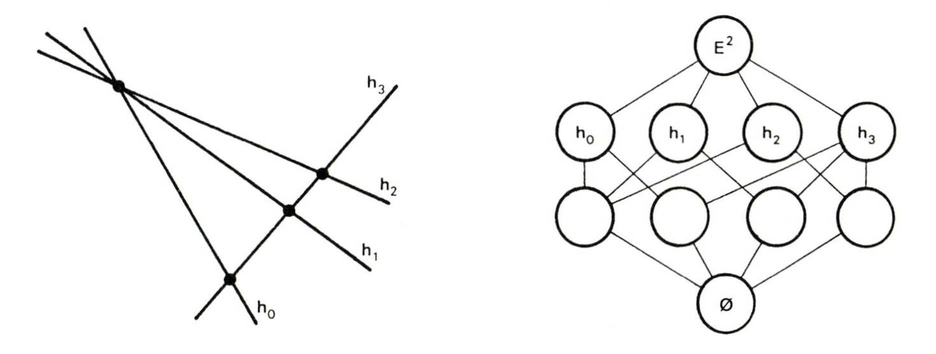

Figure 12.5. Geometric graph in two dimensions.

$\text{card}H_f = \text{card}H_e - 1$ for no predecessor f of e.

Unfortunately, $\mathcal{G}(H)$ does not allow us to determine i–degenerate subsets of P which span vertical flats. However, every i–flat can contain a line parallel to the x_j–axis for at most i values of j. In our case $i \leq d-1$, which implies that every i–degenerate subset dualizes to a $(d-i-1)$–flat at least once, if we define the direction of each coordinate–axis as the vertical direction once. This amounts to considering d instances of the problem: the jth instance exchanges the x_j– and the x_d–coordinates of the given points. We outline the algorithms below.

Algorithm 12.3 (Degenerate subsets):

for $j := 1$ **to** d **do**

Step 1: Compute P' from P by exchanging the jth and the dth coordinates of the points, define $H = \mathcal{D}(P')$, and construct $\mathcal{G}(H)$.

Step 2: To each node e of $\mathcal{G}(H)$ attach a list which stores H_e, the set of hyperplanes in H that contain e. In addition, store the cardinality of H_e at node e.

Step 3: Process all sets H_e as follows:

for each node e of $\mathcal{G}(H)$ **do**

Define $i := d - \dim e - 1$, $Q' = \mathcal{D}(H_e)$, and let Q be the corresponding subset of P. Report Q if all of the following conditions are met:

3.1: Q is i–degenerate, that is, $\text{card}Q = \text{card}H_e$ is greater than or equal to $i+2$,

3.2: Q is proper, that is, $\text{card}H_f = \text{card}H_e - 1$ for no predecessor f of e, and

3.3: Q is not reported yet, that is, each one of the first $j-1$ coordinate–axes has a parallel line in $\text{aff}Q$.

endfor

endfor.

To see that Algorithm 12.3 is correct, notice that Observations 12.5(i), 12.5(ii) and 12.4(i) justify conditions 3.1, 3.2 and 3.3 used to identify i–degenerate subsets which are maximal and proper. To further specify the algorithm, we outline an efficient method for constructing the sets H_f, for all nodes f of $\mathcal{G}(H)$. First, we present some properties of the sets.

Lemma 12.6: Let H be a set of n hyperplanes in E^d, and let $N(j)$ denote the set of j–flats in $\mathcal{G}(H)$, for $0 \leq j \leq d-1$.

(i) Each node f in $N(j)$ has at most $n-d+j$ successors.

(ii) $\sum_{f \in N(j)} \text{card}H_f \leq (d-j)\binom{n}{d-j}$.

Proof: Let f be a j-flat in $\mathcal{G}(H)$; then f is the common intersection of at least $d-j$ hyperplanes of H. A $(j-1)$-flat e is a successor of f only if there is a hyperplane h in H with $e=f\cap h$. Part (i) follows since there are only $n-d+j$ hyperplanes left.

To show part (ii), we first assume that $\mathcal{A}(H)$ is simple. In this case, there are exactly $\binom{n}{d-j}$ j-flats, each of which is equipped with a set of $d-j$ hyperplanes. So (ii) holds in this case. In the general case, let f be a j-flat contained in $k>d-j$ hyperplanes. A small perturbation of these hyperplanes makes them intersect in $\binom{k}{d-j}$ j-flats which increases the contribution to the sum in (ii) by

$$(d-j)\binom{k}{d-j}-k=k\binom{k-1}{d-j-1}-k\geq 0.$$

By a sequence of such perturbations, every arrangement can be converted to a simple one without decreasing the sum in (ii). Consequently, (ii) also holds if $\mathcal{A}(H)$ is not simple. □

By Lemma 12.6(ii), $O(n^d)$ storage suffices to represent the sets H_e, for all nodes H_e of $\mathcal{G}(H)$. To compute the sets (and analogously the lists which represent them), we proceed in the direction of decreasing dimension: H_e is constructed as the union of sets H_f, for all predecessors f of e. We outline a procedure which finds the union of sets with a total of m integers between 0 and $n-1$ in time $O(m)$. By Lemma 12.6(i), the set H_f, for f a j-flat and $j\geq 1$, is involved in less than n union operations, which implies, by Lemma 12.6(ii), that time $O(n^d)$ suffices to construct all sets. The linear time procedure makes use of an initially empty stack S and a bit-vector $B[0..n-1]$. Initially, $B[k]=0$, for $0\leq k\leq n-1$.

Procedure 12.4 (Union of sets):

for each set M **do**
  for each integer m in M **do**
    if $B[m]=0$ **then**
      Set $B[m]:=1$ and push m onto S.
    endif
  endfor
endfor;
Initialize $U:=\emptyset$, the set to be computed.
while S is not empty **do**
  Take the topmost integer number m from S, set $B[m]:=0$, and set $U:=U\cup\{m\}$.
endwhile.

Since all used bits of B are finally reset to 0, the procedure can be used repeatedly without initializing $\mathcal{B}$ except for the first time. It follows that Step 2 of Algorithm 12.3 takes time $O(n^d)$. It is not hard to see that the same is true for Step 3 which needs constant time per node and arc of $\mathcal{G}(H)$. Since Step 1 of Algorithm 12.3 takes time $O(n^d)$ by the results of Chapter 7, the main result of this section follows.

Theorem 12.7: All proper and maximal i-degenerate subsets, for all $1 \leq i \leq d-1$, of a set of n points in E^d can be reported in time and storage $O(n^d)$.

12.4. Minimum Measure Simplices

Given a finite set of points in E^d, we study the problem of identifying $d+1$ of the points such that the spanned simplex has minimum measure among all such simplices, where "measure" designates the d-dimensional generalization of "length", "area" and "volume". We start with more formal definitions of the required notions.

Let $S=\{p_0,p_1,...,p_d\}$ be a set of $d+1$ points in d dimensions, and define $S_i=\{p_0,p_1,...,p_i\}$ and $s_i=\text{conv}S_i$, for $1 \leq i \leq d$. s_i is an i-dimensional simplex and the *i-dimensional measure $\mu_i(s_i)$ of s_i* is defined recursively:

(i) $\mu_1(s_1)=d(p_0,p_1)$, that is, the one-dimensional measure is the length of the segment s_1.

(ii) For $2 \leq i \leq d$, $\mu_i(s_i)=\frac{d_i}{i}\mu_{i-1}(s_{i-1})$, with $d_i=\min\{d(p_i,x) \mid x$ a point of $\text{aff}S_{i-1}\}$; d_i is often referred to as the height of s_i.

Evidently, $\mu_d(s_d)=0$ unless S is affinely independent. Let P be a set of $n \geq d+1$ points in E^d. Any subset S of $d+1$ points of P defines a (d-dimensional) simplex $s(S)$ with measure $\mu(S)=\mu_d(s(S))$. We call $s(S)$ a *minimum simplex of P* if $\mu(S) \leq \mu(S')$, for all subsets S' of P with $\text{card}S'=d+1$.

To compute a minimum simplex of P, we assume without loss of generality that no $d+1$ points of P lie in a common hyperplane; otherwise, the measure of every minimum simplex vanishes and Algorithm 12.3 can be used to find such simplices. Call a subset T of d points of P a *base* if $b=\text{aff}T$ is non-vertical, and let p_T be a point of $P-T$ closest to b. Clearly, $\mu(T \cup \{p_T\}) \leq \mu(S)$, for all sets S with $T \subseteq S \subseteq P$ and $\text{card}S=d+1$. Intuitively, p_T is one of the first points met if hyperplane b is moved in a parallel manner upwards or downwards. The algorithm outlined below exploits the fact that every simplex defined by $d+1$ points of P has a base, that is, some d of the points are not contained in a vertical hyperplane. For each base T of P, it finds some points including the closest point, and computes the measures of the spanned simplices. Let $H = \mathcal{D}(P)$ be the dual set of hyperplanes. Every base T of P corresponds to a vertex v of the arrangement $\mathcal{A}(H)$. A point p_T can be found by moving v vertically up or down: p_T corresponds to one of the hyperplanes $h_T = \mathcal{D}(p_T)$, hit first by v. It follows that h_T contains a facet that bounds a cell c with v in bd$c$. A more formal description of the strategy follows.

Algorithm 12.5 (Minimum measure simplex):

Step 1: Construct arrangement $\mathcal{A}(H)$.

Step 2: Compute a collection of simplices including a minimum simplex as follows:

```
for each cell c in A(H) do
    for each vertex v in bdc do
        for each hyperplane h supporting a facet of c do
            unless h contains v do
                Let S contain the points that correspond to h
                and the d hyperplanes that contain v. Compute
                μ(S) and record S if it is smaller than the
                current minimum.
            endunless
        endfor
    endfor
endfor.
```

Let $deg_i(c)$ denote the number of i-faces in the boundary of cell c. The number of simplices explicitly tested for cell c is bounded from above by $deg_0(c) \cdot deg_{d-1}(c)$, and

$$\sum_{c \text{ a cell in } \mathcal{A}(H)} deg_0(c) \cdot deg_{d-1}(c) = O(n^d)$$

by Theorem 5.5. The set of vertices in the boundary of each cell c can be computed in a total time $O(n^d)$ as follows.

> In the direction of increasing dimension i, compute the set of vertices in the boundary of each i-face f by merging the sets of its incident $(i-1)$-faces.

If we modify Procedure 12.4 to do the merging, the total amount of time taken is proportional to

$$\sum_{i=0}^{d-2} 2(d-i) \left(\sum_{f \text{ an } i\text{-face of } \mathcal{A}(H)} deg_0(f) \right),$$

where $deg_0(f)$ denotes the number of vertices in f; $2(d-i)$ is the number of superfaces of each i-face since $\mathcal{A}(H)$ is simple. Obviously, the above sum is in $O(n^d)$ which implies the main result of this section.

Theorem 12.8: A minimum simplex spanned by $d+1$ of n given points in E^d, $d\geq 2$, can be computed in time and storage $O(n^d)$.

12.5. Computing Ranks: Sorting in d Dimensions?

This section investigates two d-dimensional problems which can be viewed as generalizations of sorting numbers, which is a classical one-dimensional problem. The information gained by sorting a set of numbers into an ascending sequence is essentially the collection of ranks associated with the numbers, where the rank of number a equals i if $i-1$ numbers in the set are smaller than a. Notice that the collection of ranks is sufficient to reconstruct, for each number a, the set of numbers which are smaller and the set of numbers which are greater than a. We generalize this view of sorting.

Let $P=\{p_0,p_1,\ldots,p_{n-1}\}$ be a set of n points in E^d. For each point q, not necessarily in P, we define its *rank* to be

$$\lambda(q)=\min\{\text{card}(P\cap h)|h \text{ an open half-space with } q\in \text{bd}h\}.$$

Similarly, we associate ranks with sequences of d affinely independent points. Let $S=(q_0,q_1,\ldots,q_{d-1})$ be such a sequence, and let q_d be an additional point in E^d. We write $q_i=(\psi_{i,1},\psi_{i,2},\ldots,\psi_{i,d})$, for $0\leq i\leq d$. (S,q_d) is said to have *positive orientation* if the determinant of the matrix

$$\begin{pmatrix} \psi_{0,1} & \psi_{0,2} & \cdots & \psi_{0,d} & 1 \\ \psi_{1,1} & \psi_{1,2} & \cdots & \psi_{1,d} & 1 \\ \cdots & \cdots & \cdots & \cdots & \cdots \\ \psi_{d,1} & \psi_{d,2} & \cdots & \psi_{d,d} & 1 \end{pmatrix}$$

is positive. We now define $\lambda(S)=\text{card}\{q\in P|(S,q) \text{ has positive orientation}\}$ to be the *rank of* S. Note that if sequence S' can be obtained from S by exchanging two points, then $\lambda(S')=n-\lambda(S)-m$, if m points of P belong to the hyperplane $\text{aff}\{q_0,q_1,\ldots,q_{d-1}\}$. Figure 12.6 shows a set $\{1,2,3,4,5\}$ of points in E^2 with ranks $\lambda(\{1\})=0$, $\lambda(\{2\})=0$, $\lambda(\{3\})=1$, $\lambda(\{4\})=0$, $\lambda(\{5\})=0$, and $\lambda((1,2))=3$, $\lambda((1,3))=1$, etc. Notice that in two dimensions $\lambda((i,j))$ denotes the number of points to the left of the oriented line from i to j; Figure 12.6 illustrates this fact for $i=1$ and $j=3$. Ranks of points can be used to identify non-representative members of a set (that is, points with vanishing or low rank) which is a common problem in statistics. Ranks of d-tuples of points are useful for classifying configurations combinatorially (see also Exercise 12.11).

To compute the ranks of all points in set P, we make use of the dual arrangement $\mathcal{A}(H)$, with $H=\mathcal{D}(P)$. For every point x in E^d, we define $a(x)$, $b(x)$ and $o(x)$ such that x lies below $a(x)$, above $b(x)$, and on $o(x)$ hyperplanes

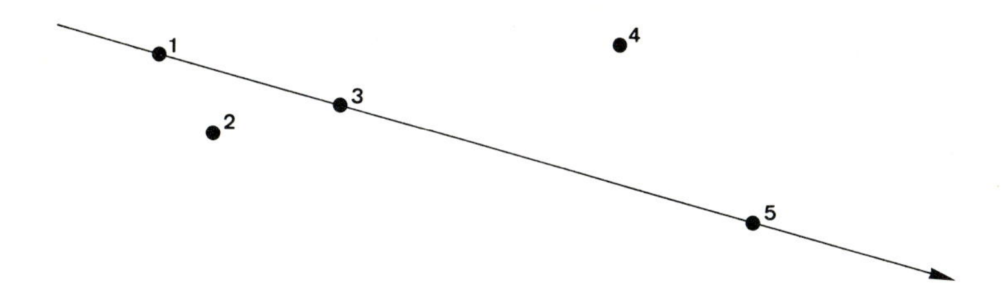

Figure 12.6. A configuration in the plane.

of H. Obviously, $a(x)+b(x)+o(x)=n$, for every point x in E^d. Ranks and the functions $a(x)$ and $b(x)$ relate to each other as follows.

Observation 12.9: (i) Let p be a point in P. Then $\lambda(p)=\min\{a(x),b(x)|x$ belongs to hyperplane $\mathcal{D}(p)\}$, unless all open half-spaces h with $p\in \mathrm{bd}h$ and $\mathrm{card}(P\cap h)=\lambda(p)$ are vertical.

(ii) Let T be a subset of P of size d and let S be a permutation of T. If T is affinely dependent then $\lambda(S)$ is not defined. Otherwise, the d dual hyperplanes intersect in a vertex v of $\mathcal{A}(H)$ (unless $\mathrm{aff}\,T$ is vertical), and $\lambda(S)=a(v)$ ($\lambda(S)=b(v)$) if (S,q) has positive orientation whenever v belongs to $\mathcal{D}(q)^-$ ($\mathcal{D}(q)^+$).

By Observation 12.9, it is easy to compute all ranks determined by non-vertical hyperplanes if $\mathcal{A}(H)$ is constructed and each face f stores the numbers $a(f)=a(x)$, $b(f)=b(x)$, and $o(f)=o(x)$, for x any point of f. To determine $\lambda(p)$ visit all faces f contained in $\mathcal{D}(p)$ and take the smallest of the associated values $a(f)$ and $b(f)$; to compute $\lambda(S)$, for S a sequence of d points, take the corresponding vertex v in $\mathcal{A}(H)$ and set $\lambda(S):=a(v)$ or $\lambda(S):=b(v)$ according to the rule given in Observation 12.9(ii). To include ranks determined by vertical hyperplanes, we exchange d^{th} and i^{th} coordinates, for $1\leq i\leq d-1$, and construct $d-1$ additional arrangements (as we did in Section 12.3). An algorithm which assigns values $a(f)$, $b(f)$, and $o(f)$ to all faces f of a given arrangement $\mathcal{A}(H)$ in E^d is outlined below. The values of the cells are assigned using breadth-first traversal starting at the topmost cell, and the values of an i-face are derived from its incident $(i+1)$-faces, for $0\leq i\leq d-1$.

Algorithm 12.6 (Ranks in arrangements):

Step 1: Determine cell c_0 above all hyperplanes in H and set $a(c_0):=0$, $b(c_0):=n$, and $o(c_0):=0$. Mark c_0 and add c_0 as the first element to an empty queue $\mathcal{Q}$.

Step 2: Compute the values of all cells as follows:

while $\mathcal{Q}$ is not empty **do**

Remove the first cell c from $\mathcal{Q}$.
for each facet f incident upon c **do**
Let c' be the other cell incident upon f.
unless c' is marked **do**
Set $a(c'):=a(c)+1$, $b(c'):=b(c)-1$, and $o(c'):=0$, mark c', and append c' to the end of $\mathcal{Q}$.
endunless
endfor
endwhile;
Step 3: Compute the values of all other faces:
for $i:=d-1$ **downto** 0 **do**
for each i-face e of $\mathcal{A}(H)$ **do**
Set $a(e)$ equal to the minimum of $\{a(f)|f$ a superface of $e\}$, set $b(e)$ equal to the minimum of $\{b(f)|f$ a superface of $e\}$, and set $o(e):=n-a(e)-b(e)$.
endfor
endfor.

In Step 1, we test each cell to determine whether it lies below a hyperplane that supports one of its facets, until a cell without such facet is found. This takes time $O(n^d)$. Step 2 obviously takes time $O(n^d)$. Its correctness follows from the fact that the breadth-first traversal visits a cell only if all adjacent cells above it have already been visited. Finally, time $O(n^d)$ suffices for Step 3 since the number of incidences between faces of $\mathcal{A}(H)$ is in $O(n^d)$ (see also Chapter 5). We conclude with the main result of this section.

Theorem 12.10: Let P be a set of n points in E^d. The rank of all points and all sequences of d points in P can be computed in time and storage $O(n^d)$.

12.6. A Vector-Sum Maximization Problem

Let $V=\{v_0,v_1,\ldots,v_{n-1}\}$ be a set of n non-zero vectors in E^d. We investigate the problem of adding and subtracting the vectors, with each vector appearing exactly once, such that the resulting vector has maximal length. More formally, let A be the set of all ordered n-tuples $(\alpha_0,\alpha_1,\ldots,\alpha_{n-1})$, termed *assignments*, with α_i either equal to $+1$ or to -1, for $0\leq i\leq n-1$. For any assignment $a=(\alpha_0,\alpha_1,\ldots,\alpha_{n-1})$, we define

$$s(a)=\sum_{i=0}^{n-1}\alpha_i v_i,$$

and let $|s(a)|$ denote the length of vector $s(a)$. To find an assignment which maximizes the length of the sum, we could explicitly compute $|s(a)|$, for all assignments a of A; but there are 2^n such assignments which makes it a rather time-consuming venture. Another method takes each vector $s(a)$ as a point $p(a)$, its endpoint, and observes that $s(a)$ can be longest only if $p(a)$ is extreme in $P=\{p(a)|a\in A\}$. We will follow this approach and compute all extreme points in P. Since conv$P$ is a zonotope (see Chapter 1 for an introductory discussion of zonotopes), the number of vertices is in $O(n^{d-1})$, and we will show how to find them in time $O(n^{d-1})$. No familiarity with zonotopes and their correspondence with arrangements is assumed in the description of the method.

For each vector v_i in V let h_i denote the hyperplane of points x with

$$\langle x,v_i\rangle=0,$$

and let h_i^+ be the half-space of points x such that

$$\langle x,v_i\rangle>0.$$

All hyperplanes in $H=\{h_i|v_i\in V\}$ contain the origin. It follows that each cell c in the arrangement $\mathcal{A}(H)$ is an unbounded cone with the apex at the origin. Now, let $m=(\mu_0,\mu_1,\ldots,\mu_{n-1})$ be an optimal assignment (that is, $|s(m)|$ is a maximum), and define

$$c(m)=\{x\in E^d|x\in h_i^+ \text{ if } \mu_i=+1 \text{ and } x\notin \mathrm{cl}h_i^+ \text{ if } \mu_i=-1, \text{ for } 0\le i\le n-1\}.$$

We show that $c(m)$ is non-empty and that it contains point $p(m)$.

Lemma 12.11: If $\mu_i=+1$ then $p(m)\in h_i^+$ and if $\mu_i=-1$ then $p(m)\notin \mathrm{cl}h_i^+$, for $0\le i\le n-1$.

Proof: Assume the existence of an index j such that $\mu_j=+1$ and $p(m)$ does not belong to h_j^+. (The other case is symmetric.) Define $q=p(m)-2v_j$ and note that $q=p(m')$, with $m'=(\mu_0,\ldots,\mu_{j-1},-\mu_j,\mu_{j+1},\ldots,\mu_{n-1})$. Since $p(m)$ lies outside h_j^+ we know that point q does not belong to h_j^+. Finally, $|s(m')|>|s(m)|$, since q and $p(m)$ lie on a line orthogonal to hyperplane h_j, h_j contains the origin, and the minimal distance of q to h_j is larger than that from $p(m)$ to h_j. This provides the required contradiction. □

By Lemma 12.11, only assignments a with $c(a)$ a cell of $\mathcal{A}(H)$ need to be checked. Furthermore, if $c(a)$ and $c(b)$ are two adjacent cells in $\mathcal{A}(H)$ and vector $s(a)$ is known, then $s(b)=s(a)-2v_i$ (where h_i is the hyperplane that separates $s(a)$ and $s(b)$, and α_i of assignment a is $+1$); so $s(b)$ can be computed in constant time. Finally, note that only one of two opposite cones $c(a)$ and $c(\overline{a})$ needs to be considered since $|s(a)|=|s(\overline{a})|$, and that a hyperplane h that avoids the origin cuts each pair of opposite cones in a bounded $(d-1)$-face or in two unbounded $(d-1)$-faces. We outline the algorithm below.

Algorithm 12.7 (Optimal assignments):

Step 1: Define hyperplane h: $x_d=1$, and for each vector v_i in V, define $g_i=h\cap h_i$ (unless h_i: $\langle x,v_i\rangle=0$ is parallel to h). Construct the arrangement $\mathcal{A}(G)$, with $G=\{g_i|h_i$ in H is not parallel to $h\}$.

Step 2: Pick a $(d-1)$-face f_0 in $\mathcal{A}(G)$, determine the assignment $a(f_0)$ such that $f_0=h\cap c(a(f_0))$, and compute $s(a(f_0))$ and $|s(a(f_0))|$. Push f_0 and $s(a(f_0))$ onto an empty stack S and mark f_0.

while S is non-empty **do**
  Take the topmost $(d-1)$-face f from S.
  for each $(d-2)$-face e incident upon f **do**
    Let f' be the other incident $(d-1)$-face of e, and let e belong to g_i.
    if f' is not marked **then**
      if f' belongs to g_i^+ **then** Set $s(a(f')):=s(a(f))+2v_i$.
        else Set $s(a(f')):=s(a(f))-2v_i$.
      endif;
      Record $|s(a(f'))|$ if it exceeds the current maximum.
      In any case, mark f' and push f' and $s(a(f'))$ onto S.
    endif
  endfor
endwhile.

Note that Algorithm 12.7 only records the length of the current longest vector-sum and not the corresponding assignment. To determine the optimal assignment, we run the algorithm again and maintain the current assignment in constant time per move, where a "move" is either a push or a pop operation involving S. The assignment is saved when we reach a vector-sum whose length matches the maximum computed in the first run. Step 1 of Algorithm 12.7 constructs a $(d-1)$-dimensional arrangement which takes time $O(n^{d-1})$ if $d-1\geq 2$, and it takes time $O(n\log n)$ if $d-1=1$. Step 2 takes time $O(n)$ for the initial $(d-1)$-face and constant time for each additional $(d-1)$-face. This yields the main result of this section.

Theorem 12.12: For $V=\{v_0,v_1,\ldots,v_{n-1}\}$ a set of n vectors in E^d, $d\geq 2$, an assignment a which maximizes the length of vector $s(a)$ can be found in $O(n\log n+n^{d-1})$ time and $O(n^{d-1})$ storage.

12.7. Exercises and Research Problems

2 **Exercise 12.1:** Let P be a set of n points, possibly with three or more collinear points and with two or more points on a common vertical line. Call Q a *convex subset of* P if $Q\subseteq P$ and $Q\cap\text{int}\,\text{conv}Q=\emptyset$. Extend Algorithm 12.1 so that it handles degenerate cases

according to the above generalization of the term "convex subset", and still runs in $O(n^3)$ time and $O(n^2)$ storage.

Problem 12.2: Find an algorithm that computes the size of a largest convex subset of n points in E^2

4 (a) in $O(n^3)$ time and $O(n)$ storage *(Hint: use the algorithm for sweeping an arrangement indicated in Exercise 7.8)*, and

5 (b) in $o(n^3)$ time.

Problem 12.3: Call a convex subset Q of a set P of n points in E^2 *empty* if $P \cap \text{conv} Q = Q$. Give an algorithm that computes the size of a largest empty convex subset of P

3 (a) in $O(n^3)$ time and $O(n^2)$ storage *(Hint: modify Algorithm 12.1 such that no non-empty triangles are added to the growing convex polygon)*,

4 (b) in $O(n^3)$ time and $O(n)$ storage, and

5 (c) in $o(n^3)$ time.

3 **Exercise 12.4:** Adjust Algorithm 12.2 (which computes the visibility graph of non-intersecting line segments) to special cases such as three or more collinear endpoints, two or more identical endpoints, two or more endpoints on a common vertical line, and segments which degenerate to points, rays, or even to lines.

Exercise 12.5: Use the approach of Section 12.2 to solve the following visibility problems in time and storage $O(n^2)$.

2 (a) For n (possibly intersecting) line segments in E^2 find a line that intersects the largest number of line segments, and

2 (b) find a line that avoids all segments and produces a best balanced separation of the set of segments.

3 (c) For n non-intersecting segments and a distinguished segment s identify all segments which are completely hidden from every point of s.

4 (d) Solve Exercises 12.5(a) through (c) in $O(n^2)$ time and $O(n)$ storage. *(Hint: use the algorithm which sweeps an arrangement (indicated in Exercise 7.8) to get the necessary improvement in storage.)*

5 **Problem 12.6:** Given a set H of n lines in E^2, decide whether there is an algorithm that sorts the vertices of the arrangement $\mathcal{A}(H)$ with respect to their x_1-coordinates in time $O(n^2)$. *(Comment: this problem is not easier than the classic problem of sorting a matrix $X+Y=(a_{ij})_{1\leq i,j\leq n}$, with $a_{ij}=x_i+y_j$, $X=(x_i)_{1\leq i\leq n}$, and $Y=(y_j)_{1\leq j\leq n}$.)*

5 **Problem 12.7:** Given n non-intersecting segments in E^2, decide upon the existence of an algorithm that finds in time $O(n^2)$ a direction (if it exists) such that the shadows of no two segments intersect if light is shed from this direction. *(Comment: such an algorithm is immediate if Problem 12.6 has a positive answer.)*

3 **Exercise 12.8:** Let P be a set of n points in E^d, $d \geq 2$. For i an integer with $0 \leq i \leq d$, we call a subset Q of at least $i+2$ points *i-spherical* if aff$Q$ is an i-flat and all points of Q are equidistant from some point not necessarily in P. Give an algorithm that reports all proper (no set $Q-\{p\}$, $p \in Q$, is i-spherical) and maximal (no i-spherical subset of P properly contains Q) i-spherical subsets Q of P in time and storage $O(n^{d+1})$. *(Hint: project P vertically onto the paraboloid U: $x_{d+1}=x_1^2+...+x_d^2$ and observe that an i-spherical subset of P maps to an i-degenerate subset of the new point set.)*

5 **Problem 12.9:** Give an algorithm which decides in time $o(n^2)$ whether or not a set of n points in E^2 contains three collinear points. Alternatively, prove $\Omega(n^2)$ as a lower

bound in some reasonable model of computation.

5 **Problem 12.10:** Give an algorithm that finds a subset Q of size four of a set P of n points in E^2 such that the area of conv$Q$ is minimal in time $o(n^4)$.

4 **Exercise 12.11:** Let P be a set of n points with indices 0 through $n-1$ in E^d. We define the *λ-matrix of P* as the matrix with entries $\lambda(i_0,i_1,...,i_{d-1})$, with $0 \leq i_j \leq n-1$ for $0 \leq j \leq d-1$, where $\lambda(i_0,i_1,...,i_{d-1})$ equals the rank $\lambda(S)$, for S the sequence of points from P with indices $i_0,i_1,...,i_{d-1}$ (see also Section 12.5). Prove that the λ-matrix of P determines whether or not (S,p) has positive orientation, for every sequence S of d points in P and p another point in P.

Exercise 12.12: Let P be a set of n points in E^2.

3 (a) Show that the locus of points x with rank at least k, for k some fixed integer, is an open convex polygon with at most n edges (see Section 12.5 for a definition of the rank of a point).

3 (b) Give an algorithm which constructs this polygon in time $O(n\sqrt{k}\log^2 n)$, $0 \leq k \leq \frac{n}{2}$. *(Hint: use the results described in Chapters 3 and 9.)*

4 **Exercise 12.13:** Given n vectors $v_0,v_1,...,v_{n-1}$ in E^d, compute an assignment $a=(\alpha_0,\alpha_1,...,\alpha_{n-1})$, $\alpha_i \in \{+1,-1\}$ for $0 \leq i \leq n-1$, which maximizes the length of

$$\sum_{i=0}^{n} \alpha_i v_i$$

in $O(n\log n + n^{d-1})$ time and $O(n)$ storage. *(Hint: use the method which sweeps a two-dimensional arrangement in quadratic time and linear storage (Exercise 7.8) as a component of the algorithm; see also Section 12.6.)*

Problem 12.14: Let P be a convex polytope with n facets of known $(d-1)$-dimensional measures in E^d. For every vector u let $\mu(u)$ be the $(d-1)$-dimensional measure of P's shadow, that is, the orthogonal projection of P onto a hyperplane with normal vector u.

4 (a) Compute u such that $\mu(u)$ is a maximum in time $O(n^{d-1})$ *(Hint: for each facet f, let v_f be the outward directed normal vector with length equal to the (d-1)-dimensional measure of f. Prove that an optimal assignment gives a direction of maximal shadow, see Section 12.6 and also Exercise 12.13.)*

5 (b) How fast can a vector v be computed such that the $(d-2)$-dimensional measure of the $((d-2)$-dimensional) boundary of P's shadow is maximized?

12.8. Bibliographic Notes

Computing largest convex subsets of finite set of points in E^2 (considered in Section 12.1) is motivated by the combinatorial investigations of Erdös, Szekeres (1935 and 1960). The first algorithm for this problem was published by Chvátal, Klincsek (1980); it runs in time $O(n^3)$ and uses $O(n^2)$ storage. In fact, Algorithm 12.1 outlined in Section 12.1 can be viewed as a modification of their method. An improvement of the storage follows from a general method of Edelsbrunner, Guibas (1986) which sweeps a two-dimensional arrangement; this solves Problem 12.2(a) and, with minor complications, also solves Problem 12.3(b).

Visibility graphs for line segments in E^2 have been studied at least since Lee (1978) who uses them for finding shortest paths that avoid the segments. He also gives an algorithm which constructs the graph for n non-intersecting line segments in time $O(n^2\log n)$. The improvement described in Section 12.2 is taken from Welzl (1985). Currently, the most efficient method for finding shortest paths in weighted graphs goes back to Dijkstra (1959); an implementation which runs in time $O(n\log n+m)$, with n the number of nodes and m the number of arcs, can be found in Fredman, Tarjan (1984). Shortest path algorithms in the presence of segments as obstacles which avoid the use of the visibility graph and gain efficiency for some cases are described by Lee, Preparata (1984a) and by Reif, Storer (1985). The former paper is the source for Problem 12.7. Problems such as those posed in Exercise 12.5 are investigated in Edelsbrunner, Overmars, Wood (1983); refinements needed to solve Exercise 12.5(d) are developed in Edelsbrunner, Guibas (1986). Information about the problem of sorting $X+Y$ (mentioned in Problem 12.6) can be found in Fredman (1976) who demonstrates the existence of an algorithm that takes $O(n^2)$ comparisons (if X and Y are vectors of length n); however, this does not imply an algorithm which takes time $O(n^2)$.

Detecting and reporting degeneracies in finite point sets became an issue of general interest in the field of computational geometry when van Leeuwen (1983) asked for an algorithm that finds three collinear points among n given points in $\boldsymbol{E}^2$ in time $o(n^2\log n)$. Soon thereafter, Chazelle, Guibas, Lee (1985) and Edelsbrunner, O'Rourke, Seidel (1986) treated the construction of arrangements and obtained, as an application, an $O(n^2)$ algorithm for the above problem. The storage of these algorithms can again be reduced by the technique given in Edelsbrunner, Guibas (1986). Although Edelsbrunner, O'Rourke, Seidel (1986) considered the problem of detecting degeneracies in arbitrary dimensions, Section 12.3 in this book treats the problem in its full generality for the first time. The geometric graph of an arrangement used to this end also has other applications: Zaslavsky (1975) exploits it in his development of a theory for counting faces in possibly degenerate arrangements in $\boldsymbol{E}^d$. Applications of finding degenerate subsets in a configuration to problems in computer vision can be found, e.g. in Li, Lavin, LeMaster (1986) (see also Exercise 12.8 for reporting spherical degeneracies). A problem closely related to finding degeneracies in d-dimensional point sets is that of finding $d+1$ points that span a simplex with minimum d-dimensional measure. The two-dimensional version of the problem was first considered by Dobkin, Munro (1985); the treatment in Section 12.4 is taken primarily from Edelsbrunner, O'Rourke, Seidel (1986).

Ranks of points in some finite set are relevant in statistics as reported, e.g. in Shamos (1976). Cole, Sharir, Yap (1987) and Edelsbrunner, Welzl (1986a) give algorithms for computing ranks in $\boldsymbol{E}^2$; the former also include a solution to Exercise 12.12(a). λ-matrices, as defined in Exercise 12.11, have been introduced by Goodman, Pollack (1983) where the exercise was solved. The method for computing ranks and λ-matrices as described in Section 12.5 is taken from Edelsbrunner, O'Rourke, Seidel (1986).

The problem of computing optimal assignments for n given vectors in $\boldsymbol{E}^d$ was communicated to the author by Jan van Leeuwen. A solution which takes $O(n^{d-1})$ time and $O(n)$ storage can be found in Edelsbrunner, Guibas (1986); this settles Exercise 12.13. Computing maximal shadows of convex polytopes in $\boldsymbol{E}^d$ reduces to the above assignment problem as demonstrated by McKenna, Seidel (1985). They thus solve Problem 12.14(a). In addition, they investigate the problem of computing a minimal shadow and give a solution along similar lines.

CHAPTER 13

VORONOI DIAGRAMS

A Voronoi diagram is a cell complex which is defined with respect to a finite set of objects in some Euclidean space. Each cell of the diagram belongs to one object of the set and contains all points for which this object is the closest, or the one with the dominant influence in some sense. The Voronoi diagram in fact expresses the proximity information of the set of objects at hand in a very explicit and computationally useful manner. We will see examples of uses of this information (in particular in Section 13.2). First, we provide a rather general definition of the notion of a Voronoi diagram which subsumes all common variants as specializations.

Let S be a finite set of subsets of $\boldsymbol{E}^d$, and for each $s \in S$ let d_s be a mapping of $\boldsymbol{E}^d$ to the positive real numbers; we call $d_s(p)$ the *distance function of* s. The set $\{p \in \boldsymbol{E}^d | d_s(p) < d_t(p),\ t \in S - \{s\}\}$ is the *Voronoi cell of* s, and the cell complex defined by the Voronoi cells of all subsets in S is called the *Voronoi diagram of* S.

There is a rich mathematical literature on Voronoi diagrams, and they find applications in many diverse areas among which are biology, visual perception, physics, and archeology. In most cases, S is a finite set of points (called *sites* in this chapter) and $d_s(p)$ is the Euclidean distance between a site s and a point p. Several independent introductions of Voronoi diagrams have lead to several alternate names: they are also known as Dirichlet tessellations, Thiessen polygons, and Blum's medial axis transform.

All Voronoi diagrams examined in this chapter follow from specializations of the generic definition given above.

(i) In Section 13.1, S is a set of sites, and $d_s(p) = d(p,s)$, for $d(p,s)$ the Euclidean distance between point p and site s. Section 13.2 demonstrates several applications of such Voronoi diagrams in $\boldsymbol{E}^2$ to problems in computational geometry.

(ii) In Sections 13.3 through 13.5, S consists of all subsets T of a finite set of sites in E^d which have some fixed cardinality k. In this case $d_T(p) = \max\{d(p,t) | t \in T\}$.

(iii) Finally, Section 13.6 generalizes both types of Voronoi diagrams to the case of weighted sites: each site s is associated with a non-negative real weight w_s, and the distance of any point p from s is given by $d_s(p) = d^2(p,s) - w_s^2$.

In all cases, the two-dimensional instances are emphasized. This is because they are very versatile which is a direct consequence of the low combinatorial and computational complexity of two-dimensional Voronoi diagrams.

13.1. Classical Voronoi Diagrams

This section investigates Voronoi diagrams for finite sets of sites (that is, points) where closeness is defined by the Euclidean distance function. A formal definition of the Voronoi diagram in this classical setting is hereafter presented.

Let S be a finite set of sites in E^d, and let $d(p,q)$ be the Euclidean distance between points p and q. For every point p in E^d, let $C(p)$ contain all sites s in S that minimize $d(p,s)$, and call two points p and q *equivalent* if $C(p) = C(q)$. Notice that every point of E^d is contained in an equivalence class imposed by the equivalence relation just defined, and that two equivalence classes have a non-empty intersection by definition. The *Voronoi diagram* $\mathcal{V}(S)$ *of* S is the cell complex whose faces are the equivalence classes of points. To demonstrate that the faces of $\mathcal{V}(S)$ have attractive geometric properties, we formalize the idea of "dominant closeness" of one site over another. For two sites s and t, we call $b_{\{s,t\}} = \{p \in E^d | d(p,s) = d(p,t)\}$ the *perpendicular bisector of* s and t, and we define $b_{(s,t)} = \{p \in E^d | d(p,s) < d(p,t)\}$. We then have the following result.

Observation 13.1: Let S be a finite set of sites in E^d, and let f be the equivalence class of points p with $C(p) = C_f$, for some subset C_f of S. Then

$$f = \left(\bigcap_{s \in C_f, t \in S - C_f} b_{(s,t)} \right) \cap \left(\bigcap_{s,t \in C_f} b_{\{s,t\}} \right).$$

By Observation 13.1, every equivalence class of points is a relatively open convex polyhedron in E^d. We call an equivalence class f a k*-face* if its dimension is k, that is, if $\dim \operatorname{aff} f = k$. We use *vertex*, *edge*, *region*, *facet*, and *cell* as synonyms for 0-face, 1-face, 2-face, $(d-1)$-face, and d-face, respectively. A k-face f and a $(k+1)$-face g are said to be *incident upon each other* if $\operatorname{cl} g$ contains f. Additionally, two cells are *adjacent* if they are incident upon a common facet. Figure 13.1 displays the Voronoi diagram of 11 sites in E^2: it consists of 11 cells or regions (one for each site), 25 edges, and 15 vertices (one edge and one vertex are not shown).

In order to argue about closeness in the Euclidean sense, we use balls which are centered at points of interest. For every point p in d dimensions, define $b(p)=\{x\in E^d | d(p,x)<\min\{d(p,s)|s\in S\}\}$; so $b(p)$ is the largest open ball centered at p which is *empty*, that is, $b(p)\cap S=\emptyset$. Obviously, $C(p)=S\cap \mathrm{bd}\, b(p)$. For each face f of $\mathcal{V}(S)$, we define $C_f=C(p)$, for p any point in f. The following result can be seen as a consequence of Observation 13.1 and the fact that the restrictions imposed by $d+1-k$ sites on the boundary of a ball let the center move freely in a k-flat, unless some $i+2$ of the $d+1-k$ sites belong to a common i-flat.

Observation 13.2: Let S be a finite set of sites in E^d and let f be a k-face of the Voronoi diagram $\mathcal{V}(S)$ of S.

(i) $\mathrm{card}\, C_f=1$ if and only if f is a cell.

(ii) $\mathrm{card}\, C_f=2$ if and only if f is a facet.

(iii) $\mathrm{card}\, C_f\geq d+1-k$, for $0\leq k\leq d-2$, and equivalently

(iv) a point p is contained in a k-face of the diagram $\mathcal{V}(S)$ only if $\mathrm{card}(S\cap \mathrm{bd}\, b(p))\geq d+1-k$, for $0\leq k\leq d-2$.

We call $\mathcal{V}(S)$ *simple* if $\mathrm{card}\, C_v=d+1$, for each vertex v of $\mathcal{V}(S)$. It is not hard to see that if $\mathcal{V}(S)$ is simple and contains at least one vertex then $\mathrm{card}(S\cap \mathrm{bd}\, b(p))=d+1-k$ if p belongs to a k-face of $\mathcal{V}(S)$, for $0\leq k\leq d$.

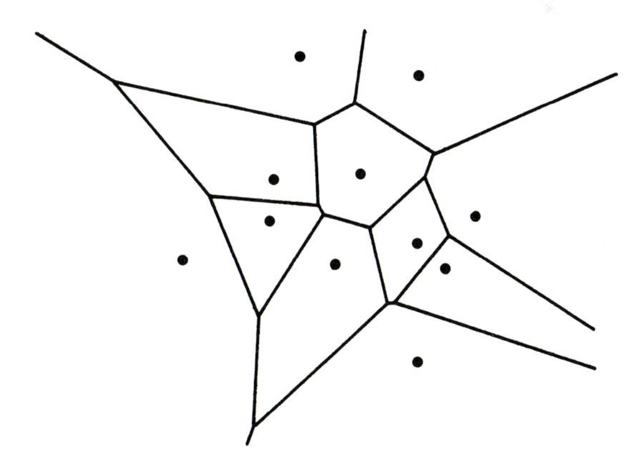

Figure 13.1. Voronoi diagram for 11 sites in the plane.

Otherwise, there must be a smallest integer number $k_0 \geq 1$ and a k_0-face f with $\text{card}\, C_f > d+1-k_0$. We then have $\text{card}\, C_g > d+1-k_0+1$, for each (k_0-1)-face g incident upon f; the minimality of k_0 implies that there is no such (k_0-1)-face. It follows that f is a complete k_0-flat, and consequently, all sites belong to a $(d-k)$-flat normal to f. This contradicts the existence of a vertex in $\mathcal{V}(S)$. The condition that any $d+1$ sites but no $d+2$ sites are equidistant from a common point is sufficient but not necessary for the simplicity of $\mathcal{V}(S)$.

The concept of simplicity will become clearer when we interpret a Voronoi diagram in E^d as a structure in a space of one higher dimensions using the method of geometric transformation. To this end, we exploit the geometric transform $\mathcal{E}$ which maps each site of S to a hyperplane in E^{d+1}. Upper bounds on the complexity of $\mathcal{V}(S)$ and algorithms that construct $\mathcal{V}(S)$ will then follow from results on arrangements and convex hulls that can be found in earlier chapters of this book. Transform $\mathcal{E}$ maps any site $s=(\sigma_1,\sigma_2,\ldots,\sigma_d)$ to the hyperplane

$$\mathcal{E}(s)\colon x_{d+1}=2\sigma_1 x_1+2\sigma_2 x_2+\ldots+2\sigma_d x_d-(\sigma_1^2+\sigma_2^2+\ldots+\sigma_d^2).$$

Let us identify E^d with the hyperplane $x_{d+1}=0$ in E^{d+1}. Then $\mathcal{E}(s)$ is the unique hyperplane that touches the paraboloid $U\colon x_{d+1}=x_1^2+x_2^2+\ldots+x_d^2$ in the point $U(s)=(\sigma_1,\sigma_2,\ldots,\sigma_d,\sigma_1^2+\sigma_2^2+\ldots+\sigma_d^2)$, which is the vertical projection of site s onto the paraboloid U. Furthermore, let $h(p)$ be the vertical projection of point p in E^d onto any non-vertical hyperplane h in E^{d+1}. Straightforward analytic calculations imply that $\mathcal{E}(s)$ encodes the Euclidean distance between a point and a site as follows.

Observation 13.3: For s a site in E^d, p a point in E^d, and $h=\mathcal{E}(s)$, we have $d^2(p,s)=d(U(p),h(p))$.

Since points $U(p)$ and $h(p)$ differ only in their $(d+1)^{\text{st}}$ coordinates, if at all, Observation 13.3 implies a strong relationship between $\mathcal{V}(S)$ and the unique cell in $\mathcal{A}(H)$ which lies above all hyperplanes in H, where H is the set $\{\mathcal{E}(s)|s\in S\}$. Notice that hyperplane $\mathcal{E}(s)$ is non-vertical, for every site s, and recall that $\mathcal{E}(s)^+$ denotes the open half-space above $\mathcal{E}(s)$.

Observation 13.4: Let S be a finite set of sites in E^d (identified with $x_{d+1}=0$ in E^{d+1}), and define $\mathcal{P}$ as the intersection of all half-spaces $\mathcal{E}(s)^+$, $s\in S$. Thus, $\mathcal{P}$ contains $U-\{U(s)|s\in S\}$.

(i) For a point p in E^d, let $\mathcal{P}(p)$ be its vertical projection onto $\text{bd}\mathcal{P}$. $\mathcal{P}(p)$ is contained in hyperplane $\mathcal{E}(s)$ if and only if $d(p,s)$ is a minimum over all sites s in S, which is equivalent to $s\in C(p)$.

(ii) For each k-face f in $\mathcal{V}(S)$, there is a k-face f' in $\text{bd}\mathcal{P}$, and vice versa, such that f is the vertical projection of f' onto E^d.

Figure 13.2 illustrates Observation 13.4 for $d=1$: four sites in E^1 (the x_1-axis) transform to four tangents to the parabola U: $x_2=x_1^2$. The faces in the boundary of the cell above all lines project onto the faces of the Voronoi diagram of the four sites. Observation 13.4(ii) implies that the number of k-faces of $\mathcal{V}(S)$ is equal to the number of k-faces of $\mathcal{P}$, for $0 \leq k \leq d$. By Corollary 6.7, we have the following upper bound on the number of k-faces of $\mathcal{V}(S)$.

Corollary 13.5: Let S be a set of n sites in $E^d, d \geq 1$. The number of k-faces of $\mathcal{V}(S)$ is in $O(n^{\min\{d+1-k, \lceil d/2 \rceil\}})$, for $0 \leq k \leq d$.

The obvious data structure to represent a Voronoi diagram is its incidence graph defined analogously as for arrangements of hyperplanes and for convex polytopes: each face of the diagram stands in one-to-one correspondence with a node of the incidence graph, and two nodes are connected by an arc if the corresponding faces are incident. To affix the diagram in space, each node of the graph that corresponds to a cell of the diagram stores the site that this cell belongs to. The following strategy for constructing the Voronoi diagram $\mathcal{V}(S)$ from the set of sites S is immediate if we recall that there are efficient methods that construct the intersection of a set of half-spaces (see Chapter 8).

Algorithm 13.1 (Voronoi diagram for sites in E^d):

Step 1: Map each site $s \in S$ to the half-space $\mathcal{E}(s)^+$ in E^{d+1} and construct the polyhedron $\mathcal{P} = \bigcap_{s \in S} \mathcal{E}(s)^+$.

Step 2: Project each face of bd$\mathcal{P}$ vertically onto $x_{d+1}=0$.

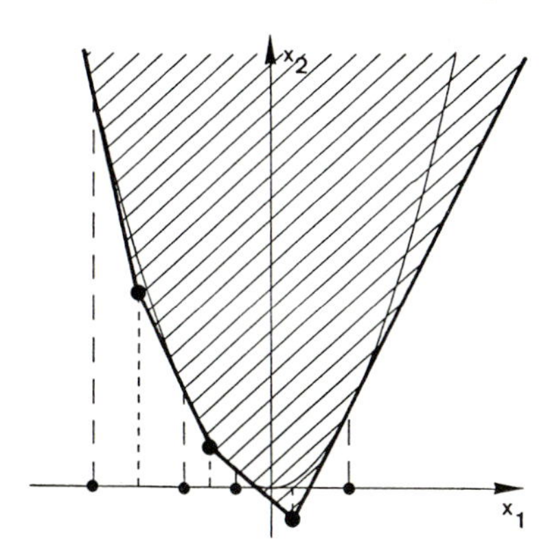

Figure 13.2. Voronoi diagram and convex polyhedron.

Notice that Step 2 is essentially void since the incidence graph of $\mathcal{P}$ is identical to the incidence graph of $\mathcal{V}(S)$. Even the geometric information stored in the facets of $\mathcal{P}$ and the cells of $\mathcal{V}(S)$ is the same, up to interpretation. If a facet f of $\mathcal{P}$ is supported by a hyperplane $\mathcal{E}(s)$ and therefore f stores $\mathcal{E}(s)$, then the corresponding cell of $\mathcal{V}(S)$ belongs to site s. Thus, the node that represents this cell is supposed to store s. However, f could have stored s in the first place, with the transformation $\mathcal{E}$ implicitly understood. By the algorithms of Chapter 8 (see Theorems 8.11 an 8.16, in particular), this implies the following result for the problem of constructing Voronoi diagrams.

Corollary 13.6: Let S be a set of n sites in E^d, $d \geq 1$. There is an algorithm which construct $\mathcal{V}(S)$ in $O(n\log n)$ time and $O(n)$ storage if $d \leq 2$, and in $O(n^{\lceil (d+1)/2 \rceil})$ time and $O(n^{\lceil d/2 \rceil})$ storage, otherwise.

13.2. Applications in the Plane

The use of Voronoi diagrams for solving geometric problems algorithmically is particularly attractive in the Euclidean plane where $O(n)$ storage suffices if the diagram is defined by n sites. Recall that a polytope with n facets in E^3 has at most $3n-6$ edges and at most $2n-4$ vertices, see Chapter 6. If the polytope is unbounded, which is the case for every polytope that corresponds to a Voronoi diagram by means of the geometric transform $\mathcal{E}$, then the number of vertices is at most $2n-5$. Intuitively, the single missing vertex is at infinity and it is an endpoint of all unbounded edges. These upper bounds, together with the lower bound implied by simple Voronoi diagrams in the plane with only three unbounded edges, lead to the following result.

Corollary 13.7: Let S be a set of $n \geq 3$ sites in the plane. The diagram $\mathcal{V}(S)$ contains n regions, at most $3n-6$ edges, and at most $2n-5$ vertices. These bounds are tight.

This section consists of six parts which demonstrate five applications of two-dimensional Voronoi diagrams to problems in computational geometry, optimization, and pattern recognition.

13.2.1. The Post Office Problem

Let S be the set of all post offices in some area on earth which can reasonably be assumed to be part of a plane. An obvious question that people living in this area may ask is

"Where is the post office closest to where I live?",

and it is certainly fortunate for the author of this book that these people will expect an immediate answer to their question. The *post office problem* models this situation, that is, it is a search problem defined as follows:

> let S be a finite set of sites in the plane; store S so that a query can be answered effectively, where a query is specified as a point and to answer the query means to determine the site closest to this point.

To solve the problem, we suggest a data structure which is based on the Voronoi diagram of S and which allows for the efficient location of a given point. By definition of the Voronoi diagram, the location of the point uniquely determines the closest sites. Both the construction and the search of the data structure rely on the algorithmic tools developed in Chapter 11. In particular, we use the layered DAG of a subdivision, which is a data structure that supports point location queries in optimal time and storage.

Algorithm 13.2 (Preprocessing):

Step 1: Construct the Voronoi diagram $\mathcal{V}(S)$ and associate each face f of $\mathcal{V}(S)$ with the set C_f of closest sites.

Step 2: Construct the layered DAG for $\mathcal{V}(S)$.

Using the data structure constructed by Algorithm 13.2, we can answer a post office query as follows.

Algorithm 13.3 (Query with given point q):

Step 1: Using the layered DAG of $\mathcal{V}(S)$, determine the region, edge, or vertex f of $\mathcal{V}(S)$ which contains q.

Step 2: Report any site in C_f.

By Corollary 13.6 and Theorem 11.12, the preprocessing takes $O(n\log n)$ time and $O(n)$ storage. The region, edge, or vertex which contains q can be determined in time $O(\log n)$ by Theorem 11.12. This implies the following complexity of the above solution.

Theorem 13.8: Let S be a set of n sites in the plane. The post office problem for S can be solved in $O(n)$ storage, $O(n\log n)$ time for preprocessing, and $O(\log n)$ time for a query.

13.2.2. Triangulating Point Sets

Triangulations are special subdivisions of the plane which have many applications due to their simple structure: all regions are triangles, except for one which is the only unbounded region of the subdivision. We formally define the notion of a triangulation with respect to a given set of sites. Let S be a finite set of sites in the plane, not all collinear. A subdivision $\mathcal{T}$ of the plane is a *triangulation of* S if

(i) compl conv$S$ is the only unbounded region of $\mathcal{T}$,
(ii) S is the set of vertices of $\mathcal{T}$, and
(iii) each bounded region of $\mathcal{T}$ is a *triangle*, that is, it is of the form int conv$\{s,t,u\}$, for s, t, and u three non-collinear sites such that no other site of S belongs to the convex hull of s, t, and u.

Figure 13.3 shows a triangulation of 11 sites in the plane. Let k denote the number of edges that belong to the boundary of the unbounded region of $\mathcal{T}$, that is, $k = \text{card}(S \cap \text{bd conv} S)$. We can draw $k-3$ simple curves, where each curve connects a unique pair of vertices, no curve intersects an edge of $\mathcal{T}$, and no two curves intersect. $\mathcal{T}$, augmented by these $k-3$ curves, is still the embedding of a planar graph, and no curve can be added without violating this property. By Euler's relation for planar graphs (see Theorem 11.1), we can therefore express the numbers of edges and regions of $\mathcal{T}$ by means of n, the number of sites, and k, the number of edges of the unbounded region.

Corollary 13.9: Let S be a set of n non-collinear sites in the plane, with $k = \text{card}(S \cap \text{bd conv} S)$. Any triangulation of S consists of n vertices, $3n-3-k$ edges, and $2n-1-k$ regions of which $2n-2-k$ are triangles.

For the sake of many applications of triangulations, it is most desirable to have nearly equilateral triangles. We therefore introduce the following definitions.

Let $\mathcal{T}$ be a triangulation of S with m triangles whose interior angles $\alpha_1, \alpha_2, \ldots, \alpha_{3m}$ are indexed such that $\alpha_i \leq \alpha_j$ if $i < j$. Since the angles of a triangle add up to π, we have

$$\sum_{i=1}^{3m} \alpha_i = m\pi.$$

We call $A(\mathcal{T}) = (\alpha_1, \alpha_2, \ldots, \alpha_{3m})$ the *equiangularity of* $\mathcal{T}$, and we write $(\alpha_1', \alpha_2', \ldots, \alpha_{3m}') < (\alpha_1, \alpha_2, \ldots, \alpha_{3m})$ if there is an index $1 \leq j \leq 3m$ such that $\alpha_j' < \alpha_j$ and $\alpha_i' = \alpha_i$, for $1 \leq i < j$. The triangulation $\mathcal{T}$ is *globally equiangular* if $A(\mathcal{T}') \leq A(\mathcal{T})$, for all triangulations $\mathcal{T}'$ of S.

We continue with the introduction of a family of triangulations, the so-called completions of the Delaunay triangulation of S, which will be shown to contain all globally equiangular triangulations of S. The *Delaunay triangulation*

$\mathcal{DT}(S)$ *of* S is in some sense dual to the Voronoi diagram $\mathcal{V}(S)$: $\mathcal{DT}(S)$ contains $\text{ri}\,\text{conv}\{s,t\}$ as an edge if and only if regions f and g in $\mathcal{V}(S)$, with $C_f=\{s\}$ and $C_g=\{t\}$, are adjacent.

If $\mathcal{V}(S)$ is simple then $\mathcal{DT}(S)$ is a triangulation, otherwise, it is not. There is an obvious incidence–preserving one–to–one correspondence between the vertices, edges, and regions of $\mathcal{V}(S)$ and the bounded regions, edges, and vertices of $\mathcal{DT}(S)$, respectively. By Observation 13.2, we have the following characterization of $\mathcal{DT}(S)$.

Observation 13.10: Let S be a finite set of sites in the plane.

(i) The relative interior of $\text{conv}\{s,t\}$ is an edge of $\mathcal{DT}(S)$ if and only if there is an empty open disc b (that is, $S\cap b=\emptyset$) with $\{s,t\}=S\cap \text{bd}\,b$.

(ii) A subset C of at least three sites of S coincides with the set of vertices of a bounded region of $\mathcal{DT}(S)$ if and only if there is an empty disc b with $C=S\cap \text{bd}\,b$.

Figure 13.3 shows the Delaunay triangulation of 11 sites in solid segments; broken segments indicate the underlying Voronoi diagram. A triangulation $\mathcal{T}$ is called a *completion of* $\mathcal{DT}(S)$ if it can be obtained from $\mathcal{DT}(S)$ by adding new edges (but no new vertices) which dissect the bounded regions of $\mathcal{DT}(S)$ with more than three edges into non–overlapping triangles. If a bounded region r is incident upon $i\geq 3$ edges, then $i-3$ mutually non–intersecting new edges can be introduced, which dissect r into $i-2$ triangles. Let n denote the number of sites in set S. Because of the close relationship between $\mathcal{DT}(S)$ and $\mathcal{V}(S)$, $O(n\log n)$

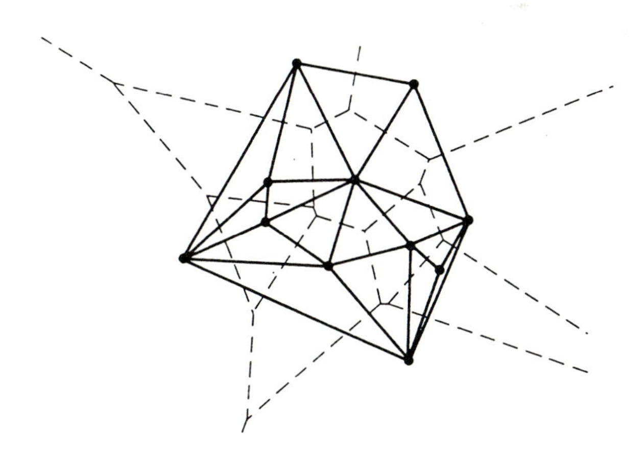

Figure 13.3. Delaunay triangulation and Voronoi diagram.

time and $O(n)$ storage suffice to construct $\mathcal{DT}(S)$ or a completion thereof. We finally show that all globally equiangular triangulations of S are completions of $\mathcal{DT}(S)$.

Theorem 13.11: Let S be a finite set of sites in the plane, not all collinear. A triangulation $\mathcal{T}$ of S is globally equiangular only if it is a completion of $\mathcal{DT}(S)$.

Proof: We proceed in three steps. First, we introduce an algorithm that produces so-called locally equiangular triangulations, which will be defined later. Second, we show that all locally equiangular triangulations are completions of $\mathcal{DT}(S)$. Finally, the assertion follows, since we will see that local equiangularity is necessary for global equiangularity.

Let C be a subset of four sites of S such that all four sites are extreme with respect to C and $S \cap \operatorname{conv} C = C$. We denote the two possible triangulations of C by $\mathcal{T}_C$ and $\mathcal{T}_C'$. Let t_1, t_2 (t_1', t_2') be the two triangles of $\mathcal{T}_C$ ($\mathcal{T}_C'$), and e (e') the edge incident upon both t_1 and t_2 (t_1' and t_2') (see Figure 13.4 which indicates e by a solid segment and e' by a broken segment). Let $b(t_1)$ be the unique open disc with the vertices of t_1 contained in $\operatorname{bd} b(t_1)$, and let s be the site in C that is not a vertex of t_1. Straightforward elementary geometry can be used to show the following result.

Fact 1: $A(\mathcal{T}_C) < A(\mathcal{T}_C')$ if $b(t_1)$ contains s, and $A(\mathcal{T}_C) > A(\mathcal{T}_C')$ if $\operatorname{cl} b(t_1)$ does not contain s (see Figure 13.4).

Let $\mathcal{T}$ be a triangulation of S which contains edge e and triangles t_1 and t_2. We call e *illegal* if $A(\mathcal{T}_C) < A(\mathcal{T}_C')$, and *to switch* e with e' means to replace e, t_1, and t_2 in $\mathcal{T}$ by edge e' and triangles t_1' and t_2'. $\mathcal{T}$ is called *locally equiangular* if it contains no illegal edge. Consider the following non-deterministic algorithm that begins with an arbitrary triangulation and produces a locally equiangular triangulation.

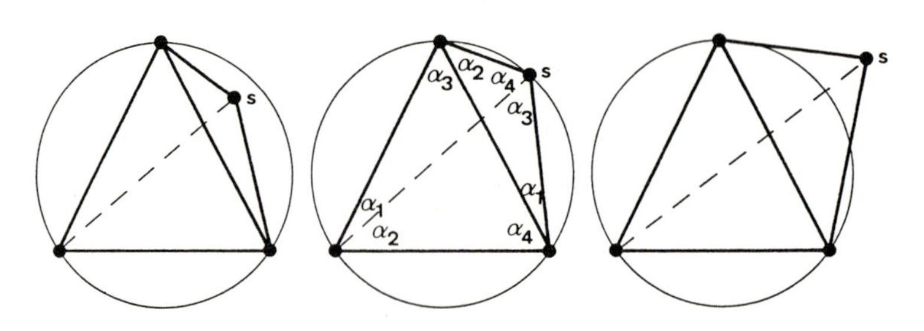

Figure 13.4. The effect of switching a diagonal.

Algorithm 13.4 (Locally equiangular triangulation):
while $\mathcal{T}$ contains an illegal edge e **do**
Switch e with e' in $\mathcal{T}$.
endwhile.

To verify that Algorithm 13.4 terminates, note that any switch increases $A(\mathcal{T})$ and that there are only finitely many triangulations of S. Obviously, the algorithm halts when a locally equiangular triangulation $\mathcal{T}$ is reached. We show that $\mathcal{T}$ has to be a completion of $\mathcal{DT}(S)$.

Assume the contrary. By definition of $\mathcal{DT}(S)$, there is a triangle t_1 in $\mathcal{T}$ with a site s contained in $b(t_1)$. Since $\mathcal{T}$ is a triangulation, s is not contained in $\mathrm{cl}t_1$. Let e be the edge of t_1 such that the triangle $t^* = \mathrm{conv}(e \cup \{s\})$ does not intersect t_1. Of all such triplets triangle, site, edge, let (t_1, s, e) maximize the angle in t^* at vertex s. Since e is not an edge of the unbounded region of $\mathcal{T}$, there is a triangle $t_2 \neq t_1$ in $\mathcal{T}$ incident upon e; let v be its vertex which is not an endpoint of e. By the local equiangularity of $\mathcal{T}$, $b(t_1)$ does not contain v (see Fact 1), which implies $t_2 \neq t^*$. Let $b(t_2)$ be the open disc with the vertices of t_2 in its boundary. The circle $\mathrm{bd}t_2$ intersects $\mathrm{bd}t_1$ in the two endpoints of e, and since v belongs to $\mathrm{bd}t_2$ but is not contained in $b(t_1)$, $b(t_2)$ contains the part of $b(t_1)$ which is separated from t_1 by edge e. It follows that s belongs to $b(t_2)$. Let e_2 be the edge of t_2 such that $\mathrm{conv}(e_2 \cup \{s\}) \cap t_2 = \emptyset$. The triplet (t_2, s, e_2) defines a greater angle than (t_1, s, e), which is a contradiction.

We conclude that every locally equiangular triangulation of S is a completion of $\mathcal{DT}(S)$. The assertion follows since every globally equiangular triangulation is also locally equiangular. □

Notice that the smallest angles of any two completions of the Delaunay triangulation of S are the same. It follows that every completion of the Delaunay triangulation of S maximizes the minimum angle α_1.

13.2.3. Delaunay Triangulations from Convex Hulls

Instead of indirectly using Voronoi diagrams, there is a more direct strategy for computing the Delaunay triangulation $\mathcal{DT}(S)$ of a finite set S of sites in E^2: $\mathcal{DT}(S)$ can be derived from the convex hull of the set $P = \mathcal{U}(S) = \mathcal{D}(\mathcal{E}(S))$ of points in E^3. Recall that $\mathcal{E}$ maps a site in E^2 to a plane in E^3, and that $\mathcal{D}$ maps a plane in E^3 to a point in E^3. This method yields, as a byproduct, the so-called furthest-site Delaunay triangulation of S which is in some sense complementary to $\mathcal{DT}(S)$.

Recall that $P = \{\mathcal{U}(s) = (\sigma_1, \sigma_2, \sigma_1^2 + \sigma_2^2) \mid s = (\sigma_1, \sigma_2) \text{ in } S\}$. Instead of resorting to the duality between conv$P$ and the intersection of half-spaces $\mathcal{E}(s)^+$, $s \in S$, we use direct arguments to establish the correspondence between conv$P$ and $\mathcal{DT}(S)$. For $b=\{p \in E^2 | d(p,c) \leq \rho,\ c=(\gamma_1,\gamma_2)$ and $\rho \geq 0\}$ a closed disc, we extend $\mathcal{E}$ to yield plane $\mathcal{E}(b)$: $x_3=2\gamma_1 x_1+2\gamma_2 x_2-(\gamma_1^2+\gamma_2^2-\rho^2)$. Intuitively, $\mathcal{E}$ maps b to the unique plane in E^3 which intersects the paraboloid U in the vertical projection of bd$b$ onto U. Straightforward analytic calculations show that $\mathcal{E}$ encodes the relative position of a site and a disc in E^2.

Observation 13.12: Let s be a site in E^2, let $p=U(s)$ be the vertical projection of s onto U in E^3, and let b be a closed disc in E^2. Then s belongs to int$b$, bd$b$, or compl$b$ if and only if p belongs to $\mathcal{E}(b)^-$, $\mathcal{E}(b)$, or $\mathcal{E}(b)^+$, respectively.

Call a k-face f of conv$P$ a *lower* (*upper*) *k-face*, for $0 \leq k \leq 2$, if there is a non-vertical plane h which contains f and clh^+ (clh^-) contains conv$P$. Note that facets contained in vertical planes are neither lower nor upper facets, and that the only facet of conv$P$, when dim aff$P=2$, is a lower and an upper facet unless it lies in a vertical plane. Edges and vertices might well be lower and upper edges and vertices at the same time. Let f be a lower facet with vertex-set $U(C)$, C subset of S, and let h be the unique plane that supports f. Obviously, $h \cap P=U(C)$ and $h^- \cap P=\emptyset$ which implies, by Observations 13.10(ii) and 13.12, that there is a bounded region r in $\mathcal{DT}(S)$ with vertex-set C. By definition of P, r is the vertical projection of f onto the plane $x_3=0$ identified with E^2. The inverse is also obvious which yields the following relationship between conv$P$ and $\mathcal{DT}(S)$.

Observation 13.13: Let S be a finite set of sites in E^2 (identified with the plane $x_3=0$ in E^3), and define $P=U(S)$. For each k-face f in $\mathcal{DT}(S)$ there is a lower k-face f' of conv$P$, and vice versa, such that f is the vertical projection of f' onto E^2, for $0 \leq k \leq 2$.

We now define the *furthest-site Delaunay triangulation* $\mathcal{DT}_f(S)$ *of* S: edge $e=$ri conv$\{s,t\}$, for s and t sites in S, is an edge of $\mathcal{DT}_f(S)$ if there is an open disc b with bd$b \cap S=\{s,t\}$ and $b \cap S=S-\{s,t\}$. We conclude that s and t are extreme in S. Consequently, region $r=$int conv$C$, C subset of ext$S$, is a region of $\mathcal{DT}_f(S)$ if and only if there is an open disc b with bd$b \cap S=C$ and $b \cap S=S-C$. Note that, in general, $\mathcal{DT}_f(S)$ is not a triangulation of S; rather, it is a subdivision that can be refined to a triangulation of ext$S$. Figure 13.5 shows $\mathcal{DT}_f(S)$, for S the set of sites displayed in Figure 13.3. By the same reasoning as above, we have the following relationship between conv$P$ and $\mathcal{DT}_f(S)$.

Observation 13.14: Let S be a finite set of sites in E^2 (identified with

$x_3=0$ in E^3), and define the set $P=U(S)$ of points in E^3. For each k-face f in $\mathcal{DT}_f(S)$, there is an upper k-face f' of conv$P$, and vice versa, such that f is the vertical projection of f' onto E^2, for $0 \leq k \leq 2$.

We have thus learned that $\mathcal{DT}(S)$ and $\mathcal{DT}_f(S)$ can both be obtained by projecting faces of conv$P$ vertically onto the plane $x_3=0$, where P is the vertical projection of S onto the paraboloid U. The following algorithm for constructing $\mathcal{DT}(S)$ and $\mathcal{DT}_f(S)$ is now obvious.

Algorithm 13.5 (Delaunay triangulations):

Step 1: Compute $P=U(S)=\{U(s)|s \in S\}$ and construct the incidence graph of conv$P$.

Step 2: For each face of conv$P$ decide whether it is a lower face, an upper face, a lower and upper face, or neither a lower nor an upper face.

Step 3: Project each lower (upper) face of conv$P$ vertically onto $x_3=0$ which yields $\mathcal{DT}(S)$ ($\mathcal{DT}_f(S)$).

Steps 2 and 3 are straightforward and take time proportional to $n=$card$S$. Step 1 takes $O(n\log n)$ time and $O(n)$ storage by Theorem 8.16 which thus implies the following theorem.

Theorem 13.15: Let S be a set of n sites in E^2. There is an algorithm that constructs $\mathcal{DT}(S)$ and $\mathcal{DT}_f(S)$ in $O(n\log n)$ time and $O(n)$ storage.

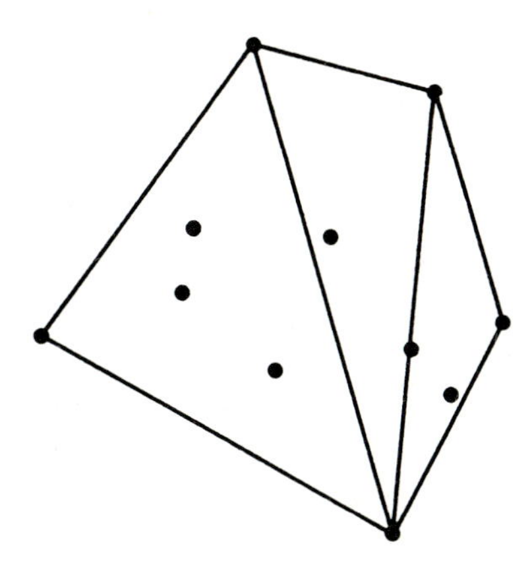

Figure 13.5. Furthest-site Delaunay triangulation.

13.2.4. Finding Closest Neighbors

Let S be a finite set of sites in the Euclidean plane, and define $C(s)=\{t$ in $S-\{s\}|d(s,t)\leq d(s,u)$, u in $S-\{s\}\}$. A site t in $C(s)$ is called a *closest neighbor of s*. We consider the problem of computing $C(s)$, for each site s in S. The Delaunay triangulation of S turns out to lead to an efficient algorithm for this problem. The algorithm to be described is based on the following observation.

Lemma 13.16: Let S be a finite set of sites in the plane, let s be a site in S, and let t be a closest neighbor of s. Sites s and t are adjacent in $\mathcal{DT}(S)$, that is, they are endpoints of a common edge.

Proof: Note that s and t are adjacent in $\mathcal{DT}(S)$ if there is an empty open disc whose boundary intersects S in exactly the two sites, s and t. Define $b=\{p\in E^2|d(p,(s+t)/2)<d(s,t)/2\}$. Trivially, sites s and t lie on $\text{bd}b$, and $d(s,p)<d(s,t)$ for every point $p\neq t$ in $\text{cl}b$. Since t is a nearest neighbor of s, we conclude that b is empty and that $\{s,t\}=\text{bd}b\cap S$. □

The algorithm suggested by Lemma 13.16 proceeds as follows.

Algorithm 13.6 (Closest neighbors):

Step 1: Construct $\mathcal{DT}(S)$.

Step 2: For each site $s\in S$, find the sites adjacent to s in $\mathcal{DT}(S)$ which are closest to s.

Step 2 takes time proportional to $n=\text{card}S$ since $\mathcal{DT}(S)$ defines at most $3n-6$ pairs of adjacent sites. By Theorem 13.15, we have the following result.

Theorem 13.17: Let S be a set of n sites in the plane. There is an algorithm that finds the closest neighbors of each site s in S in $O(n\log n)$ time and $O(n)$ storage.

It is worthwhile to note that the same strategy can be used to identify all pairs $\{s,t\}$ of sites with s in $C(t)$ and t in $C(s)$, or to find all pairs $\{s,t\}$ with $d(s,t)=\min\{d(u,v)|u,v\in S\}$.

13.2.5. Minimum Spanning Trees

Suppose that a collection of locations has to be connected by streets in a way that minimizes the total costs and that allows for travel to each other location. We assume that each location is a site in the plane and that the cost of connecting two locations by a street is the distance between the two locations. The minimum spanning tree solves this optimization problem because a desired network of streets must be a spanning tree of the collection of locations. First, we present formal definitions of the concepts.

Let S be a finite set of sites in the Euclidean plane. An undirected (free) tree $\mathcal{T}=(N,A)$ spans S if $N=S$. We call $c(\{s,t\})=d(s,t)$ the *weight* or *length* or *cost of* arc $\{s,t\}$, and define

$$c(\mathcal{T})=\sum_{\{s,t\}\in A} c(\{s,t\})$$

as the *cost of* $\mathcal{T}$. The tree $\mathcal{T}$ is a *minimum spanning tree of* S if $c(\mathcal{T})\leq c(\mathcal{T}')$, for every tree $\mathcal{T}'$ that spans S. We describe an algorithm for constructing a minimal spanning tree of S below, prove its correctness, and finally show how $\mathcal{DT}(S)$ can be used to speed up the construction. The algorithm works incrementally, that is, it adds arcs, one at a time, to a growing tree $\mathcal{T}^*$ which spans a subset S^* of S. For each site s in $S-S^*$, it maintains a closest site c_s in S^*, that is, $d(s,c_s)\leq d(s,t)$, for all sites t in S^*.

Algorithm 13.7 (Minimum spanning tree):

Initially, $S^*=\{u_0\}$, for u_0 an arbitrary site of S, $A=\emptyset$, and $c_s=u_0$, for each site s in $S-\{u_0\}$.

while $S^*\neq S$ **do**

Determine a site s in $S-S^*$ with $d(s,c_s)\leq d(t,c_t)$, for all sites t in $S-S^*$, set $A:=A\cup\{\{s,c_s\}\}$, and set $S^*:=S^*\cup\{s\}$.

for all sites t in $S-S^*$ **do**

if $d(t,s)<d(t,c_t)$ **then**

Set $c_t:=s$.

endif

endfor

endwhile.

The body of the while–loop of Algorithm 13.7 is executed $n-1$ times, where $n=\text{card}S$. Each execution takes time $O(n)$ to find a shortest arc connecting a site in S^* with a site not in S^*, and to update c_t, for all sites t in $S-S^*$. The straightforward implementation of Algorithm 13.7 thus runs in $O(n^2)$ time and $O(n)$ storage. We prove that Algorithm 13.7 is indeed correct.

Lemma 13.18: Let S be a finite set of sites in the plane. Algorithm 13.7 constructs a minimum spanning tree of S.

Proof: Since each arc added to $\mathcal{T}^*$ connects a site in S^* with a site not in S^*, We know that Algorithm 13.7 constructs a tree $\mathcal{T}=(S,A)$; so $\mathcal{T}$ spans S. Let $A=\{a_1,a_2,...,a_{n-1}\}$, with the arcs indexed such that $A_i=\{a_1,a_2,...,a_i\}$ is the set of arcs (and S_i is the set of sites or nodes) of $\mathcal{T}^*$ immediately after the ith execution of the body of the while-loop, for $1 \leq i \leq n-1$. Assume that $\mathcal{T}$ is not minimal and let $\mathcal{T}'=(S,A')$ be a tree with $c(\mathcal{T}')<c(\mathcal{T})$ that maximizes j such that $A_j \subseteq A'$. We have $j<n-1$ since $A \neq A'$. The graph $\mathcal{G}=(S,A' \cup \{a_{j+1}\})$ contains a unique cycle, and there is at least one arc $a=\{s,t\} \neq a_{j+1}$ in this cycle, with s in S_j and t in $S-S_j$, which is not an arc of $\mathcal{T}$. Otherwise, $\mathcal{T}$ contains this cycle, but this is also not possible since $\mathcal{T}$ is a tree. Now, we have $c(a_{j+1}) \leq c(a)$ which follows from the strategy for choosing a new arc applied by Algorithm 13.7. Consequently, $\mathcal{T}''=(S,A''=A' \cup \{a_{j+1}\}-\{a\})$ is a tree which spans S, and which satisfies $c(\mathcal{T}'') \leq c(\mathcal{T}')$ and $A_{j+1} \subseteq A''$; a contradiction to the choice of $\mathcal{T}'$. □

Note that Algorithm 13.7, as described, is non-deterministic as it is free to break ties arbitrarily. In fact, a reasoning similar to the one used in the proof of Lemma 13.18 shows that, for any minimum spanning tree of S, there is a tie-breaking sequence that leads Algorithm 13.7 to construct it. We use this to prove that all minimum spanning trees of S are subgraphs of the graph that contains an arc connecting two sites if $\mathcal{DT}(S)$ contains an edge connecting the two sites.

Lemma 13.19: Let S be a finite set of sites in the plane. A minimum spanning tree of S contains an arc $\{s,t\}$ only if riconv$\{s,t\}$ is an edge of $\mathcal{DT}(S)$.

Proof: Assume the existence of a minimum spanning tree $\mathcal{T}$ with an arc $\{s,t\}$ such that s and t are not adjacent in $\mathcal{DT}(S)$. By Observation 13.10(i), the closed disc $b=\{p \in E^2 | d(p,(s+t)/2) \leq d(s,t)/2\}$ contains at least a third site u, and thus $\max\{d(u,s),d(u,t)\}<d(s,t)$. It follows that $\{s,t\}$ cannot be added as an arc to $\mathcal{T}^*$ by Algorithm 13.7: if s and u are in S^* and t belongs to $S-S^*$ then $c_t \neq s$ since $d(t,u)<d(t,s)$; if s is contained in S^* and t and u belong to $S-S^*$ then $d(u,c_u)<d(t,c_t)$ if $c_t=s$. □

Algorithm 13.7 can thus be improved if the only arcs considered as candidates for inclusion in A each connect two sites adjacent in $\mathcal{DT}(S)$. To allow for an efficient determination of a site s in $S-S^*$ with $d(s,c_s) \leq d(t,c_t)$, for all sites t in $S-S^*$, we organize all sites t of $S-S^*$ in a priority queue $\mathcal{Q}$ which discriminates according to $d(t,c_t)$. Recall that $\mathcal{Q}$ supports the following operations in time $O(\log m)$ if it stores m elements: determine a smallest element and remove it from $\mathcal{Q}$, insert a new element into $\mathcal{Q}$, and delete an element from $\mathcal{Q}$. If c_t changes during the computation, which can happen if an adjacent site s of t is added to S^*, then t is deleted from $\mathcal{Q}$ and reinserted with a new c_t; these two manipulations are attributed to the edge riconv$\{s,t\}$ in $\mathcal{DT}(S)$. Since the number of edges in $\mathcal{DT}(S)$ is linear in the number of vertices, the number of manipulations in $\mathcal{Q}$ is in $O(n)$, with $n=\text{card}S$, which implies the following result.

Theorem 13.20: Let S be a set of n sites in the plane. There is an algorithm that takes $O(n\log n)$ time and $O(n)$ storage for constructing a minimum spanning tree of S.

13.2.6. Shapes of Point Sets

This section introduces a generalization of the convex hull of a finite set of sites in the Euclidean plane. The generalization leads to a parametrized family of subdivisions which seems to capture the intuitive notions of the "fine shape" and the "crude shape" of a set of sites. This is done by showing straight edges connecting sites as well as area of "foreground" juxtaposed against a complementary "background".

The generalization of the convex hull is based on the concept of an α-*disc*, with α any real number. This is

a closed disc of radius $1/\alpha$, if $\alpha > 0$,
a closed half-plane, if $\alpha = 0$, and
the closed complement of a disc of radius $-1/\alpha$, if $\alpha < 0$.

For a finite set S of sites in E^2, an α-disc b is *full* if b contains S. We define the α-*hull of* S as the intersection of all full α-discs. Sample members of the continuous family of α-hulls of S are

the entire plane, for α sufficiently large,
the smallest enclosing disc,
conv$S$, for $\alpha = 0$, and
S itself, for α sufficiently small.

Figure 13.6 shows three α-hulls of the same set of sites, one for $\alpha > 0$ and two for $\alpha < 0$. All members of the family satisfy the following simple relationship.

Observation 13.21: Let S be a finite set of sites in E^2. If $\alpha_1 < \alpha_2$ then the α_1-hull of S is contained in the α_2-hull of S.

A discrete family of so-called α-shapes of S is obtained by straightening the boundaries of the α-hulls. Before introducing α-shapes, we need some additional definitions. Let α be any real number. A site s in S is said to be α-*extreme* if, for any sufficiently small real number $\epsilon > 0$, there is a full $(\alpha + \epsilon)$-disc b with $\text{bd}b \cap S = \{s\}$. The role of ϵ in our definition of extremeness will become more transparent when we consider pairs of sites: it takes care of the special case of an α-disc that contains more than two sites on its boundary. An α-disc b is said *to*

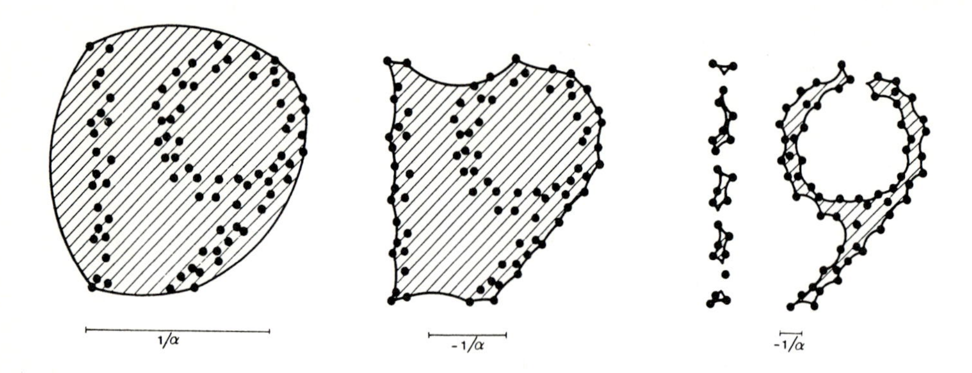

Figure 13.6. Three α–hulls of the same set of sites.

lean on an ordered pair (s,t) of sites if s and t lie on the boundary of b and an infinitesimal movement of b normal to and towards the left of vector $t-s$ puts s and t outside b, see Figure 13.7. Notice that $-2/d(s,t) < \alpha < 2/d(s,t)$, for every α–disc which leans on (s,t). Furthermore,

the center of b is to the left of $t-s$, if $\alpha > 0$,
the interior of half–plane b is to the left of $t-s$, if $\alpha = 0$, and
the center of compl$b$ is to the right of $t-s$, if $\alpha < 0$.

Now, an ordered pair (s,t) of sites is said to be α–*exposed* if, for any sufficiently small real number ϵ, there is a full $(\alpha+\epsilon)$–disc b that leans on (s,t). Figure 13.7 illustrates the concept of α–exposedness for negative and positive values of α. Note that s is 0–extreme if and only if it is in ext$S$, and that (s,t) is 0–exposed if and only if ri conv$\{s,t\}$ is an edge of conv$S$ and no site lies to the right of the line aff$\{s,t\}$ directed from s to t. The following straightforward observations turn out to be crucial for our development of an algorithm that constructs shapes of a set of sites in E^2.

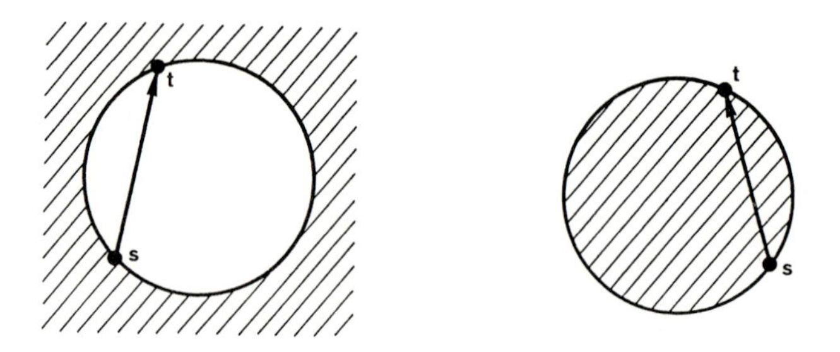

Figure 13.7. α–exposed pairs (s,t).

Observation 13.22: Let S be a finite set of sites in E^2, and let s and t be two sites in S.

(i) The pair (s,t) is α–exposed only if s and t are both α–extreme.

(ii) If s is α–extreme and $\alpha \geq 0$, then there is another site u in S such that (s,u) is α–exposed.

(iii) There is a real number M_s such that s is α–extreme if and only if $\alpha < M_s$.

(iv) There are real numbers $m_{(s,t)}$ and $M_{(s,t)}$ such that (s,t) is α–exposed if and only if $m_{(s,t)} \leq \alpha < M_{(s,t)}$; this includes the possibility that $m_{(s,t)} \geq M_{(s,t)}$.

The *α–shape of S* is a subdivision of E^2 that contains a site s of S as a vertex if s is α–extreme. In addition, it contains an edge $\text{riconv}\{s,t\}$ if (s,t) or (t,s) is α–exposed. A *region of* the α–shape is a maximal connected subset of E^2 reduced by all edges and vertices of the α-shape. We say that a region r is *activated by* an incident edge $e = \text{riconv}\{s,t\}$ if (s,t) is α–exposed, (t,s) is not α–exposed, and r lies to the left of e directed from s to t. Now, r is *active*, or "foreground", if it is activated by all its bounding edges. Otherwise, r is *inactive* or "background". Figure 13.8 shows the α–shapes that correspond to the α–hulls shown in Figure 13.6. We note that a region r of the α–shape of S is activated by an incident edge if and only if r is activated by all its incident edges. This is intuitively clear, since an inactive region r' can be thought of as being carved out of the foreground by moving a full $(\alpha+\epsilon)$–disc. Each pair of sites it touches "deactivates" r', and it touches all pairs of sites that define an edge of r'. Thus, it leaves no edge that can activate r'. To construct the α–shape of S, for a given real number α, we take advantage of the following relationship between α–shapes and Delaunay triangulations.

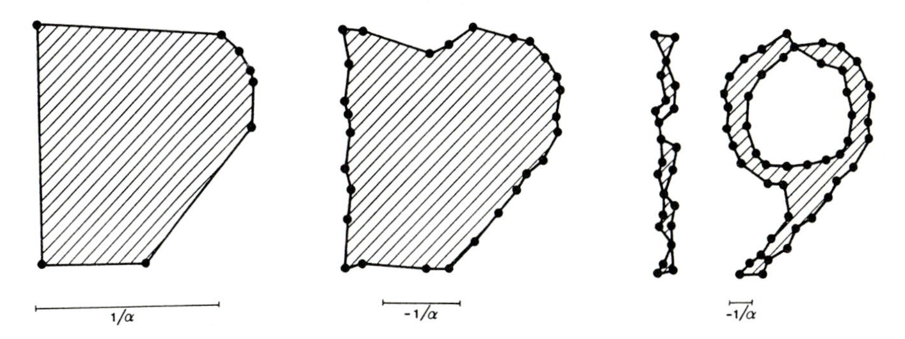

Figure 13.8. Three α–shapes for the same set of sites.

Observation 13.23: Let S be a finite set of sites in E^2. For $\alpha<0$ ($\alpha\geq 0$), e is an edge of the α–shape of S only if e is also an edge of $\mathcal{DT}(S)$ ($\mathcal{DT}_f(S)$).

Observation 13.23 is an immediate consequence of Observation 13.10(i) and the definitions of $\mathcal{DT}_f(S)$ and the α–shape. We elaborate on this relationship which will finally allow us to derive the value of the numbers M_s, $m_{(s,t)}$, and $M_{(s,t)}$ mentioned in Observation 13.22. For purposes of uniformity, we exploit a geometric transformation which identifies E^2 with the plane $x_3=0$ in E^3 and which maps each site $s=(\sigma_1,\sigma_2)$ to its vertical projection $U(s)=(\sigma_1,\sigma_2,\sigma_1^2+\sigma_2^2)$ on the paraboloid U: $x_3=x_1^2+x_2^2$. Define $U(S)=\{U(s)|s\in S\}$ and write $\mathcal{P}$ for the convex hull of the set $U(S)$ of points in E^3. The reader who is not familiar with this transformation and its properties may find it useful to refer to Section 13.2.3 which demonstrates the close relationship between $\mathcal{P}$ and the Delaunay triangulations $\mathcal{DT}(S)$ and $\mathcal{DT}_f(S)$ of the set S of sites in E^2.

Let h be a closed half–space in E^3 such that neither h nor $\mathrm{cl\,compl}h$ contains the paraboloid U. Let $b(h)$ be the vertical projection of $h\cap U$ onto the plane $x_3=0$, and define $\alpha(h)$ such that $b(h)$ is an $\alpha(h)$–disc. By Observation 13.12, $b(h)$ is a full $\alpha(h)$–disc if and only if h contains the polytope $\mathcal{P}$. For each facet f of $\mathcal{P}$, let h_f be the closed half–space which contains $\mathcal{P}$ such that $\mathrm{bd}h_f$ contains f. If $\dim\mathrm{aff}\mathcal{P}=2$, in which case $\mathcal{P}$ has only a single facet f, there are two such half–spaces for f and we treat f like two facets, one for each half–space. Based on the values $\alpha(h_f)$, for all facets f of $\mathcal{P}$, we develop rules which allow us to decide for which values of α a site is α–extreme and for which values of α a directed pair of sites is α–exposed.

We examine the directed pairs of sites first. Let $e=\mathrm{ri\,conv}\{U(s),U(t)\}$ be an edge of $\mathcal{P}$ and let f be an incident facet. We call f the *left (right) facet of* (s,t) if f is to the left (right) of the line $\mathrm{aff}e$ directed from s to t in plane $\mathrm{aff}f$, viewed from outside h_f. We associate (s,t) with the two half–spaces defined by its incident facets and with the two half–spaces h, with $s,t\in\mathrm{bd}h$, which realize the smallest and the largest values of $\alpha(h)$. In particular, we define

$$h_{(s,t),L}=h_f,\text{ with } f \text{ the left facet of } (s,t),$$
$$h_{(s,t),R}=h_g,\text{ with } g \text{ the right facet of } (s,t),$$
$$h_{(s,t),m}\text{ such that } \alpha(h_{(s,t),m})=-2/d(s,t),\text{ and}$$
$$h_{(s,t),M}\text{ such that } \alpha(h_{(s,t),M})=2/d(s,t).$$

Note that $\mathrm{bd}h_{(s,t),m}=\mathrm{bd}h_{(s,t),M}=\mathcal{E}(b)$, for b the closed disc with center $(s+t)/2$ and radius $d(s,t)/2$ and $\mathcal{E}$ the geometric transform discussed in Section 13.2.3. So $h_{(s,t),m}$ and $h_{(s,t),M}$ are well defined which is also true for $h_{(s,t),L}$ and $h_{(s,t),R}$, unless $\mathrm{card}S<2$, in which case $\mathcal{P}$ has no facet and the computation of α-shapes is trivial. We now introduce the set $H_{(s,t)}$ of all half–spaces which contain $\mathcal{P}$ and which have sites s and t in their boundaries. Obviously, $h_{(s,t),L}$ and $h_{(s,t),R}$ belong to $H_{\{s,t\}}$. Furthermore, $H_{\{s,t\}}$ contains both $h_{(s,t),m}$ and $h_{(s,t),M}$ only if $\dim \operatorname{aff} P = 2$ and the boundaries of all four half-spaces defined for (s,t) coincide. We can now assign values to the numbers $m_{(s,t)}$ and $M_{(s,t)}$ (see Observation 13.22(iv)) which delimit the interval of values of α such that (s,t) is α-exposed. The values of $m_{(s,t)}$ and $M_{(s,t)}$ depend on whether or not $H_{\{s,t\}}$ contains the half-spaces $h_{(s,t),m}$ and $h_{(s,t),M}$.

Lemma 13.24: Let S be a finite set of at least three sites in E^2, define $P = \operatorname{conv} U(S)$, and let $e = \operatorname{ri} \operatorname{conv}\{U(s), U(t)\}$ be an edge of P. The ordered pair (s,t) of sites is α-exposed if and only if $m_{(s,t)} \leq \alpha < M_{(s,t)}$, with

(i) $m_{(s,t)} = \alpha(h_{(s,t),m})$ and $M_{(s,t)} = \alpha(h_{(s,t),L})$ if $h_{(s,t),m} \in H_{\{s,t\}}$,

(ii) $m_{(s,t)} = \alpha(h_{(s,t),R})$ and $M_{(s,t)} = \alpha(h_{(s,t),M})$ if $h_{(s,t),M} \in H_{\{s,t\}}$,

and

(iii) $m_{(s,t)} = \alpha(h_{(s,t),R})$ and $M_{(s,t)} = \alpha(h_{(s,t),L})$ if $H_{\{s,t\}}$ contains neither $h_{(s,t),m}$ nor $h_{(s,t),M}$.

Proof: Let $A_{\{s,t\}}$ contain all closed half-spaces h with e in $\operatorname{bd} h$, including those that do not contain P. Recall that $b(h)$ is an $\alpha(h)$-disc in the plane defined by the equation $x_3 = 0$ with s and t in $\operatorname{bd} b(h)$. To argue about $A_{\{s,t\}}$, we use a plane g normal to line $\operatorname{aff} e$ and endow it with a Cartesian system of coordinates with the origin at $g \cap \operatorname{aff} e$. For a point (or vector) v in g with unit distance from the origin, we define the half-space $h(v)$: $\langle v, x \rangle \geq \langle v, U(s) \rangle$ in $A_{\{s,t\}}$. Let C be the circle of all points v in g with unit distance from the origin, and let v_L, v_R, v_m, and v_M be the points on C such that $h(v_L) = h_{(s,t),L}$, $h(v_R) = h_{(s,t),R}$, $h(v_m) = h_{(s,t),m}$, and $h(v_M) = h_{(s,t),M}$. Obviously, $v_m = -v_M$ in any case, and $v_L = -v_R$ only if $\dim \operatorname{aff} P = 2$. Figure 13.9 displays the circle C and the points v_L, v_R, v_m, and v_M. The view taken in Figure 13.9 is normal to g such that $U(t)$ lies further in the background than $U(s)$; it follows that the counterclockwise angle from v_R to v_L does not exceed π. We identify two intervals of points v on C such that (s,t) is $\alpha(h(v))$-exposed if and only if v belongs to both intervals.

1. Half-space $h(v)$ contains P if and only if point v belongs to the interval from v_R to v_L.
2. Disc $b(h(v))$ leans on (s,t) if and only if point v belongs to the interval from v_m to v_M.

Figure 13.9 displays the four different counterclockwise sequences that can be realized by points v_L, v_R, v_m, and v_M: v_R, v_m, v_L, v_M in (a), v_R, v_M, v_L, v_m in (b), v_R, v_L, v_M, v_m in (c), and v_R, v_L, v_m, v_M in (d). We notice that the sequences v_R, v_m, v_M, v_L and v_R, v_M, v_m, v_L are impossible since the angle from v_R to v_L cannot be larger than π. The intersections of the intervals identified in 1 and 2 are

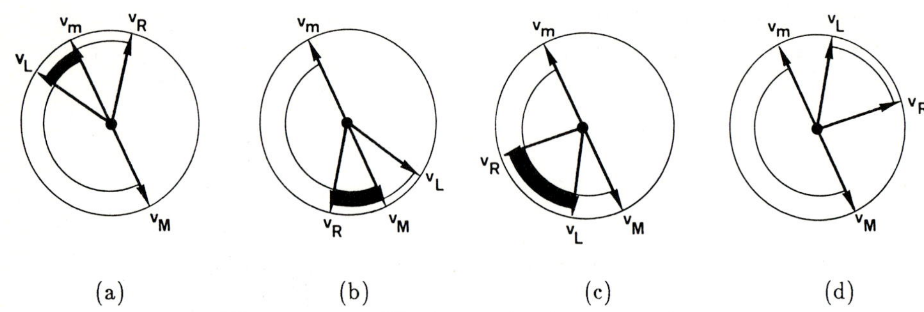

Figure 13.9. The intervals of α–exposedness.

thus from v_m to v_L in (a), from v_R to v_M in (b), from v_R to v_L in (c), and empty in (d). The assertion thus follows since $\alpha(h(v))$ is monotonically increasing from v_m to v_M, and since the definition of α–exposedness ensures that the intervals of α–exposedness are closed at their lower bound and open at their upper bound. □

To calculate the interval of values for α such that a site is α–extreme requires fewer cases than for α–exposed ordered pairs.

Lemma 13.25: Let S be a finite set of at least three sites in the plane, and define $P = \text{conv}\, U(S)$. Site s is α–extreme if and only if $\alpha < M_s$, where M_s is the maximum of $\{M_{(s,t)} | t \in S - \{s\}$ and $M_{(s,t)} > m_{(s,t)}\}$.

Proof: By Observation 13.22(i), s is α–extreme if $\alpha < M_s$, with M_s as defined above. As such, assume that s is also α–extreme for some $\alpha \geq M_s$. By definition, there is a full $(\alpha+\epsilon)$–disc b with $\{s\} = \text{bd}\, b \cap S$, provided $\epsilon > 0$ is small enough. However, as b pivots in the counterclockwise direction about s, the boundary of b encounters another site t; otherwise, there is no real number β, and no site u in S, such that (s,u) is β–exposed. However, (s,t) is thus α–exposed which contradicts the assumption. □

Lemmas 13.24 and 13.25 suggest the following algorithm for computing α–shapes of S:

Algorithm 13.8 (α–shape):

Step 1: Map each site s in S to the point $U(s)$ in E^3 and construct $P = \text{conv}\, U(S)$.

Step 2: For each facet f of P, compute $\alpha(h_f)$.

Step 3: For each edge $\text{riconv}\{U(s), U(t)\}$ of P, compute the two *activity intervals* $I_{(s,t)} = [m_{(s,t)}, M_{(s,t)})$ and $I_{(t,s)} = [m_{(t,s)}, M_{(t,s)})$ (see Lemma 13.24).

Step 4: For each vertex $U(s)$ of P, compute the *activity interval*

$I_s = (-\infty, M_s)$ (see Lemma 13.25).

Step 5: Determine all activity intervals that contain α and project the corresponding edges and vertices vertically onto the plane given by $x_3 = 0$. For each region created, test an edge to determine whether it is "foreground" or "background".

Theorem 8.16 implies that Step 1 takes $O(n \log n)$ time and $O(n)$ storage, with $n = \text{card}S$. We argue that time $O(n)$ suffices for Steps 2 through 5. Step 2 involves trivial calculations that take constant time for each of at most $2n-4$ facets. Each of the at most $3n-6$ edges can be treated in constant time in Step 3, since a half-space h with e in bd$h$ contains P if and only if it contains a third vertex of both incident facets of e. This allows us to discriminate between the three cases of Lemma 13.24. In Step 4, a vertex needs time proportional to the number of incident edges. The sum of these numbers taken over all vertices of P is equal to twice the number of edges of P, which implies that it is linear in n. Finally, Step 5 checks each one of at most $7n-12$ intervals in constant time. We remark that the upper bounds on the number of edges and activity intervals hold unless $n < 3$.

Note that Steps 1 through 4 of the algorithm do not depend on the choice of α. To formalize this idea, we call P the *α-spectrum of S* if it is augmented with the values $\alpha(h_f)$ and the activity intervals as described in Steps 2 through 4.

Theorem 13.26: Let S be a set of n sites in E^2.

(i) There is an algorithm that constructs the α-spectrum of S in $O(n \log n)$ time and $O(n)$ storage.

(ii) For a given real number α, the α-shape of S can be computed in time $O(n)$ from the α-spectrum.

13.3. Higher-Order Voronoi Diagrams

This section introduces a generalization of ordinary Voronoi diagrams for a finite set of sites to so-called higher-order Voronoi diagrams. The higher-order Voronoi diagrams of a finite set S of sites in E^d form a discrete one-parametric family of cell complexes, where the parameter k assumes all integer values between 1 and card$S-1$. Intuitively, the diagram for parameter k associates each subset of S which has cardinality k with the cell of points p such that the k sites in the subset are the k closest sites of p. We will see that the majority of subsets of size k define empty cells, unless k is very small or very large. We begin with a formal definition of higher-order Voronoi diagrams.

Let S be a set of n sites in E^d, and, for $1 \leq k \leq n-1$, let S_k be the collection of all subsets T of S with $\text{card}\, T = k$. For every point p in E^d, we define the distance from p to T as $d_T(p) = \max\{d(p,s) | s \in T\}$, and we let $C_k(p)$ denote the subset of S_k that contains all sets T with minimum $d_T(p)$. If $\text{card}\, C_k(p) = 1$ then the only set in $C_k(p)$ contains the k sites of S which are closest to p. Recall the definitions of the hyperplane

$$b_{\{s,t\}} = \{p \in E^d | d(p,s) = d(p,t)\}$$

and of the half-space

$$b_{(s,t)} = \{p \in E^d | d(p,s) < d(p,t)\},$$

for s and t two sites in S. By the following result, the set of points p with identical $C_k(p)$ defines the relative interior of a convex polyhedron in E^d.

Observation 13.27: Let S be a set of n sites in E^d, let k be a positive integer smaller than n, and let C_f be a subset of S_k. We define I as the intersection and J as the union of all sets of C_f, and we let f be the set of all points p with $C_k(p) = C_f$. Then

$$f = \left(\bigcap_{s \in J, t \in S-J} b_{(s,t)} \right) \cap \left(\bigcap_{s \in I, t \in J-I} b_{(s,t)} \right) \cap \left(\bigcap_{s,t \in J-I} b_{\{s,t\}} \right).$$

We call two points p and q *k-equivalent* if $C_k(p) = C_k(q)$. By Observation 13.27, every equivalence class f is a convex polyhedron, and we call f an *i-face* if $\dim \text{aff} f = i$. For $1 \le k \le n-1$, the *k^{th}-order Voronoi diagram* $\mathcal{V}_k(S)$ *of* S consists of all i-faces, $0 \le i \le d$, defined by the concept of k-equivalence. Note that $\mathcal{V}_1(S)$ coincides with the ordinary Voronoi diagram $\mathcal{V}(S)$. Most of the terminology introduced in Section 13.1 for ordinary Voronoi diagrams can be extended to k^{th}-order Voronoi diagrams. Figure 13.10 shows the 2nd- and the 3rd-order Voronoi diagrams for the set of 11 sites also shown in Figure 13.1. To generalize Observation 13.2, we define, for every point p in E^d, the open ball $b_k(p) = \{x \in E^d | d(p,x) < d_T(p),\ T \in C_k(p)\}$; so, $b_k(p)$ is the largest open ball centered at p that contains at most $k-1$ sites of S. Obviously, each set T in $C_k(p)$ contains $S \cap b_k(p)$, and each site in $S \cap \text{bd}\, b_k(p)$ belongs to at least one set T in $C_k(p)$. For each face f of $\mathcal{V}_k(S)$, we define $C_f = C_k(p)$, for p any point of f.

Observation 13.28: Let S be a finite set of sites in E^d and let f be an i-face of $\mathcal{V}_k(S)$.

(i) $\text{card}\, C_f = 1$ if and only if f is a cell.

(ii) $\text{card}\, C_f = 2$ if and only if f is a facet.

(iii) $\text{card}\, C_f \ge d+1-i$, for $0 \le i \le d-2$, and equivalently

(iv) a point p is contained in an i-face of the diagram $\mathcal{V}_k(S)$ only if $\text{card}(S \cap \text{bd}\, b_k(p)) \ge d+1-i$, $0 \le i \le d-2$.

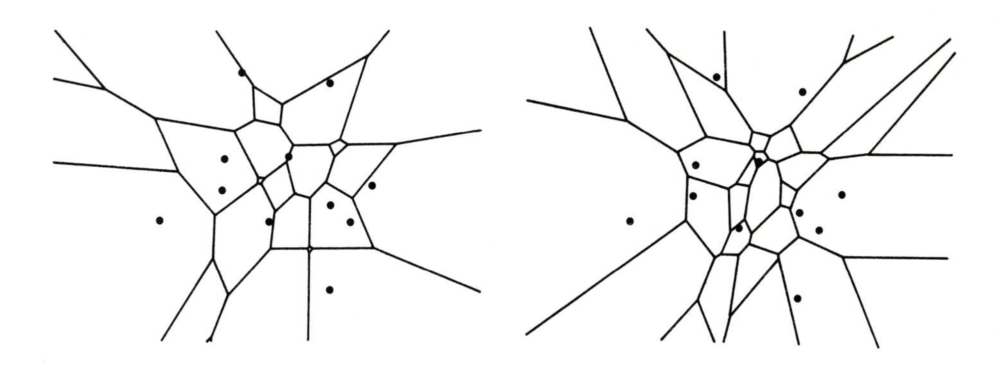

Figure 13.10. 2nd– and 3nd–order Voronoi diagrams.

We use the geometric transform $\mathcal{E}$ introduced in Section 13.1 to relate k^{th} order Voronoi diagrams in E^d with certain structures in arrangements of hyperplanes in E^{d+1}. The transformation identifies E^d with the hyperplane $x_{d+1}=0$ in $d+1$ dimensions, and it maps each site $s=(\sigma_1,\sigma_2,...,\sigma_d)$ to the hyperplane $\mathcal{E}(s)$: $x_{d+1}=2\sigma_1x_1+2\sigma_2x_2+...+2\sigma_dx_d-(\sigma_1^2+\sigma_2^2+...+\sigma_d^2)$ in E^{d+1} which touches the paraboloid U: $x_{d+1}=x_1^2+x_2^2+...+x_d^2$ in the vertical projection $U(s)$ of s onto U. By Observation 13.3, which forms the basis of the relationship between ordinary Voronoi diagrams and intersections of half–spaces, we have the following more general relationship between higher–order Voronoi diagrams of S and certain structures in the arrangement $\mathcal{A}(\mathcal{E}(S))$.

Observation 13.29: Let S be a finite set of sites in E^d (identified with $x_{d+1}=0$ in E^{d+1}), define the set of hyperplanes $H=\mathcal{E}(S)$, and let $F_k(H)$ contain all cells in the arrangement $\mathcal{A}(H)$ which lie below k and above $n-k$ hyperplanes of H.

(i) For each cell (that is, each d–face) c in $\mathcal{V}_k(S)$ there is a cell (that is, $(d+1)$–face) c' in $F_k(H)$, and vice versa, such that c is the vertical projection of c' onto E^d.

(ii) For each i–face f in $\mathcal{V}_k(S)$, $0\leq i\leq d-1$, there is an i–face f' in the boundary of at least two cells in $F_k(H)$, and vice versa, such that f is the vertical projection of f' onto E^d.

Figure 13.11 illustrates Observation 13.29 for $d=1$: the four sites in E^1 are the same as shown in Figure 13.2; the 2nd–order Voronoi diagram is obtained by a vertical projection of the shaded cells. Notice that there is no overlap in the vertical projection of any two cells in $F_k(H)$, and that a face belongs to the common boundary of at least two such cells if and only if it lies below at most $k-1$ and above at most $n-k-1$ hyperplanes of H. It follows that a face is in the

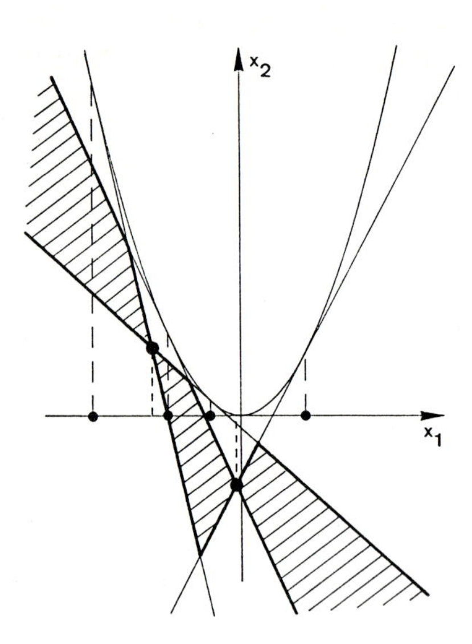

Figure 13.11. 2nd–order Voronoi diagram via projection.

common boundary of at least two cells in $F_k(H)$ if and only if it is in the intersection of the k–level and the $(k+1)$–level of the arrangement $\mathcal{A}(H)$; see Chapter 3 for a definition of levels in arrangements. The boundary of all cells in $\mathcal{V}_k(S)$ is therefore the vertical projection onto E^d of the intersection of the k–level and the $(k+1)$–level of $\mathcal{A}(H)$. Recall that $a(p)$, $o(p)$, and $b(p)$ denote the numbers of hyperplanes h in H with point p in h^-, in h, and in h^+, respectively. Similarly, $a(f)=a(p)$, $o(f)=o(p)$, and $b(f)=b(p)$, for f any face in $\mathcal{A}(H)$ and any point p in f.

The one–to–one correspondence between the cells of $\mathcal{V}_k(S)$ and the cells c of $\mathcal{A}(H)$ which satisfy $a(c)=k$ implies an upper bound on the number of cells of $\mathcal{V}_k(S)$ in terms of the number of k–sets of a related set of points in E^{d+1}. This can be seen as follows: associate each cell c in $F_k(H)$ with the set of hyperplanes h in H for which $c \subseteq h^-$; if we now dualize the hyperplanes of H to points using the dual transform $\mathcal{D}$ introduced in Section 1.4, then Observation 1.5 implies that each such set of k hyperplanes corresponds to a unique k–set of the point set $P=\mathcal{D}(H)=U(S)$. Since $e_k^{(d+1)}(n)$ denotes the maximum number of k–sets of any set of n points in E^{d+1}, $\mathcal{V}_k(S)$ contains no more than $e_k^{(d+1)}(n)$ cells, for S a set of n sites in E^d. This bound is loose, even in the asymptotic sense (at least for $d=1$ or 2), since P is a rather special set of points, that is, because each point of P is extreme. The next section elaborates on this observation and develops a tighter bound on the number of cells of higher–order Voronoi diagrams in two dimensions.

The relationship between higher–order Voronoi diagrams in E^d and arrangements in E^{d+1} can be used for the design of an algorithm which constructs higher–order Voronoi diagrams. The results of Chapter 7 on constructing arrangements suggest the following strategy for constructing all k^{th}–order Voronoi diagrams of S, for $1 \leq k \leq n-1$, where $n = \text{card}S$:

Algorithm 13.9 (All higher–order Voronoi diagrams):

Step 1: Map each site s in S to the hyperplane $\mathcal{E}(s)$ and construct the arrangement $\mathcal{A}(H)$, with $H = \mathcal{E}(S)$.

Step 2: For each face f in $\mathcal{A}(H)$, compute $a(f)$ and $b(f)$.

Step 3: Compute each higher–order Voronoi diagram in turn as follows:

for $k := 1$ **to** $n-1$ **do**

Project the cells c, with $a(c) = k$, and the i–faces f $(i \leq d-1)$, with $a(f) \leq k-1$ and $b(f) \leq n-k-1$, vertically onto the hyperplane $x_{d+1} = 0$; this yields $\mathcal{V}_k(S)$.

endfor.

By Theorem 7.6, the arrangement $\mathcal{A}(H)$ can be constructed in time $O(n^{d+1})$, if H is a set of n hyperplanes in E^{d+1}. The computation of functions $a(f)$ and $b(f)$, for all faces f of $\mathcal{A}(H)$, takes time $O(n^{d+1})$ if we use the same strategy as applied in Algorithm 12.6 outlined in Chapter 12. Finally, Step 3 distinguishes between the nodes of the incidence graph of $\mathcal{A}(H)$ which represent faces of $\mathcal{V}_k(S)$ and the nodes which do not; each node can thus be tackled in constant time as Step 2 provides the necessary preprocessing. The subgraph of these nodes is an appropriate presentation of $\mathcal{V}_k(S)$. To affix the diagram in space, each facet f stores the two sites s and t such that f belongs to the perpendicular bisector $b_{\{s,t\}}$ of s and t. This implies the following result.

Theorem 13.30: Let S be a set of n sites in E^d. There is an algorithm that constructs $\mathcal{V}_k(S)$, for all k from 1 to $n-1$, in time and storage $O(n^{d+1})$.

If $\mathcal{V}_k(S)$, for some fixed k, is to be constructed, the expensive computation of $\mathcal{A}(H)$ can be avoided by the methods presented in Chapter 9 (see Algorithm 9.2). We will return to this result in Section 13.5 which discusses the case $d = 2$.

13.4. The Complexity of Higher–Order Voronoi Diagrams

This section offers asymptotically tight bounds on the number of faces of kth–order Voronoi diagrams in the Euclidean plane. Let S be a set of n sites in E^2 and define $H = \mathcal{E}(S)$. To derive an upper bound on the number of faces of $\mathcal{V}_k(S)$, for $1 \leq k \leq n-1$, we can assume that arrangement $\mathcal{A}(H)$ in E^3 is simple. As such, every vertex of $\mathcal{A}(H)$ belongs to exactly three planes, every edge of $\mathcal{A}(H)$ is common to exactly two planes, and no two lines in $\mathcal{A}(H)$ are parallel. Observation 13.29 can be reformulated for this case as follows:

> for each vertex v, edge e, and region r of $\mathcal{V}_k(S)$, there is a vertex v' of $\mathcal{A}(H)$ with $k-2 \leq a(v') \leq k-1$, an edge e' of $\mathcal{A}(H)$ with $a(e')=k-1$, and a cell c' of $\mathcal{A}(H)$ with $a(c')=k$, and vice versa, such that v, e, and r are the vertical projections of v', e', and c' onto the plane given by the equation $x_3=0$, respectively.

We let $f_i(H,k)$ be the number of i-faces f in $\mathcal{A}(H)$ with $a(f)=k$, for $0 \leq i \leq 3$. Notice that $\mathcal{V}_k(S)$ contains $f_0(H,k-1)+f_0(H,k-2)$ vertices, $f_1(H,k-1)$ edges, and $f_3(H,k)$ regions. By the special structure of $\mathcal{A}(H)$ (each plane in H supports a facet f with $a(f)=0$), $f_i(H,k)$ cannot grow as fast with the cardinality of H as the corresponding numbers in general three-dimensional arrangements (see Theorems 3.11 and 3.3). We give upper and lower bounds on $f_i(H,k)$, $i=0,1,3$, after examining the special structure of $\mathcal{A}(H)$.

Let c be a cell in $\mathcal{A}(H)$ and write k for $a(c)$. We define the *upper* (*lower*) *skeleton of* c to contain all edges e in bd$c$ with $a(e)=k-2$ ($a(e)=k$) together with all incident vertices (see Figure 13.12). Notice that the upper (lower) skeleton of c projects vertically onto some edges and vertices of $\mathcal{V}_{k-1}(S)$ ($\mathcal{V}_{k+1}(S)$), if it exists. The upper or lower skeleton t of any cell c in $\mathcal{A}(H)$ defines, in a natural way, an undirected graph $\mathcal{G}_t(N,A)$: each vertex of t is a node in N, and an arc $\{v,w\}$ is in A if t contains an edge with endpoints v and w. For convenience, we let each unbounded edge have two endpoints, one at infinity. We prove the following result on the structure of the graph $\mathcal{G}_t$.

Lemma 13.31: Let S be a set of n sites in E^2, and define $H=\mathcal{E}(S)$.

(i) For t the upper or lower skeleton of any cell c in $\mathcal{A}(H)$, with $1 \leq a(c) \leq n-1$, $\mathcal{G}_t$ is a possibly empty connected graph without cycles.

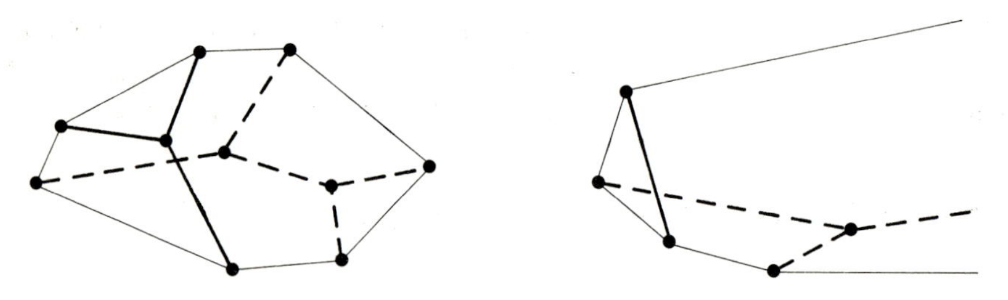

Figure 13.12. Upper and lower skeletons.

(ii) $\mathcal{G}_t$ is empty if and only if $a(c)=1$ and t is the upper skeleton of c, or if $a(c)=n-1$ and t is the lower skeleton of c.

Proof: By virtue of the construction of H, each plane h in H touches the paraboloid U: $x_3=x_1^2+x_2^2$ in a point p; note that $a(p)=0$. Let f be a facet of $\mathcal{A}(H)$ contained in h, let e be an edge incident upon f, and let $g \neq h$ be the plane in H such that e belongs to $g \cap h$. If the half-space g^+ above g contains f then $a(e)=a(f)$, and $a(e)=a(f)-1$ if f is contained in the half-space g^- below g. Since g^+ contains point p, it contains f if and only if p and f lie on the same side of $g \cap h$ in h. Observe that f is a convex region in plane h, and that we can draw two lines through point p which are tangent to f, unless f contains p or all intersections of h with other planes in H separate f and p. The two points where the tangents touch the boundary of f divide the edges incident upon f into two categories. We have $a(e)=a(f)-1$, for all edges e that lie on the path which separates p from f, and $a(e)=a(f)$, for all other edges e. Consequently, $a(e)=a(f)$, for all edges e incident upon f, if and only if f contains p which is equivalent to $a(f)=0$. Notice that f is thus the only facet f' of a cell c such that $a(c)=1$ and $a(f')=0$. Similarly, $a(e)=a(f)-1$, for all edges e incident upon f, if and only if $a(f)=n-1$; in this case, f is the only facet f' of a cell c such that $b(c)=1$ and $b(f')=0$. This proves part (ii) of Lemma 13.31.

Without loss of generality, let t be a non-empty upper skeleton of some cell c with $1 \leq a(c) \leq n-1$. A cycle in $\mathcal{G}_t$ implies the existence of a facet f such that f is below the planes spanned by any facet of c which shares an edge with f. Therefore, $a(f)>0$ and $a(e)=a(f)-1$, for all edges e incident upon f (see Figure 13.13(a)). By the above argument, f is thus the only facet of a cell $\bar{c}$ such that $a(f)=a(\bar{c})=n-1$. This contradicts $a(c)=a(\bar{c})+1 \leq n-1$. If $\mathcal{G}_t$ is not connected, then t falls into several connected components and there is a facet f incident upon edges of at least two components (see Figure 13.13(b)). Then the edges e incident upon f with $a(e)=a(f)-1$ do not induce a connected path. This is again a contraction to the above argument. Analogously, we can prove the same for lower skeletons. □

We note that $\mathcal{G}_t$ is a tree even if t is the upper skeleton of the bottommost cell c, that is, $a(c)=n$. Trivially, $\mathcal{G}_t$ is empty if t is the upper skeleton of the topmost cell or the lower skeleton of the bottommost cell. Therefore, $\mathcal{G}_t$ is a non-empty graph with cycles only if t is the lower skeleton of the topmost cell. In this case, t corresponds to the boundary of the regions in $\mathcal{V}(S)$. The remainder of this section exploits the fact that the lower and upper skeletons of all cells c in $\mathcal{A}(H)$ induce trees, unless $a(c)=0$. Define $\boldsymbol{u}(H,k)$ as the number of unbounded cells c in $\mathcal{A}(H)$ with $a(c)=k$, for $1 \leq k \leq n-1$. We prove the following relation between the functions $\boldsymbol{f}_0$, $\boldsymbol{f}_1$, $\boldsymbol{f}_3$, and $\boldsymbol{u}$, which count faces of $\mathcal{A}(H)$. For convenience, we omit H as an argument of these functions as it will be

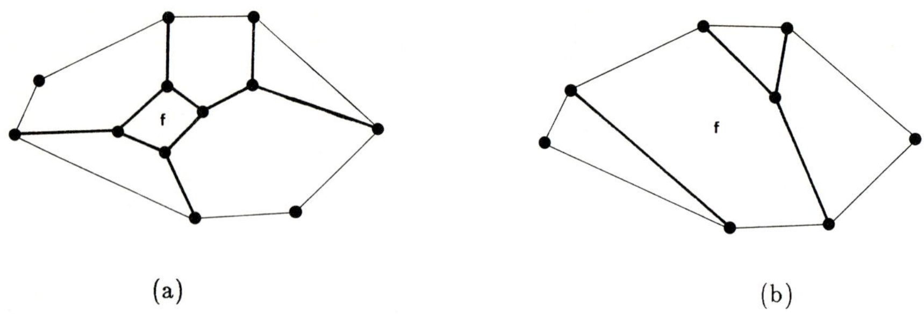

(a) (b)

Figure 13.13. Impossible upper and lower skeletons.

understood from the context throughout the following discussion.

Lemma 13.32: Let S be a set of n sites in E^2 such that the arrangement $\mathcal{A}(H)$ is simple, where $H = \mathcal{E}(S)$. Then

(i) $f_1(k-1) = 3f_3(k) - u(k) - 3$, and

(ii) $f_0(k-1) = \sum_{i=1}^{k} (-1)^{k-i}(2f_3(i) - u(i) - 2)$, for $1 \leq k \leq n-1$.

Proof: The vertical projection onto the plane given by the equation $x_3 = 0$ of all vertices v with $a(v) = k-1$ or $k-2$, all edges e with $a(e) = k-1$, and all cells c with $a(c) = k$ yields the kth-order Voronoi diagram $\mathcal{V}_k(S)$. There are $u(k)$ unbounded regions and edges in $\mathcal{V}_k(S)$, and, since $\mathcal{A}(H)$ is simple, each vertex is incident upon exactly three edges. To be able to use Euler's relation for planar graphs (see Theorem 11.1), we assume that all unbounded edges of $\mathcal{V}_k(S)$ have one common endpoint at infinity. Thus, there are $f_0(k-1)+f_0(k-2)+1$ vertices, $f_1(k-1)$ edges, and $f_3(k)$ regions in $\mathcal{V}_k(S)$, which implies that

$$f_0(k-1) + f_0(k-2) - f_1(k-1) + f_3(k) = 1.$$

The next formula is obtained by counting the number of incidences between the edges and the vertices of $\mathcal{V}_k(S)$ in two different ways: an edge contributes two incidences, the vertex at infinity contributes $u(k)$ incidences, and every other vertex contributes three incidences. This leads to

$$2f_1(k-1) = 3f_0(k-1) + 3f_0(k-2) + u(k).$$

Part (i) of Lemma 13.32 follows by a combination of both formulas, as does

$$f_0(k-1) + f_0(k-2) = 2f_3(k) - u(k) - 2.$$

Part (ii) of the assertion is now a consequence of the trivial relation

$$f_0(k-1) = (f_0(k-1) + f_0(k-2)) - (f_0(k-2) + f_0(k-3)) + \ldots \pm f_0(0).$$

□

The following identity leads to a recurrence relation for $f_3(k)$ which finally produces the main result of this section. We define $f_0(k)=f_1(k)=0$ if $k<0$.

Lemma 13.33: Let S be a set of n sites in E^2 such that the arrangement $\mathcal{A}(H)$ is simple, where $H=\mathcal{E}(S)$. Then $f_3(k)=f_1(k-2)-2f_0(k-3)$, for $1 \leq k \leq n-1$.

Proof: For a cell c in $\mathcal{A}(H)$, let e_c be the number of edges e in its upper skeleton; so $a(e)=a(c)-2$. Let v_c be the number of vertices v in the boundary of c such that $a(v)=a(c)-3$; these vertices are exactly the vertices incident upon three edges of the upper skeleton t of c. By Lemma 13.31, the corresponding graph $\mathcal{G}_t$ is a tree with v_c inner nodes, each of which has degree 3. It follows that $e_c=2v_c+1$. Now

$$f_1(k-2)=\sum_{a(c)=k} e_c=2f_0(k-3)+f_3(k),$$

which implies the assertion. □

We derive the recurrence relation

$$f_3(k)=2f_3(k-1)-f_3(k-2)-\boldsymbol{u}(k-1)+\boldsymbol{u}(k-2)-2$$

by substituting the identities of Lemma 13.32 into those of Lemma 13.33 and by adding $f_3(k)$ and $f_3(k-1)$. To initialize the iteration, we note that $f_3(1)=n$ and $f_3(2)=3n-3-\boldsymbol{u}(1)$, by construction of H. Using induction, it is now straightforward to establish the following result.

Theorem 13.34: Let S be a set of n sites in E^2 with $\mathcal{A}(H)$ simple, where $H=\mathcal{E}(S)$. Then

$$f_3(k)=(2k-1)n-(k^2-1)-\sum_{i=1}^{k-1}\boldsymbol{u}(i).$$

It is worth noting that Lemmas 13.31 through 13.33 and Theorem 13.34 also hold for more general arrangements $\mathcal{A}(H)$. The planes in H need not touch the paraboloid U; it suffices that each plane supports a facet f of $\mathcal{A}(H)$ with $a(f)=0$.

The bounds on the number of faces in $\mathcal{V}_k(S)$ that follow from our calculations are stated below. Note that arrangement $\mathcal{A}(H)$ is simple if and only if no three sites of S are collinear and no four are cocircular. A region r in $\mathcal{V}_k(S)$ is unbounded if and only if the k sites closest to any point in r define a k-set of S, $1 \leq k \leq n-1$. Chapter 3 writes $\boldsymbol{e}_k(S)$ for the number of k-sets of S; so $\boldsymbol{u}(k)=\boldsymbol{e}_k(S)$ if no three sites are collinear. Theorem 13.34 and Lemma 13.32 now yield formulas for the number of regions, edges, and vertices of $\mathcal{V}_k(S)$ in terms of the number of k-sets of S.

Corollary 13.35: Let S be a set of n sites in E^2 with no three sites collinear and no four sites cocircular. The kth–oder Voronoi diagram $\mathcal{V}_k(S)$ consists of

$$(2k-1)n-(k^2-1)-\sum_{i=1}^{k-1} e_i(S) \text{ regions,}$$

$$(6k-3)n-3k^2-e_k(S)-3\sum_{i=1}^{k-1} e_i(S) \text{ edges, and}$$

$$(4k-2)n-2k^2-e_k(S)-2\sum_{i=1}^{k-1} e_i(S) \text{ vertices, for } 1 \leq k \leq n-1.$$

By Corollary 3.15, $\sum_{i=1}^{k-1} e_i(S) \leq (k-1)n$ if $k-1 < n/2$, with the equality satisfied if all sites in S are extreme. Since

$$\sum_{i=1}^{n-1} e_i(S) = n(n-1),$$

if no three sites are collinear, and $e_i(S) = e_{n-i}(S)$, we also have

$$\sum_{i=1}^{k-1} e_i(S) = n(n-1) - \sum_{i=1}^{n-k} e_i(S) \geq (k-1)n$$

if $k > n/2$, and the bound is again tight. These results imply optimal bounds on the number of regions in $\mathcal{V}_k(S)$.

Corollary 13.36: Let S be a set of n sites in E^2.

(i) If $k > n/2$ then $\mathcal{V}_k(S)$ contains at most $kn-k^2+1$ regions, and this is tight.

(ii) If no three sites are collinear, no four sites are cocircular, and if $k-1 < n/2$, then $\mathcal{V}_k(S)$ contains at least $kn-k^2+1$ regions, and this is tight.

Interestingly enough, the number of regions of $\mathcal{V}_k(S)$ is thus exactly $kn-k^2+1$ if n is odd, if $k=(n+1)/2$, and if no three sites in S are collinear and no four sites are cocircular.

13.5. Constructing Higher–Order Voronoi Diagrams

This section briefly addresses the issue of computing the kth–order Voronoi diagram, for some fixed value of k. Let S be a set of n sites in E^2 identified with $x_3=0$ in E^3, and let H be the set of planes in E^3 obtained by application of the transform $\mathcal{E}$. By Corollary 13.35, the kth-order Voronoi diagram $\mathcal{V}_k(S)$ of S, for $1 \leq k \leq n-1$, consists of $O(k(n-k))$ regions, edges, and vertices. By Observation 13.29, the edges and vertices of $\mathcal{V}_k(S)$ are in one–to–one correspondence with the edge and vertices of $\mathcal{A}(H)$ that belong to the k-level and the $(k+1)$-level of $\mathcal{A}(H)$. These edges, however, define a skeleton in $\mathcal{A}(H)$ which can be constructed in time $O(\sqrt{n}\log n)$ per edge (see Chapter 9, Theorem 9.2 and Theorem 9.13). To commence the construction, we first need to determine an initial vertex of $\mathcal{V}_k(S)$. To this end, we make use of the fact that a vertex v of $\mathcal{A}(H)$ belongs to both the k-level and the $(k+1)$-level of $\mathcal{A}(H)$ if and only if $a(v) \leq k-1$ and $b(v) \leq n-k-1$. Since the algorithm in Chapter 9 simulates a perturbation that eliminates all degeneracies, we can assume that $\mathcal{A}(H)$ is simple.

Algorithm 13.10 (Initial vertex of $\mathcal{V}_k(S)$):

- **Step 1:** Determine two sites s and t which appear as consecutive vertices of conv$S$. Compute planes $g = \mathcal{E}(s)$ and $h = \mathcal{E}(t)$, and line $l = g \cap h$.
- **Step 2:** Determine plane f in $H-\{g,h\}$ that intersects line l in point p with $a(p) = k-1$.
- **Step 3:** Compute the intersection v of planes f, g, and h; v is a vertex of the skeleton that projects onto the boundary of all regions in $\mathcal{V}_k(S)$.

Note that the choice of planes g and h in Step 1 guarantees that the unbounded edges e and $\bar{e}$ on l satisfy $a(e)=0$ and $b(\bar{e})=0$. It follows that l contains a vertex v_0 with $a(v_0)=0$ and a vertex v_{n-3} with $b(v_{n-3})=0$. We have $a(v_{n-3}) = n-3$ since $a(v)+b(v) = n-3$ for every vertex v. Consequently, $a(v_i) = i$, for $0 \leq i \leq n-3$, if $v_0, v_1, \ldots, v_{n-3}$ is the sorted sequence of intersections of l and all planes in $H-\{g,h\}$. There are standard algorithms for computing the kth-smallest member of some collection of n elements, drawn from a totally ordered set, in time $O(n)$. Algorithm 13.10 can be implemented so that it takes time $O(n)$ if we use such an algorithm as a subroutine to find the kth vertex on line l. We conclude with the following result on the complexity of constructing k^{th}-order Voronoi diagrams in E^2.

Theorem 13.37: Let S be a set of n sites in E^2. For each integer $1 \leq k \leq n-1$, there is an algorithm which constructs the kth-order Voronoi diagram $\mathcal{V}_k(S)$ of S in $O(k(n-k)\sqrt{n}\log n)$ time and $O(k(n-k))$ storage.

Application of kth-order Voronoi diagrams use the fact that, for a point p, the region of $\mathcal{V}_k(S)$ that contains p determines the k closest sites of p. If p belongs to an edge or vertex of $\mathcal{V}_k(S)$ then the kth-closest site is not unique. If p

13.8. Bibliographic Notes

In the mathematical literature, Voronoi diagrams appeared as early as in Dirichlet (1850) and in Voronoi (1907, 1908); they were mainly interested, however, in diagrams for regular distributions of sites. Problems about packings and coverings of space by balls and other convex figures belong to the main applications of Voronoi diagrams (see Fejes Tóth, (1972) and Rogers (1964)). Applications to other areas of science can be found in Bowyer (1981) for biology, Fairfield (1979) for visual perception, Brostow, Dussault, Fox (1978) for physics, and Hodder, Orton (1976) for archaeology. Shamos, Hoey (1975) introduced Voronoi diagrams to computer science and also gave an algorithm that constructs the diagram for n sites in E^2 in time $O(n\log n)$ if the Euclidean metric is used for measuring distances. Recently, Fortune (1986) gave an $O(n\log n)$ time algorithm which computes the two-dimensional Voronoi diagram for n points in the Euclidean plane. His approach can easily be implemented and generalized to more complicated objects like line segments and circles. The relationship between Voronoi diagrams in E^d and particular convex polytopes in E^{d+1} was first exploited by Brown (1979) to design algorithms that work in arbitrary dimensions; the complexity mentioned in Corollary 13.6 is achieved by the convex hull algorithms of Preparata, Hong (1977), if $d=2$, and of Seidel (1981), if $d\geq 3$ (see also Chapter 8). Using the same relationship, Paschinger (1982) and Seidel (1982) independently derived the tight upper bounds on the number of faces of Voronoi diagrams given in Exercise 13.2. The particular transformation of Section 13.1, which relates Voronoi diagrams in E^d with substructures of arrangements in E^{d+1} using the paraboloid U: $x_{d+1}=x_1^2+x_2^2+...+x_d^2$, was first used in Edelsbrunner, Seidel (1986); its advantage over the transformation based on the unit-sphere is that it maps sites with integer coordinates to hyperplanes with integer coefficients.

A considerable number of applications of Voronoi diagrams for sets of sites in E^2 can be found in Shamos (1978); his collection includes the problems discussed in Sections 13.2.1 through 13.2.5. The post office problem addressed in Section 13.2.1 was first mentioned in Knuth (1973). Lipton, Tarjan (1977) gave the first optimal solution to this problem, and the more practical solution suggested in this book uses the techniques of Edelsbrunner, Guibas, Stolfi (1986) (see also Chapter 11). Edelsbrunner, Maurer (1985) demonstrate a solution to the post office problem using the techniques developed in Chapter 9, and thus solve Exercise 13.3. Delaunay triangulations of sets of sites in E^2 are named after the Russian mathematician Boris Delone (see Delaunay (1934), written in memory of Georges Voronoi, and Rogers (1964)). Lawson (1972) discussed applications of locally equiangular triangulations and also formulated the algorithm used in the proof of Theorem 13.11. Sibson (1978) was the first to observe that locally equiangular triangulations in E^2 are necessarily completions of Delaunay triangulations. A generalization of Sibson's result to triangulations of a set of non-intersecting line segments can be found in Lee, Lin (1986). The algorithm for computing closest neighbors presented in Section 13.2.4 is taken from Shamos, Hoey (1975). Lee, Preparata (1978) demonstrate that a linear time algorithm for this problem exists if the sites are the vertices of a convex polygon; this solves Exercise 13.8. A proof of Exercise 13.9(a) can be found in Day, Edelsbrunner (1984); they reduce the problem to finding so-called Hadwiger numbers of congruent discs. The results on Hadwiger numbers for congruent balls (see Schütte (1953)) and for general d-dimensional convex objects (see Grömer (1961)) imply solutions to Exercises 13.9(b) and (c). An algorithm that finds the closest neighbors of each site in a given set of size n in time $O(n\log n)$ can be found in Vaidya (1986); this solves Exercise 13.9(d). He exploits the result of Exercise 13.9(c). Exercise 13.10(a) is solved by Edelsbrunner, Skiena (1986) who extend an argument given in Avis (1984) where the problem is solved for even integer numbers n. Solutions to Exercise 13.10(d) and (e) are given in Avis,

Erdös, Pach (1986) and Edelsbrunner, Skiena (1986). An algorithm that finds all furthest neighbors of n vertices of a convex polygon in time $O(n)$, and thus solves Exercise 13.10(c), can be found in Aggarwal, Klawe, Moran, Shor, Wilber (1986). Solutions to Exercises 13.11(a) through (c) appear in Sutherland (1935) and Shamos (1975). The upper bound on the number of diametrical pairs of a finite set of points in three dimensions can be found in Heppes (1956) (see also Moser, Pach (1986)). The algorithm for constructing minimum spanning trees, as outlined in Section 13.2.5, was developed independently by Prim (1957) and Dijkstra (1959). Other strategies are described in Yao (1975) and Cheriton, Tarjan (1976), where a solution to Exercise 13.13 can be found. Shamos, Hoey (1975) were the first to notice the relationship between minimum spanning trees and Delaunay triangulations of planar point sets. The Gabriel graph (Exercise 13.14) and the relative neighborhood graph (Exercise 13.15) of a point set in $\boldsymbol{E}^2$ are supergraphs of every minimum spanning tree and subgraphs of the graph induced by the Delaunay triangulation. This was realized by Matula, Sokal (1980) and Toussaint (1980). Supowit (1983) gives a solution to Exercise 13.15(b). The concepts of α-hulls and α-shapes were introduced in Edelsbrunner, Kirkpatrick, Seidel (1983); they also offer a solution to Exercise 13.21. Jung (1901, 1910) solves Exercise 13.16(b) and thus shows that the α-hull of a point set S in $\boldsymbol{E}^2$ with diameter 1 is not the entire plane, unless α is larger than $\sqrt{3}$. Contour diagrams of planar point sets, as defined in Exercise 13.19, are used in Moss (1967). An application of α-shapes to range search can be found in Chazelle, Edelsbrunner (1985a).

The generalization of Voronoi diagrams to so-called k^{th}-order Voronoi diagrams was introduced in Shamos, Hoey (1975). Our development of upper and lower bounds on the complexity of higher-order Voronoi diagrams borrows from Lee (1982). Lee (1982) also describes the first algorithm for planar point sets; it constructs the kth-order diagram from the $(k-1)^{\text{st}}$-order diagram. An improvement to the approach taken in Section 13.5 can be found in Chazelle, Edelsbrunner (1985b). The simultaneous construction of all higher-order diagrams as well as the encoding of the lists of closest sites (see Section 13.5) is demonstrated in Edelsbrunner, O'Rourke, Seidel (1986).

There are various generalizations of the classical concept of a Voronoi diagram that have been described in the literature. We will mention only a few of these papers. Lee, Drysdale (1981) associated with each site a weight which influences the distance of a point from this site as an additive term. An optimal algorithm that constructs the Voronoi diagram for a finite set of such weighted sites in the plane can be found in Fortune (1986). Aurenhammer, Edelsbrunner (1984) considered the case of multiplicative weights and present an optimal algorithm for the two-dimensional case. Ash, Bolker (1986) offer an extensive mathematical analysis of Voronoi diagrams under various weighting schemes. Algorithms for the particular scheme addressed in Section 13.6 can be found in Imai, Iri, Murota (1985) and Aurenhammer (1987). Edelsbrunner, Seidel (1986) give a rather general framework for Voronoi diagrams subsuming all generalizations encountered in this chapter. They also give a solution to Exercise 13.27.

CHAPTER 14

SEPARATION AND INTERSECTION IN THE PLANE

This chapter presents applications of the combinatorial results in Part I and the computational results in Part II to separation and intersection problems in Euclidean spaces. A typical problem in this category, which will not be discussed in this chapter, however, is the linear separability of two point sets in E^d. This is the question whether or not there is a hyperplane such that two given point sets are contained in different closed half-spaces defined by the hyperplane. This problem has been briefly discussed in Chapter 10, and we have seen that there is an algorithm which finds a separating hyperplane in time linear in the total number of points, if it exists.

For several independent reasons, we will restrict our attention to problems in two dimensions only. One of these reasons is that in three dimensions there are already more open problems than there are results. Some extensions of our two-dimensional results to three and higher dimensions are indicated in the exercise section of this chapter.

The organization of this chapter is as follows: Section 14.1 describes an algorithm that constructs a ham-sandwich line for two separated sets of points in E^2. The algorithmic techniques employed are those described in Chapter 10. In Section 14.2, we use ham-sandwich cuts for the design of a data structure that supports so-called line queries. A combination of this data structure and arrangements of lines preprocessed for point location queries (see Chapters 7 and 11) is used to solve a self-dual intersection problem in Section 14.3. Finally, Section 14.4 demonstrates that the data structure of Section 14.2.1 can be used to answer range queries that are more general than line queries without loss of efficiency.

14.1. Constructing Ham–Sandwich Cuts in Two Dimensions

According to Chapter 4, a line l is a *bisector of* a finite set P of points in E^2 if each open half–plane defined by l contains at most half of the points in P. Consequently, every bisector of P contains a point of P, if card$P$ is odd. Line l is called a *ham–sandwich cut of* two point sets P and Q in E^2 if it is a bisector of both P and Q. By Theorem 4.7, any two finite sets in E^2 have a common bisector and therefore a ham–sandwich cut.

The main purpose of this section is to offer an optimal algorithm that determines a ham–sandwich cut for two separated sets of points in E^2. We describe the geometric fundamentals for this algorithm in Section 14.1.1. The main structure of the algorithm follows the prune–and–search strategy outlined in Chapter 10; as a matter of fact, the algorithm is an adaptation of the line searching strategy of Section 10.4.3. We give the details of the computations in Sections 14.1.2 and 14.1.3. Section 14.1.4 describes the overall algorithm which cleverly combines the procedures described in Sections 14.1.2 and 14.1.3.

14.1.1. Ham–Sandwich Cuts and Duality

The algorithm developed in the subsequent sections finds a ham–sandwich cut of two finite point sets P and Q in optimal time, provided P and Q are separated by a line. The dual explanation of this algorithm is more intuitive which leads us to translate the concepts introduced above into dual space. For the dualization, we use the transform $\mathcal{D}$ which maps a point to a non–vertical line and vice versa. Specifically, it maps a point $p=(\pi_1,\pi_2)$ to the line

$$\mathcal{D}(p)\colon\ x_2=2\pi_1x_1-\pi_2,$$

and it maps a non–vertical line $l\colon x_2=\lambda_1x_1+\lambda_2$ to the point

$$\mathcal{D}(l)=(\frac{\lambda_1}{2},-\lambda_2).$$

Transform $\mathcal{D}$ is its own inverse and it preserves incidence and above–below relationships between points and lines (see Section 1.4).

By application of $\mathcal{D}$, sets P and Q are mapped to sets of non–vertical lines $G=\mathcal{D}(P)$ and $H=\mathcal{D}(Q)$. For $A\in\{G,H\}$, we call a point p a *bisector of* A if $p\in l^+$, for at most half of the lines l in A, and $p\in l^-$, for at most half of the lines l in A. Furthermore, we call a point p a *ham–sandwich cut of* G and H if it is a bisector of G and of H.

Observation 14.1: Let P and Q be two sets of m and n points in E^2 and define $G=\mathcal{D}(P)$ and $H=\mathcal{D}(Q)$. A point p is a ham–sandwich cut of G

and H if and only if line $\mathcal{D}(p)$ is a non-vertical ham-sandwich cut of point sets P and Q.

Following the exposition of Chapter 3, we define, for each point p in E^2, the integer functions $a_A(p)$, $o_A(p)$, and $b_A(p)$ such that

$$p \in l^-, \text{ for } a_A(p) \text{ lines } l \text{ of } A,$$
$$p \in l, \text{ for } o_A(p) \text{ lines } l \text{ of } A, \text{ and}$$
$$p \in l^+, \text{ for } b_A(p) \text{ lines } l \text{ of } A,$$

where A is either equal to G or equal to H. The *k-level $\mathcal{L}_k$ of* the arrangement $\mathcal{A}(A)$ is defined as the set of points p such that

$$a_A(p) < k \quad \text{and} \quad a_A(p) + o_A(p) \geq k,$$

for $1 \leq k \leq \text{card}A$ and $A \in \{G,H\}$. Notice that a point p is a bisector of A if and only if it lies between the $\lfloor \frac{\text{card}A}{2} \rfloor$-level and the $\lceil \frac{\text{card}A}{2} \rceil$-level of $\mathcal{A}(A)$, boundaries included. Thus, the problem of finding a ham-sandwich cut of G and H reduces to finding a point in the intersection of these regions defined for G and H.

For convenience, we will assume that m and n, the cardinalities of G and H, are odd. This simplifies the above intersection of closed regions to the intersection of the median levels of $\mathcal{A}(G)$ and $\mathcal{A}(H)$, where the *median level of* $\mathcal{A}(A)$ is the k-level of $\mathcal{A}(A)$, for $k = \frac{\text{card}A + 1}{2}$ and $A \in \{G,H\}$. The following result implies that the above assumption is justified.

Observation 14.2: Let A be a non-empty set of lines in E^2, let the cardinality of A be even, and let A' be a set obtained by removing an arbitrary line from A. A bisector of A' is also a bisector of A.

Notice that the transformation of the ham-sandwich problem into dual space leaves vertical ham-sandwich cuts in primal space without correspondence in dual space. As a result of this slight deficiency, there are pairs of sets of lines without ham-sandwich cut. By the following result, it is trivial to find a ham-sandwich cut for the point sets corresponding to such sets of lines.

Lemma 14.3: Let P and Q be two sets of points in E^2 and define $G = \mathcal{D}(P)$ and $H = \mathcal{D}(Q)$. If G and H have no ham-sandwich cut, then there is a vertical line which is a ham-sandwich cut of P and Q.

Proof: By Theorem 4.7, there is at least one line l which is a ham-sandwich cut of P and Q. By Observation 14.1, we have a one-to-one correspondence between the ham-sandwich cuts of G and H, which are points, and the non-vertical ham–sandwich cuts of P and Q, which are lines. If G and H have no ham–sandwich cut then P and Q have no non–vertical ham–sandwich cut, which leaves as the only possibility that P and Q have a vertical ham–sandwich cut. □

The algorithm to be described in the following sections will construct a ham–sandwich cut for two sets of points which are separated by a line. This additional restriction will play an important role in the algorithm; it guarantees that the ham–sandwich cut is essentially unique. To make this more concrete, we assume that the vertical line l_0: $x_1=0$ separates point sets P and Q, that all points of P have non–positive x_1–coordinates, and that all points of Q have non–negative x_1–coordinates. As a consequence, all lines in $G=\mathcal{D}(P)$ have non–positive slope and all lines in $H=\mathcal{D}(Q)$ have non–negative slope (see Figure 14.1). Now, any level of $\mathcal{A}(G)$ and of $\mathcal{A}(H)$ is a piecewise linear function from x_1 to x_2. By the separation of slopes of the lines in G and H, any level of $\mathcal{A}(G)$ is monotonically non–increasing and any level of $\mathcal{A}(H)$ is monotonically non–decreasing (see Figure 14.1). This implies the following result.

Lemma 14.4: Let G and H be two sets of non–vertical lines in E^2 such that $m=\text{card}G$ and $n=\text{card}H$ are odd and the lines in G (H) all have non–positive (non–negative) slope. Then the intersection of any level of $\mathcal{A}(G)$ with any level of $\mathcal{A}(H)$ is a possibly degenerate line segment.

As a matter of fact, the intersection of a level of $\mathcal{A}(G)$ and a level of $\mathcal{A}(H)$ is, in general, a point, and only in degenerate cases is this intersection empty or a horizontal line segment.

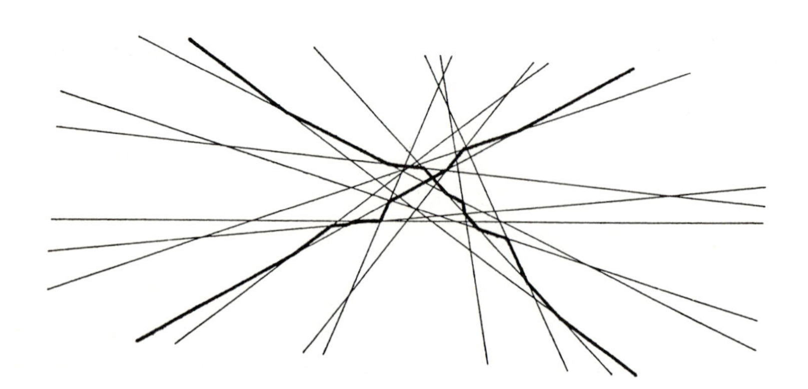

Figure 14.1. The intersection of two median levels.

14.1.2. Testing a Line

Let G and H be two sets of m and n lines in E^2 such that all lines in G (H) have non-positive (non-negative) slope, and let $k_G \leq m$ and $k_H \leq n$ be two positive integer numbers. In this section, we demonstrate that the following problem can be solved in time $O(m+n)$:

> given a test line t, decide whether t contains a point of the intersection of the k_G-level $\mathcal{L}_{k_G}(G)$ of $\mathcal{A}(G)$ and the k_H-level $\mathcal{L}_{k_H}(H)$ of $\mathcal{A}(H)$, and, if not, on which side of t the two levels intersect.

Notice that the above problem is more general than the problem of testing on which side of a line t the ham-sandwich cuts of G and H lie. For convenience, we abbreviate $\mathcal{L}_{k_G}(G)$ as $\mathcal{L}_G$ and $\mathcal{L}_{k_H}(H)$ as $\mathcal{L}_H$. In addition, we define $s = \mathcal{L}_G \cap \mathcal{L}_H$; by Lemma 4.4, s is a possibly degenerate line segment. We will see that we can assume that s is non-empty which implies that $\mathcal{L}_G$ intersects a vertical line l above or on (below or on) $\mathcal{L}_H$ if l is sufficiently far to the left (right).

When we test a line t as specified above, we distinguish four cases depending on whether t is vertical, t is horizontal, t has positive slope, or t has negative slope. We discuss the first three cases in this sequence; the fourth case is symmetric to the third and will not be treated explicitly.

If line t is vertical then we compute the intersections of t with all lines in G and H and we determine the k_G^{th} intersection p_G from the top with a line in G and the k_H^{th} intersection p_H from the top with a line in H. By definition of $\mathcal{L}_G$ and $\mathcal{L}_H$, we have

$$p_G = t \cap \mathcal{L}_G \quad \text{and} \quad p_H = t \cap \mathcal{L}_H.$$

A straightforward case-analysis can now be used to decide our problem:

Case 1.1: $p_G = p_H$. Then this point belongs to s which implies that t contains a point of the intersection between the two relevant levels.
Case 1.2: p_G lies above p_H. Then s lies to the right of line t.
Case 1.3: p_G lies below p_H. Then s lies to the left of line t.

If line t is horizontal then t can be parallel to some of the lines in G and H which complicates matters slightly. To decide this case, we determine numbers m_1, m_2, and m_3, where m_1 is the number of horizontal lines in G that lie above t, m_2 is the number of horizontal lines in G that lie below t, and m_3 is the number of non-horizontal lines in G. In addition, we determine the analogous numbers n_1, n_2, and n_3 for the lines in H. Obviously, $m_1+m_2+m_3=m$, if t does not coincide with any line in G, and $m_1+m_2+m_3=m-1$, otherwise. The analogous statement holds for the lines in H. There are three cases to be distinguished:

Case 2.1: $k_G \leq m_1$ or $k_H \leq n_1$. In this case, either all of $\mathcal{L}_G$ or all of $\mathcal{L}_H$ lies above line t (see Figure 14.2(a) which shows G and $\mathcal{L}_G$ for the case $k_G = 3$ and $m_1 = 4$).

Case 2.2: $k_G \geq m - m_2 + 1$ or $k_H \geq n - n_2 + 1$. In this case, either all of $\mathcal{L}_G$ or all of $\mathcal{L}_H$ lies below line t.

Case 2.3: $m_1 < k_G < m - m_2 + 1$ and $n_1 < k_H < n - n_2 + 1$. Both $\mathcal{L}_G$ and $\mathcal{L}_H$ intersect line t. The intersection s_G of t with $\mathcal{L}_G$ is the $(k_G - m_1)^{\text{th}}$ intersection from the right with a non-vertical line of G, if t does not coincide with any line of G, and s_G is the closed line segment between the $(k_G - m_1 - 1)^{\text{st}}$ intersection and the $(k_G - m_1)^{\text{th}}$ intersection from the right, otherwise. Here, we count an intersection point once for each non-vertical line that contains it. Analogous rules hold for the intersection s_H between line t and $\mathcal{L}_H$. There are three subcases to be distinguished:

Case 2.3.1: $s_G \cap s_H \neq \emptyset$. Thus, line t contains s, the intersection between $\mathcal{L}_G$ and $\mathcal{L}_H$.

Case 2.3.2: s_G lies to the left of s_H. Then s lies below line t. (This case is illustrated in Figure 14.2(b).)

Case 2.3.3: s_G lies to the right of s_H. Then s lies above line t.

Matters are more complicated if the test line t is non-vertical and non-horizontal. This is because t may intersect one of $\mathcal{L}_G$ and $\mathcal{L}_H$ in more than one connected component (see Figure 14.3). Fortunately, line t intersects $\mathcal{L}_G$ in exactly one point whenever the slope of t is positive, and it intersects $\mathcal{L}_H$ is a

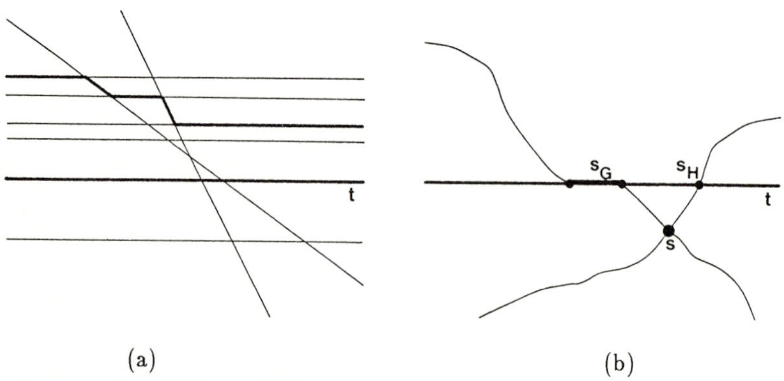

Figure 14.2. Testing a horizontal line.

unique point if the slope of t is negative. Below, we discuss the case when the slope of t is positive; the case when the slope of t is negative can be handled symmetrically.

In a first step, we compute the intersections of t with all lines in set G and we determine the k_G^{th} intersection from the right; this is the intersection of t with $\mathcal{L}_G$ (see Figure 14.3). Call this point p_G and let $\hat{t}$ be the vertical line through point p_G. Since $\mathcal{L}_G$ is a monotonically non–increasing function it lives only in the closed quadrant above t and to the left of $\hat{t}$ and in the closed quadrant below t and to the right of $\hat{t}$. To decide which side of t contains s, we therefore test the vertical line $\hat{t}$ as described above. The result for $\hat{t}$ determines the result for t as follows:

Case 3.1: Line $\hat{t}$ contains a point of s. In this case, line t contains this point too.
Case 3.2: $\mathcal{L}_G$ and $\mathcal{L}_H$ intersect to the left of $\hat{t}$. Then s lies above t. (This case is illustrated in Figure 14.3.)
Case 3.3: $\mathcal{L}_G$ and $\mathcal{L}_H$ intersect to the right of $\hat{t}$. Then s lies below t.

Since the ith point of a multiset of points on a line can be determined in time linear in the cardinality of the multiset, all operations described above can be performed in time $O(m+n)$.

14.1.3. Finding Test Lines and Pruning

We have seen in Section 14.1.2 how a given line can be tested in time linear in the number of lines in sets G and H. The problem addressed in this section is to

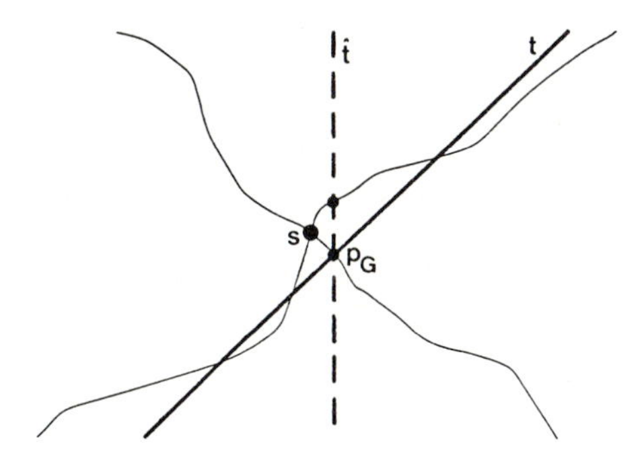

Figure 14.3. Testing a line which is neither vertical nor horizontal.

choose effective test lines, where a test line is effective if it leads to the elimination of a constant fraction of the lines in G and H. We will describe the mechanics of eliminating lines at the end of this section. More specifically, our goal is to determine two test lines such that we can drop at least one eighth of the lines in G and H without affecting the solution, which is the intersection of two given levels $\mathcal{L}_G$ in $\mathcal{A}(G)$ and $\mathcal{L}_H$ in $\mathcal{A}(H)$. The strategy for determining two such test lines is the same as the one described in Section 10.4.3; only a few minor details are different. For reasons of completeness, we present the strategy which is illustrated in Figure 10.6.

Procedure 14.1 (Searching lines in the plane):

Step 1: Determine the median line l_{med} in an ordering by slope of the lines in $L = G \cup H$. Partition L into the set L_1 of lines that are less steep than l_{med}, the set L_2 of lines parallel to l_{med}, and the set L_3 of lines that are steeper than l_{med}.

Step 2: Match each line in L_1 with a unique line in L_3, if available. This leaves $|\text{card}L_1 - \text{card}L_3|$ lines of either L_1 or L_3 unmatched. Let L_{pair} be the set of pairs of matched lines.

Step 3: Project the intersection point of each pair of lines in L_{pair} vertically onto the x_1-axis and let $\hat{P}$ be the multiset of points obtained.

Step 4: Determine the median point in $\hat{P}$ and let $\hat{t}$ be the vertical line through this point. Test line $\hat{t}$. This finishes the overall algorithm if $\hat{t}$ contains a point of the intersection between $\mathcal{L}_G$ and $\mathcal{L}_H$, or it determines the open half-plane $\hat{t}^{pos}$ bounded by $\hat{t}$ that contains $\mathcal{L}_G \cap \mathcal{L}_H$.

Step 5: Construct a multiset $\overline{P}$ on line $\hat{t}$ as follows: for each pair of lines in L_{pair} whose intersection does not belong to half-plane $\hat{t}^{pos}$, add the projection of this point onto $\hat{t}$ to $\overline{P}$, where the projection is done along the direction determined by line l_{med}, and for each line in L_2 (which is parallel to l_{med}), add the intersection with $\hat{t}$ to $\overline{P}$.

Step 6: Determine the median point in $\overline{P}$ and let $\overline{t}$ be the line parallel to l_{med} that contains this point. Test line $\overline{t}$. This either finishes the overall algorithm, or it determines the open half-plane $\overline{t}^{pos}$ bounded by $\overline{t}$ that contains $\mathcal{L}_G \cap \mathcal{L}_H$.

The half-planes $\hat{t}^{pos}$ and $\overline{t}^{pos}$ will be used to eliminate lines from G and H: if a line l in G or H does not intersect the open quadrant $q = \hat{t}^{pos} \cap \overline{t}^{pos}$, then l cannot contain any point of the intersection between $\mathcal{L}_G$ and $\mathcal{L}_H$. Thus, we can drop line l from its set. Notice that dropping line l from either G or H changes the structure of the levels in the corresponding arrangement. Assume without loss of generality that l belongs to G and define $G' = G - \{l\}$. If quadrant q lies below line l, then the relevant part of the k_G-level of $\mathcal{A}(G)$ (that is, the part that lies in q) is contained in the (k_G-1)-level of $\mathcal{A}(G')$; we therefore decrease k_G by one. If q lies above line l, then k_G remains unchanged. It is important to realize that the intersection between the two current levels does not change, although the levels themselves change. However, they change in a way that does not introduce new intersections.

It is not hard to see that all operations taken to compute the two test lines $\hat{t}$ and $\bar{t}$ can be finished in time linear in the cardinalities of sets G and H. Furthermore, we can prove the following result on the effectiveness of the choice of test lines.

Lemma 14.5: Let L be a set of ℓ lines in E^2 and let q be the open quadrant as computed in Procedure 14.1. At least $\frac{\ell}{8}$ of the lines in L avoid q.

Proof: For $i=1,2,3$, define $\ell_i=\text{card}L_i$, for L_i as constructed in Procedure 14.1, and assume without loss of generality that $\ell_1 \leq \ell_3$, that is, the number of lines in L which are less steep than line l_{med} is at most as large as the number of lines in L which are steeper than l_{med}. Since l_{med} is the median in a sorted order of L, we also have $\ell_1+\ell_2 \geq \ell_3$. Since $\ell_1 \leq \ell_3$, we have $\text{card}L_{pair}=\ell_1$, which implies

$$\text{card}\hat{P}=\ell_1.$$

At least half of the pairs of lines in L_{pair} do not intersect inside the half-plane $\hat{t}^{pos}$ constructed from multiset $\hat{P}$. Consequently,

$$\text{card}\bar{P} \geq \frac{\ell_1}{2}+\ell_2.$$

At least half of the points in multiset $\bar{P}$ do not lie in half-plane $\bar{t}^{pos}$, and each such point either corresponds to a pair of lines where at least one line avoids quadrant $q=\hat{t}^{pos} \cap \bar{t}^{pos}$ or to a line parallel to l_{med} which avoids q. It follows that there are at least

$$\frac{\ell_1+2\ell_2}{4} \geq \frac{\ell}{8}$$

lines in L which avoid q. □

14.1.4. The Overall Algorithm

In this section, we will see how the computations described in Sections 14.1.2 and 14.1.3 fit together to an efficient algorithm that computes ham-sandwich cuts. As mentioned earlier, the general structure of the algorithm is what we call prune-and-search in Chapter 10. The input to the algorithm consists of two sets G and H of m and n lines in E^2 with the property that m and n are odd and that all lines in G have non-positive slope and all lines in H have non-negative slope. Equivalently, the x_2-axis separates the sets of points that correspond to G and H by means of the dual transform $\mathcal{D}$. The algorithm aims at intersecting the k_G-level of arrangement $\mathcal{A}(G)$ with the k_H-level of arrangement $\mathcal{A}(H)$, where initially $k_G = \frac{m+1}{2}$ and $k_H = \frac{n+1}{2}$. During the algorithm, we remove lines from sets G and H which requires that numbers k_G and k_H of the current levels be updated. Below, we give a description of the algorithm.

Algorithm 14.2: (Ham-sandwich cut in the plane):

SEARCH: Find two line $\hat{t}$ and $\bar{t}$ and test them as described in Section 14.1.3. Testing $\hat{t}$ and $\bar{t}$ either leads to the discovery of a ham-sandwich cut of G and H which ends the algorithm, or it yields an open quadrant q determined by $\hat{t}$ and $\bar{t}$ that contains the intersection between the k_G-level of $\mathcal{A}(G)$ and the k_H-level of $\mathcal{A}(H)$.

PRUNE: Eliminate all lines from G and H that avoid quadrant q. In addition, if quadrant q lies below line $\bar{t}$ then set $k_G := k_G - 1$ for each line eliminated from G and set $k_H := k_H - 1$ for each line eliminated from H.

RECUR: Repeat Algorithm 14.2 for the new sets G and H and the new numbers k_G and k_H which indicate the relevant levels of $\mathcal{A}(G)$ and $\mathcal{A}(H)$.

It is interesting to see that Algorithm 14.2 does without extra computations that take care of particularly small sets G and H. This is possible since Lemma 14.5 guarantees that the SEARCH step either finds a ham-sandwich cut or the PRUNE step gets rid of at least one line. In the end, that is, when $\mathrm{card}\, G = \mathrm{card}\, H = 1$, the choice of line $\hat{t}$ in the SEARCH step finishes the computation right there.

To analyze the time required by Algorithm 14.2, we first realize that one iteration takes time linear in the sizes of the current sets G and H. Here, we do not take into account the amount of time spent in the recursive call. Let ℓ_i be the sum of the cardinalities of sets G and H at the beginning of the ith iteration of Algorithm 14.1. Thus, we have $\ell_1 = m + n$ and, by Lemma 14.5,

$$\ell_i \leq \frac{7\ell_{i-1}}{8},$$

for $i \geq 2$. Consequently, the time $T(\ell)$ required by Algorithm 14.2 to find a ham-sandwich cut for a total of ℓ lines satisfies the recurrence relation

$$T(\ell) = T(\frac{7\ell}{8}) + O(\ell),$$

which implies $T(\ell) = O(\ell)$. The duality results of Section 14.1.1 allow us to finally conclude the main result of this section.

Theorem 14.6: Let P and Q be two sets of m and n points in E^2 such that P and Q can be separated by a straight line. Then there is an algorithm that constructs a ham-sandwich cut of P and Q in time $O(m+n)$.

14.2. Answering Line Queries

In this section, we describe two solutions to a search problem defined for points and lines in E^2.

> A *line query* comes as a line l and it asks that all points of a given finite point set P that lie on l be reported. The *line search problem* is the problem of designing a data structure for P that allows us to answer line queries.

As usual, we measure the performance of a solution by the amount of time it takes to answer a line query, the amount of storage required by the data structure, and the amount of time needed to construct the data structure. Section 14.2.1 presents a solution which minimizes the amount of storage required; it makes use of the existence of a ham-sandwich cut of two finite sets of points in the plane (see Section 4.2) and of Algorithm 14.2 which computes such lines. In Section 14.2.2, we discuss a solution to the line search problem which optimizes the time needed to answer a line query; it resorts to the construction of arrangements of lines and to a data structure for answering point location queries.

14.2.1. The Ham-Sandwich Tree

The so-called ham-sandwich tree is a binary tree which stores a finite set of points in E^2. Each node ν of the tree stores a directed line *line*(ν), a possibly empty linear array *points*(ν), and two pointers *left*(ν) and *right*(ν) to the children of ν. Let P be a set of n points in E^2, let $\vec{l}$ be a directed line whose undirected version l is a bisector of P, and let l^{pos} and l^{neg} be the open half-planes to the left and the right of $\vec{l}$, respectively. The *ham-sandwich tree* $\mathcal{T}(P)$ *of* P (and $\vec{l}$) can now be defined recursively.

> If $P = \emptyset$ then $\mathcal{T}(P)$ is the empty tree. Otherwise, the root ρ of $\mathcal{T}(P)$ stores *line*(ρ) $= \vec{l}$, the array *points*(ρ) stores the sorted sequence of points

3. combine the solutions of the subproblems to a solution of the initial problem.

Several variations of this scheme exist and there is no commonly accepted definition of how tolerantly the term divide-and-conquer should be used.

In this section, we apply the divide-and-conquer paradigm to the construction of the stabbing region of a given set $S=\{s_1,s_2,\ldots,s_n\}$ of not necessarily vertical or disjoint segments in E^2. Below, we give a formal description of the global strategy.

Algorithm 15.1 (Divide-and-conquer):
if card$S=1$ **then**
$\mathcal{S}(S)=\mathcal{D}(s)$, for s the only segment in S.
else
DIVIDE: Set $k:=\lfloor n/2\rfloor$ and compute sets $S_1:=\{s_1,s_2,\ldots,s_k\}$ and $S_2:=\{s_{k+1},s_{k+2},\ldots,s_n\}$.
RECUR: Construct $\mathcal{S}(S_1)$ and $\mathcal{S}(S_2)$ recursively.
MERGE: Construct $\mathcal{S}(S)=\mathcal{S}(S_1)\cap\mathcal{S}(S_2)$.
endif.

Algorithm 15.1 is a fairly clean application of the divide-and-conquer paradigm, in the sense that its structure is little influenced by particularities of the problem. For instance, the DIVIDE step partitions S into two subsets S_1 and S_2, where the only condition is that S_1 and S_2 be roughly equally large. Such an arbitrary partition is not always possible or desirable as indicated by Algorithm 8.9 which uses divide-and-conquer to construct convex hulls in E^3.

The crucial part of almost every divide-and-conquer algorithm is its MERGE step which combines the solutions of the subproblems to a solution of the main problem. In our case, it is not hard to compute the intersection of the stabbing regions $\mathcal{S}(S_1)$ and $\mathcal{S}(S_2)$ in time $O(n)$, where $n=\text{card}S_1+\text{card}S_2$. To this end, we exploit the fact that every vertical line intersects at most two edges of $\mathcal{S}(S_i)$, for $i=1,2$. Thus, a vertical line can be swept from left to right computing $\mathcal{S}(S_1)\cap\mathcal{S}(S_2)$ in constant time per vertex and intersection of two edges. By Theorem 15.4, this number is in $O(n)$ which implies that the MERGE step of Algorithm 15.1 can be performed in time $O(n)$. If $T(n)$ denotes the amount of time needed by Algorithm 15.1 to construct the stabbing region of n segments, then we have

$$T(n)=2T(\frac{n}{2})+O(n)=O(n\log n),$$

which implies the following result.

Theorem 15.5: Let S be a set of n not necessarily vertical or disjoint segments in E^2. The stabbing region $\mathcal{S}(S)$ can be constructed in $O(n\log n)$ time and $O(n)$ storage.

We add a few remarks on the implementation of Algorithm 15.1 and, in particular, of its MERGE step. It is true that the vertical sweep line intersects at most two edges of $\mathcal{S}(S_i)$ at a time, for $i=1,2$. Thus, it intersects up to four edges at each point in time which already gives rise to a considerable number of cases to be distinguished. It is certainly less painful to implement the construction of $\mathcal{S}(S)$ in three phases, two of which are symmetric.

1. Replace each double wedge $\mathcal{D}(s)$ by its upper boundary, that is, its upper left and upper right ray. This yields n piecewise linear functions f_i, $1 \leq i \leq n$, from x_1 to x_2. Next, we compute the *lower envelope* $f_l(x_1) = \min\{f_i(x_1) | 1 \leq i \leq n\}$ using a straightforward modification of Algorithm 15.1.
2. Symmetrically, compute the analogously defined upper envelope f_u of the lower boundaries of all double wedges.
3. In a last sweep, compute the boundary of $\mathcal{S}(S)$ which is the set of points below or on f_l and above or on f_u.

15.5. Incremental Construction

A set of objects is said to be processed *incrementally* if we proceed step by step performing the computations for one object at a time. If the algorithm is to construct some structure from the set, then this strategy builds the structure for the first $i-1$ objects and then adds the ith object, that is, the structure is updated so that it represents the first i objects of the set.

At first sight, this paradigm may seem too simplistic to justify a section of this chapter devoted to describing the obvious. Nevertheless, this paradigm can lead to optimal algorithms if some properties of the problem at hand make it possible to add single objects efficiently, as we have seen in Chapters 7 and 8. Chapter 7 describes the incremental construction of an arrangement of hyperplanes which is optimal in time and in storage. In this algorithm, the order in which the hyperplanes are added is immaterial and one hyperplane can be added in optimal time. Chapter 8 demonstrates a convex hull algorithm which works incrementally and is optimal in even dimensions. This algorithm takes advantage of a particular ordering of the points, and the time needed to add two different points varies considerably. However, due to the geometry of the problem it is possible to show that the time spent for "expensive" points is made up by a suitable number of "inexpensive" points, so that the total amount of time needed turns out to be optimal. These examples illustrate that incremental algorithms may or may not be sensitive to particular orderings of the objects, and they may or may not distribute the amount of total work equally among all objects.

In this section, we demonstrate an optimal algorithm for constructing the stabbing region for a set S of n vertical line segments in E^2. It is true that this problem can be solved by intersecting $2n$ half-planes and thus can be reduced to computing convex hulls in two dimensions (see Sections 8.1 and 8.3). Due to the special situation – the half-planes come in pairs bounded by parallel lines – it is possible to design a slightly more elegant algorithm for the problem.

Below, we give a formal description of the algorithm which assumes that no two segments of the set $S=\{s_1,s_2,...,s_n\}$ lie on a common vertical line, and that the vertical line containing s_i is to the left of the vertical line containing s_j if $i<j$. If two segments s and t lie on a common vertical line, then we replace them by $s\cap t$ since $s\cap t$ intersects every non-vertical transversal of $\{s,t\}$. If $s\cap t=\emptyset$ then S has no non-vertical transversal. For each $1\leq i\leq n$, we define $S_i=\{s_1,s_2,...,s_i\}$.

Algorithm 15.2 (Incremental construction):
 Initial step: Compute $\mathcal{S}(S_2):=\mathcal{D}(s_1)\cap\mathcal{D}(s_2)$.
 Iteration: Update the stabbing region as follows:
  for $i:=3$ **to** n **do**
   $\mathcal{S}(S_i):=\mathcal{S}(S_{i-1})\cap\mathcal{D}(s_i)$.
  endfor.

To represent a stabbing region $\mathcal{S}(S_{i-1})$, we use two deques $\mathcal{Q}_t$ and $\mathcal{Q}_b$ which store the upper and the lower boundary of the convex polygon $\mathcal{S}(S_{i-1})$. Here, the upper (lower) boundary of $\mathcal{S}(S_{i-1})$ consists of all edges such that the interior of $\mathcal{S}(S_{i-1})$ lies below (above) the lines that contain them. We will discuss $\mathcal{Q}_t$ and its maintenance in further details; the maintenance of $\mathcal{Q}_b$ works symmetrically.

Each node ν of $\mathcal{Q}_t$ represents an edge of $\mathcal{S}(S_{i-1})$ which is contained in the line dual to the lower endpoint of a segment s_j, with $j\leq i-1$. To compute this line in constant time from ν, it suffices to equip ν with a pointer to s_j. When we update $\mathcal{Q}_t$ such that it represents the upper boundary of $\mathcal{S}(S_i)$, we take advantage of the fact that the slope of the lines that bound the stripe $\mathcal{D}(s_i)$ is greater than the slopes of all lines containing edges of $\mathcal{S}(S_{i-1})$ (see Figure 15.5).

To describe the update process, we let $\nu_{t,l}$ and $\nu_{t,r}$ be the nodes of $\mathcal{Q}_t$ that represent the leftmost and the rightmost edges of the upper boundary of $\mathcal{S}(S_{i-1})$, respectively, and we define nodes $\nu_{b,l}$ and $\nu_{b,r}$ analogously for $\mathcal{Q}_b$. For a node ν of $\mathcal{Q}_t$, we denote the line that contains the edge represented by ν as $line(\nu)$, and we let $succ(\nu)$ ($pred(\nu)$) represent the next edge to the right (left), if it exists.

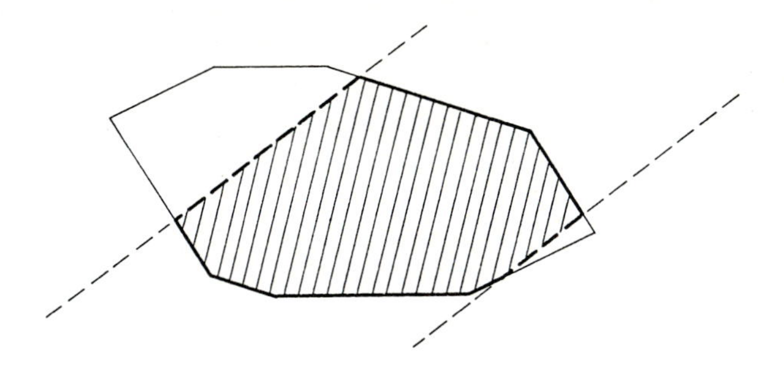

Figure 15.5. Updating the stabbing region.

Note that the leftmost vertex v_l of $S(S_{i-1})$ is thus the intersection between lines $line(\nu_{t,l})$ and $line(\nu_{b,l})$, and that $line(\nu_{t,r}) \cap line(\nu_{b,r})$ is the rightmost vertex of $S(S_{i-1})$.

Procedure 15.3 (Adding a vertical segment):

Let p_t and p_b be the upper and the lower endpoint of segment s_i, respectively, and set $l_b := D(p_t)$ and $l_t := D(p_b)$.

Case 1: Vertex v_l lies below line l_b or vertex v_r lies above line l_t. In this case, $D(s_i)$ avoids $S(S_{i-1})$, we set $S(S_i) := \emptyset$, and we halt since this implies that $S(S)$ is empty.

Case 2: $D(s_i) \cap S(S_{i-1}) \neq \emptyset$.

Step 2.1: Update the left end as follows:

Case 2.1.1: Vertex v_l lies in $D(s_i)$, that is, it lies between lines l_b and l_t including the boundary. In this case, the left end does not have to be updated.

Case 2.1.2: Vertex v_l lies above line l_t. Set $\nu := \nu_{t,l}$ and $\nu' := succ(\nu)$.

while ν' exists and $line(\nu) \cap line(\nu')$ lies above or on l_t **do**

Remove ν from $\mathcal{Q}_t$ and set $\nu := \nu'$ and $\nu' := succ(\nu)$.

endwhile;

Create a new node ν'' and set $succ(\nu'') := \nu$, $pred(\nu) := \nu''$, and $line(\nu'') := l_t$.

Step 2.2: Update the right end symmetrically. The only difference is that no new node is added after the iteration which removes an edge at a time.

Although a single segment s_i can take time $\Omega(n)$ when it is added, we have the following result.

Theorem 15.6: Algorithm 15.2 takes time $O(n)$ to construct the stabbing region of a presorted set of n vertical line segments in the plane.

Proof: Except for a constant number of steps, each step of Procedure 15.3 removes a node from $\mathcal{Q}_t$. The result follows, since each node can be removed at most once and a total of at most $2n$ nodes are created. □

15.6. Prune–and–Search

We use the prune–and–search paradigm in Chapter 10 to solve linear programs and in Chapter 14 to find ham–sandwich lines. The idea of the paradigm is to perform just enough computations to detect a constant fraction of the data as irrelevant and to remove it. After deleting the constant fraction from the data, the algorithm simply recurs. The total amount of time spent by the algorithm turns out to be asymptotically the same as the amount of time for one iteration. This is because the amount of data involved in the sequence of iterations decreases geometrically.

It is clear from this description that only problems that produce little output, as opposed to creating an elaborate geometric structure, qualify as candidates for the application of the prune–and–search paradigm. Such a candidate is problem 3 stated in Section 15.1 as follows:

> determine whether or not a given finite set of vertical segments admits a transversal, and, if it does, construct one.

By duality, this problem is equivalent to deciding whether or not the stabbing region of the set is empty, and, if it is non–empty, computing a point of it. Thus, the problem reduces to a two–dimensional linear program with $2n$ constraints.

To avoid being redundant, we do not give the details of a solution (those can be found in Sections 10.2 and 10.3), but rather restrict ourselves to a high–level description of the algorithm which also brings out the structure of the prune–and–search paradigm in a formal way. To this end, we let $S=\{s_1,s_2,\ldots,s_n\}$ be the set of vertical line segments. The algorithm below can be implemented in time $O(n)$ which does not leave enough room to presort the segments. The algorithm will work on the sets P_t and P_b of upper and lower endpoints, respectively, rather than on the segments themselves. It can be interpreted as the dual of a simplified version of Algorithm 10.2.

Algorithm 15.4 (Prune–and–search):

FIND_TEST: Pair the points in P_t and, separately, in P_b and let η_{med} be the median slope of any line through two paired points.

BISECT: Decide whether there is a line h with slope η_{med} such that $P_t \subseteq \text{cl}h^+$ and $P_b \subseteq \text{cl}h^-$; in this case, we are done. Otherwise, decide whether the slope of such a line needs to be greater or less than η_{med}. There is also the possibility that neither greater nor smaller slopes improve the situation in which case we exit with a negative answer.

PRUNE: Without loss of generality, assume that η_{med} is too small a slope. Then eliminate the appropriate point from each pair whose connecting line has slope less than or equal to η_{med}.

RECUR: Repeat Algorithm 15.4 for the new sets P_t and P_b.

This proves the following result which is also a consequence of Theorem 10.2 and the duality results described in Sections 10.2 and 15.2.

Theorem 15.7: Let S be a set of n vertical line segments in E^2. The existence of a transversal of S can be decided in time $O(n)$, and, if one exists, a transversal of S can be found in the same amount of time.

15.7. The Locus Approach

The *locus approach* is a rather high–level and intuitive idea which aims at establishing a one–to–one correspondence between solutions to a problem and points in some space. The advantage of such a correspondence is that it is intuitively easier to work with points than with other objects and that there is a number of sophisticated tools available that solve problems formulated for points. In a sense, this is what we did in Section 15.2 where we map segments to double wedges and lines to points. Then we define the concept of a stabbing region such that a point corresponds to a transversal if and only if it belongs to the stabbing region.

We illustrate these ideas by means of the *transversal search problem* stated as problem 4 in Section 15.1:

> store a set S of n not necessarily vertical or disjoint line segments and, for a given query line l, decide whether or not l is a transversal of S.

The developments in Sections 15.2 through 15.4 suggest constructing the stabbing region $\mathcal{S}(S)$ of S which costs $O(n)$ storage and $O(n\log n)$ time (see Theorem 15.5). For a non–vertical line l, we compute the point $p = \mathcal{D}(l)$ and determine whether or not p belongs to $\mathcal{S}(S)$. If it does, then l is a transversal, and if it does not belong to $\mathcal{S}(S)$, then l is not a transversal. The special case of vertical lines can be easily handled by computing the intersection of the vertical projections of all segments in S. A vertical line l is a transversal if and only if its vertical projection, which is a point, belongs to this interval.

To decide whether or not a point p lies in $\mathcal{S}(S)$, we can use the layered DAG (see Chapter 11) which represents a planar subdivision so that a query point can be located efficiently, that is, in logarithmic time. Since $\mathcal{S}(S)$ is a rather special subdivision (see Lemma 15.3), we can also use binary search along the horizontal direction first and then compare point p with at most two edges of $\mathcal{S}(S)$. The data structure which supports this strategy consists of two linear arrays, one for the edges that bound $\mathcal{S}(S)$ from above and one for the edges that bound $\mathcal{S}(S)$ from below. Both methods imply the following result.

Theorem 15.8: A set S of n not necessarily vertical or disjoint line segments in the plane can be stored in a data structure such that time $O(\log n)$ suffices to decide whether or not a given line is a transversal of S. This data structure takes $O(n)$ storage and $O(n \log n)$ time for its construction.

The locus approach is particularly useful for solving search problems. It suggests to apply a geometric transformation that maps every query object to a point in some space (preferably the plane) and to partition this space such that the answer to a query is determined by the face of the partition that contains the point corresponding to the query object. Obviously, this interpretation of the locus approach brings about a strong connection to geometric transformations and to solutions for the point location problem (see Chapter 11). These general ideas gave birth to concepts like the Voronoi diagram (see Chapter 13), the solution to the line search problem described in Section 14.2.2, etc.

The subdivision defined by the stabbing region $\mathcal{S}(S)$ of a set S of line segments is rather special and it is not necessary to use sophisticated data structures to locate points in it. A better illustration of the full power of the locus approach can be given if we solve problem 5 stated in Section 15.1:

> store a set S of n not necessarily vertical or disjoint line segments and, for a given query line l, determine the number of segments in S that intersect l.

Using the ideas of Section 15.2, we map each segment of S to a double wedge defined by two lines. Let H be the set of at most $2n$ lines obtained, and let $\mathcal{A}(H)$ be the arrangement defined by H. It is easy to see that any two points p and q of a common region or edge of $\mathcal{A}(H)$ correspond to lines $\mathcal{D}(p)$ and $\mathcal{D}(q)$ that intersect the same segments of S. Thus, we can associate each face f of $\mathcal{A}(H)$ with a unique subset S_f of S such that a non-vertical line l intersects exactly the

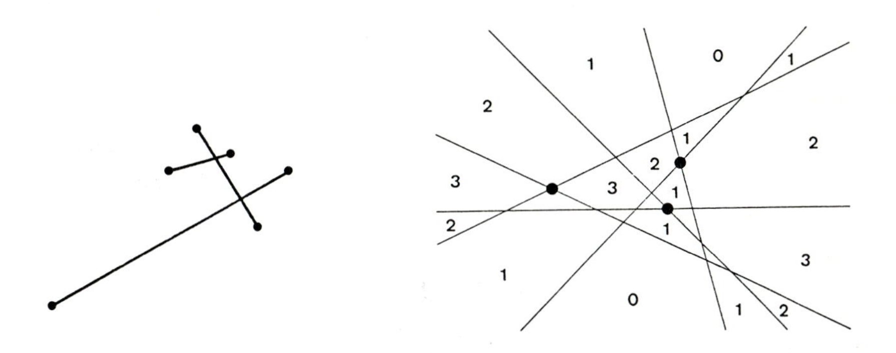

Figure 15.6. Three line segments and the corresponding arrangement.

segments in S_f if point $\mathcal{D}(l)$ belongs to face f.

To solve the above search problem, we construct the arrangement $\mathcal{A}(H)$, and we store with each face f of $\mathcal{A}(H)$ the cardinality of S_f. The construction of $\mathcal{A}(H)$ can be done in time $O(n^2)$, and the cardinalities of associated sets can be computed in an additional time $O(n^2)$ using depth–first– or breadth–first–search in the incidence graph of $\mathcal{A}(H)$ (see Chapter 7). An example of such an arrangement is given in Figure 15.6 which shows the dual set of line segments and the cardinalities associated with regions of $\mathcal{A}(H)$ where space permits it. If $\mathcal{A}(H)$ is stored in a layered DAG, we can perform point location queries in logarithmic time which implies the following result.

Theorem 15.9: A set S of n not necessarily vertical or disjoint line segments in the plane can be stored in a data structure such that time $O(\log n)$ suffice to determine how many segments of S a given query line intersects. The structure takes storage and time $O(n^2)$ for its construction.

15.8. Dynamization by Decomposition

For certain computations, it is necessary to use so–called *dynamic* solutions to a search problem. This is a data structure that allows us to perform an intermixed sequence of insertions, deletions, and queries, where an *insertion* adds a new object to the set represented by the data structure and a *deletion* removes an object from this set. We assume that the set of objects is initially empty and grows as insertions are performed. The performance of a dynamic solution to a search problem is measured by the amount of storage required by the data structure and by the amount of time needed to perform a sequence of insertions, deletions, and queries. An example of a computation which uses a dynamic data structure can be found in Chapter 9: it describes an algorithm that constructs arbitrary skeletons in arrangements using a dynamic solution to the so-called penetration search problem. The modification of a *static* (that is, non-dynamic) data structure to a dynamic one is commonly called the *dynamization of* the static data structure.

To date, the predominant paradigms for dynamization either use balanced trees that replace linear arrays, or they maintain systems of static data structures. We will show how the second paradigm can be used to dynamize the static solution to the transversal search problem described in Section 15.7. The dynamization method presented below works in general for so-called *decomposable search problems*, that is, for search problems that satisfy the following condition:

> let the set S of objects be arbitrarily partitioned into two sets S_1 and S_2, let q be an arbitrary query object, and let a_1 and a_2 be the answers to the queries for S_1 and q and for S_2 and q, respectively; then the answer a for S and q can be computed in constant time from a_1 and a_2.

Clearly, the transversal search problem is decomposable, since a line l is a transversal of a set S of segments if and only if it is a transversal of both sets S_1 and S_2, where S_1 and S_2 define an arbitrary partition of S.

Recall that the static data structure described in Section 15.7 solves the transversal search problem for a set S of n segments in $O(n)$ storage, $O(n\log n)$ time for construction, and $O(\log n)$ time for a query, where n is the cardinality of S. To improve the forthcoming notation, we let $\mathcal{R}(S)$ denote this data structure representing the segments in S. If we use $\mathcal{R}(S)$ in a dynamic environment, we can reflect an insertion or a deletion by rebuilding the entire data structure. Thus, we have a solution that takes time $O(\log n)$ per query and time $O(n\log n)$ per insertion or deletion. It is clear from this observation that the difference between static and dynamic data structures exists only on an intuitive level of understanding and has to do with the performance of the data structure for different tasks. The following dynamization method realizes a trade-off between both functions and thus can be used to improve the $O(n^2\log n)$ time needed for a sequence of n insertions, deletions, and queries.

Intuitively, the method below works by partitioning the current set S into a number of roughly equally large subsets. Let k be the number of subsets and let $S_1, S_2, \ldots, S_k$ be subsets of S with the following properties:

(i) $S_1 \cup S_2 \cup \ldots \cup S_k = S$,
(ii) $S_i \cap S_j = \emptyset$ if $i \neq j$, and
(iii) $\mathrm{card} S_i \leq \lceil \frac{n}{k} \rceil$, for $1 \leq i \leq k$.

Notice that condition (iii) implies that $\mathrm{card}S_i \leq \lceil \frac{\mathrm{card}S}{k} \rceil$, since $\mathrm{card}S \leq n$ as there can be at most n insertions in a sequence of n requests.

The dynamic data structure for the transversal search problem is a system of static data structures such that each set S_i is represented by $\mathcal{R}_i = \mathcal{R}(S_i)$. A static data structure $\mathcal{R}_i$ is called a *block*, and it is said to be *saturated* if the cardinality of S_i is equal to $\lceil n/k \rceil$; it is *non–saturated* if this cardinality is less than $\lceil n/k \rceil$. In addition to the k blocks, we maintain a dictionary of all segments in S and a list of non–saturated blocks. With each segment s represented in the dictionary we store the integer number i such that S_i contains s. Next, we sketch the procedures for querying this dynamic system of static data structures, for inserting a segment into S, and for deleting a segment from S.

Procedure 15.5 (Answering a transversal query):

Set *answer* :=**true** and let l be the query line.
for i:=1 **to** k **do**
  if l is not a transversal of S_i **then**
    answer :=**false**
  endif
endfor.

By Theorem 15.8, a transversal query for set S_i takes time $O(\log\frac{n}{k})$. Since we pose a query for each i between 1 and k, we need time $O(k\log\frac{n}{k})$ for a transversal query for S.

Procedure 15.6 (Inserting a segment):

Let s be the new segment and use the dictionary to decide whether or not s is already contained in S. If it is not yet an element of S, then perform the following steps:

Step 1: Find a non–saturated block $\mathcal{R}_i$ and rebuild it such that it represents $S_i := S_i \cup \{s\}$. If the new block $\mathcal{R}_i$ is saturated, then remove it from the list of non–saturated blocks.

Step 2: Insert s into the dictionary and equip the node that stores s with the integer number i.

Each dictionary operation takes time $O(\log n)$, and the reconstruction of $\mathcal{R}_i$ costs time $O(\frac{n}{k}\log\frac{n}{k})$.

Procedure 15.7 (Deleting a segment):

Let s be the segment to be removed. First, we search in the dictionary for segment s. If we find s, then we get the integer number i such that S_i contains s. In this case, we perform the following steps:

Step 1: Delete s from the dictionary.

Step 2: Set $S_i := S_i - \{s\}$ and reconstruct $\mathcal{R}_i$ such that it represents the new set S_i. Add $\mathcal{R}_i$ to the list of non-saturated blocks if it was saturated before the deletion of s.

The cost for the deletion of s is time $O(\frac{n}{k}\log\frac{n}{k})$ to rebuild $\mathcal{R}_i$ and $O(\log n)$ to remove s from the dictionary. The manipulation involving the list of non-saturated blocks takes only constant time. It is clear that the amount of storage for the system of static data structures is in $O(n)$ which implies the following result if we define $k = \lfloor\sqrt{n}\rfloor$.

Theorem 15.10: A sequence of n request starting with an initially empty set S can be executed in $O(n^{3/2}\log n)$ time and $O(n)$ storage, where a request is an insertion of a line segment into S, a deletion of a line segment from S, or a transversal query.

15.9. Exercises and Research Problems

2 **Exercise 15.1:** (a) Prove that if every three of a finite set of vertical line segments admit a transversal then there is a transversal for all line segments in the set. *(Hint: dualize and apply Helly's theorem, that is, Theorem 4.2.)*

4 (b) Prove that if every four of a set of $n \geq 6$ congruent non-intersecting circles admit a transversal, then there is a transversal for all circles in the set.

4 (c) Prove that for every positive integer number n there are n segments in E^2 such that all subsets of size $n-1$ admit a transversal but the set itself does not.

3 **Exercise 15.2:** Prove that the stabbing region of n line segments with a total number of m endpoints is at most $8m-4$. *(Hint: show first that the stabbing region consists of at most $m+1$ convex polygons and then apply the same argument as that used in the proof of Theorem 15.4.)*

3 **Exercise 15.3:** Let S be a set of n non-intersecting line segments in E^2. A transversal t of S defines a permutation of S that is unique up to reversion. Prove that there are at most n different permutations defined by S and its transversals. *(Hint: prove that each convex polygon of the stabbing region of S, defined in Section 15.2, corresponds to a unique permutation and that the two unbounded polygons of the stabbing region correspond to the same permutation.)*

3 **Exercise 15.4:** Consider the following problem arising in computer graphics. Let P be the set of grid points in E^2 whose integer coordinates vary between 0 and m, and assume that a particular algorithm represents a non-vertical line by the closest point of each column that intersects the line. Given a subset of P of size n, decide in time $O(n)$ whether or not this subset represents a line as specified above. *(Hint: replace each point by a vertical line segment such that the point is used to represent the line if and only if the corresponding line segment intersects the line.)*

2 **Exercise 15.5:** (a) Define *Ackermann's function* $A(n)=A_n(n)$ as follows: $A_k(1)=2$, for $k\geq 1$, $A_1(n)=A_1(n-1)+2$, for $n\geq 2$, and $A_k(n)=A_{k-1}(A_k(n-1))$, for $k\geq 2$. Prove $A_k(2)=4$, for all k, $A_1(n)=2n$, $A_2(n)=2^n$, and $A_3(n)=2^{A_3(n-1)}$, for all $n\geq 2$.

2 (b) Define $\alpha(n)=\min\{m|A(m)\geq n\}$. Verify that $\alpha(n)\leq 3$, for $n\leq 16$, that $\alpha(n)\leq 4$, for n at most a "tower" of 65536 2's, and that $\alpha(n)$ goes to infinity when n does so.

Exercise 15.6: A sequence $a_1,a_2,\ldots,a_\ell$, with $a_i\in\{1,2,\ldots,n\}$ for $1\leq i\leq \ell$, is called a *Davenport–Schinzel sequence with parameter* k, (a $DS(k)$*–sequence*, for short) if there are no $k+2$ indices $i_1<i_2<\ldots<i_{k+2}$ such that $a_{i_1}\neq a_{i_2}$ and $a_{i_j}=a_{i_{j-2}}$, for $1\leq j\leq k$.

2 (a) Prove that $\ell\leq 2n-1$ if $k=2$.

4 (b) Prove that the maximal length of any $DS(3)$–sequence is in $\Theta(n\alpha(n))$ (see Exercise 15.5(b) for a definition of $\alpha(n)$).

2 (c) Let a *V–function* be a continuous function from x_1 to x_2 that consist of two rays emanating from a common point. Let F be a collection of n V–functions $f_1,f_2,\ldots,f_n$, and call $\min\{f_i(x_1)|1\leq i\leq n\}$ the *lower envelope of* F. Prove that we obtain a $DS(3)$–sequence if we read the edges of the lower envelope of F from left to right and represent each edge by the index of the V–function that contains it.

3 (d) Prove that the $DS(3)$–sequence obtained from the collection F of V–functions, as described in (c), consist of at most $4n-2$ numbers. *(Hint: compare this claim with Theorem 15.4.)*

2 (e) Let a *U–function* be a continuous function from x_1 to x_2 that consists of two rays whose endpoints are connected by a line segment, and let a *uniform collection of* U–functions be a set of U–functions such that the left rays of any two U–functions are parallel, the same is true for their right rays, and the slopes of the lines that contain the line segments of the U–functions are strictly between the slope of all left rays and the slope of all right rays. Let F be a uniform collection of n U–functions. Prove that the transformation from F to a sequence of integer numbers between 1 and n (as described in (c)) yields a $DS(3)$–sequence.

4 (f) Prove that the maximum length of a sequence of integer numbers obtained as described in (e) is in $\Omega(n\alpha(n))$.

Exercise 15.7: Let P be a set of n possibly overlapping convex polygons with a total of n edges in the plane.

3 (a) Define the stabbing region of P by use of the dual transform $\mathcal{D}$ (see Section 15.2) and show that it is bounded by $\Theta(n\alpha(n))$ edges in the worst case. *(Hint: show that the boundary of the stabbing region of P lies in the union of the lower envelope of a uniform collection of at most n U-functions and the upper envelope of another uniform collection of at most n U-functions.)*

2 (b) Give an algorithm that constructs the stabbing region of P in $O(n\alpha(n)\log n)$ time and $O(n\alpha(n))$ storage. *(Hint: generalize the divide-and-conquer algorithm of Section 15.4.)*

3 **Exercise 15.8:** Let S be a set of n line segments in E^2 and let r be a region in the subdivision defined by S. Prove that the number of edges of r is in $O(n\alpha(n))$. *(Hint: a connected piece of r's boundary can be encoded as a cyclic sequence of integers which gives a $DS(4)$–sequence if it is disconnected at an arbitrary position; this $DS(4)$–sequence can be turned into a $DS(3)$–sequence if we label each edge on a single line segment by one of at most three different numbers.)*

3 **Exercise 15.9:** Modify the incremental construction of the stabbing region for a sorted sequence of n vertical line segments so that the algorithm runs in time per line segment

$O(\log n)$ and in total time $O(n)$.

4 **Exercise 15.10:** (a) Design a data structure that stores n line segments in $O(n \log\log n)$ storage such that a transversal query can be answered in time $O(\log n)$ and a line segment can be inserted and deleted in time $O(n)$.

4 (b) Solve the problem in (a) for a set of n vertical line segments in $O(n)$ storage, $O(\log n)$ query time, and $O(\log^2 n)$ time per insertion and deletion of a line segment.

4 (c) Let S be an initially empty set of line segments. Describe an algorithm that answers the transversal queries in a given sequence of n requests in time $O(n \log^2 n)$, where a request is either an insertion of a line segment into S, a deletion of a line segment from S, or a transversal query. *(Remark: we assume that the entire sequence is known beforehand and we do not require that the queries are answered in any particular order.)*

4 **Exercise 15.11:** Describe a data structure that stores n line segments in $O(n)$ storage and allows us to compute the number of line segments that intersect a given query line in time $O(n^{\alpha})$, for $\alpha = \log_2 \frac{1+\sqrt{5}}{2} < 0.695$.

1 **Exercise 15.12:** (a) Extend the definition of the stabbing region of a set of line segments from two to three dimensions. *(Hint: map the endpoints of a line segment to planes, map a plane to a point, and define the double wedge of a line segment such that a plane intersects the line segment if and only if its corresponding point is contained in the double wedge.)*

4 (b) Prove that the number of facets in the boundary of the stabbing region for n line segments in three dimensions is in $O(n^2)$.

15.10. Bibliographic Notes

The algorithmic methods dealing with transversals for vertical line segments (see Section 15.5) are taken from O'Rourke (1981), and both the combinatorial and the algorithmic results on transversals for not necessarily vertical line segments (see Sections 15.2, 15.3, and 15.4) are taken from Edelsbrunner, Maurer, Preparata, Rosenberg, Welzl, Wood (1982). Algorithms that compute transversals for rectangles, circles, and other objects are described in Edelsbrunner (1985) and in Atallah, Bajaj (1987). Pach, Sharir (1987) give a solution to Exercise 15.12(b) using two-dimensional arrangements and a divide-and-conquer argument.

Trivial algorithms that decide the existence of transversals for a finite set of objects follow from theorems of the type

> "if every k objects of the set have a common transversal, then all objects in the set have a transversal".

Such theorems hold for sets of vertical line segments ($k=3$, see Exercise 15.1(a)) and for sets of at least six congruent and non-intersecting circles ($k=4$, see Grünbaum (1958) who proves Exercise 15.1(b)). Surprisingly, no such constant k exists for arbitrary line segments in the plane (see Lewis (1980) who proves Exercise 15.1(c)). Additional results on related problem can be found in Hadwiger, Debrunner (1959). A solution to Exercise 15.3 which considers transversals for non-intersecting segments in the plane can be found in Katchalski, Lewis, Zaks (1985).

Exercises 15.6(c) and (e) reveal the close relationship between the boundary of stabbing regions of line segments and Davenport-Schinzel sequences defined in Exercise 15.6.

Davenport, Schinzel (1965) were the first to define such sequences as they came up in investigations of differential equations. Davenport–Schinzel sequences depend on two parameters: n, the size of the alphabet, and k, which is such that the longest alternating scattered subsequence is of length $k+1$. The historically first non-trivial upper bound on the maximum length of Davenport–Schinzel sequences was derived by Szemerédi (1974). He showed that the length of any Davenport–Schinzel sequence for n letters and any constant k is in $O(n \log^* n)$, where $\log^* n$ is the inverse of the function $A_3(n)$ defined in Exercise 15.5. For the case $k=3$, Hart, Sharir (1986) improved this result to $O(n\alpha(n))$, where $\alpha(n)$ is the inverse of Ackermann's function $A(n)$ defined in Exercise 15.5 (see also Ackermann (1928)). Solutions to Exercises 15.6(f) and 15.8 can be found in Wiernik (1986) and in Pollack, Sharir, Sifrony (1987).

The remainder of this section offers a few historical remarks regarding the algorithmic design paradigms described in this chapter. The use of geometric transforms is widespread in the geometric literature. We mention only Martin (1982) as a general source, Goodman, Pollack (1982b) for a detailed investigation of duality in two dimensions, and Brown (1980) for the introduction of geometric transforms as a standard tool in the design of algorithms for geometric problems. One of the first mentions of divide-and-conquer as a design paradigm for algorithms can be found in Aho, Hopcroft, Ullman (1974) long after the method was in general use. One of the roots of the paradigm can be traced back to Bellman (1957) who popularized a method generally known as "dynamic programming". The design paradigm referred to in this book as prune-and-search was independently developed by Dyer (1984) and by Megiddo (1983b). It can be seen as a generalization of the linear time algorithm that finds the median of an unsorted set drawn from some totally ordered domain described in Blum, Floyd, Pratt, Rivest, Tarjan (1972). The idea referred to in Section 15.7 as the locus approach is rather general and comes in many flavors. For example, it is a popular approach to path finding methods (see Lozano-Perez, Wesley (1979) who call the idea the "configuration space" approach). Many applications of the locus approach to problems in computational geometry can be found in Preparata, Shamos (1985). A solution to Exercise 15.11 using techniques which are fundamentally different from the locus approach is offered in Dobkin, Edelsbrunner (1984). A comparison of this result with Theorem 15.9 illustrates that the locus approach has the tendency to lead to very fast algorithms that sometimes need a lot of storage, however. The dynamization of static data structures is a much investigated problem in data structuring which has given rise to a variety of interesting methods. Most of these methods can be classified as either balancing techniques using trees or as decomposition techniques that maintain systems of static data structures. One of the main sources for such techniques is the book of Overmars (1983). The particular decomposition method described in Section 15.8 can be found in Maurer, Ottmann (1979) and in van Leeuwen, Wood (1980). Special dynamization methods that solve Exercises 15.10(a) through 15.10(c) are offered in Gowda, Kirkpatrick (1980), in Overmars, van Leeuwen (1981), and in Edelsbrunner, Overmars (1985).

REFERENCES

Ackermann, W. Zum Hilbertschen Aufbau der reellen Zahlen. *Math. Ann.* **99** (1928), 118–133.

Aggarwal, A., Klawe, M.M., Moran, S., Shor, P. and Wilber, R. Geometric applications of a matrix sorting algorithm. *In* "Proc. 2nd Ann. ACM Sympos. Comput. Geom. 1986", 285–292.

Aho, A.V., Hopcroft, J.E. and Ullman, J.D. *The Design and Analysis of Computer Algorithms.* Addison-Wesley, Reading, Mass., 1974.

Alexanderson, G.L. and Wetzel, J.E. Arrangements of planes in space. *Discrete Math.* **34** (1981), 219–240.

Alexandroff, P. and Hopf, H. *Topologie I.* Grundlehren der math. Wiss. 45, Julius Springer, Berlin, 1935.

Alexandrov, A.D. *Konvexe Polyeder.* Akademie-Verlag, Berlin, 1958.

Alon, N. and Györi, E. The number of small semispaces of a finite set of points in the plane. *J. Combin. Theory Ser.A* **41** (1986), 154–157.

Alon, N. and Kalai, G. A simple proof of the upper bound theorem. *European J. Combin.* **6** (1985), 211–214.

Appel, K. and Haken, W. Every planar map is four colourable. *Illinois J. Math.* **21** (1977), 429–567.

Ash, P.F. and Bolker, E.D. Generalized Dirichlet tessellations. *Geom. Dedicata* **20** (1986), 209–243.

Atallah, M. and Bajaj, Ch. Efficient algorithms for common transversals. *Inform. Process. Lett.* (1987), to appear.

Aurenhammer, F. Power diagrams: properties, algorithms, and applications. *SIAM J. Comput.* **16** (1987), 78–96.

Aurenhammer, F. and Edelsbrunner, H. An optimal algorithm for constructing the weighted Voronoi diagram in the plane. *Pattern Recognition* **17** (1984), 251–257.

Avis, D. The number of furthest neighbour pairs of a finite planar set. *Amer. Math. Monthly* **91** (1984), 417–420.

——. On the partitionability of point sets in space. *In* "Proc. 1st ACM Sympos. on Comput. Geom. 1985", 116–120.

Avis, D. and ElGindy, H. Triangulating simplicial point sets in space. *Discrete Comput. Geom.* **2** (1987), 99–111.

Avis, D., Erdös, P. and Pach, J. Repeated distances in space. Manuscript, School Comput. Sci., McGill Univ., Montreal, Quebec, 1986.

Bárány, I. and Füredi, Z. Empty simplices in Euclidean space. Rep. 689, School Operations Res., Cornell Univ., Ithaca, NY, 1986.

Barnette, D. The minimum number of vertices of a simple polytope. *Israel J. Math.* **10** (1971), 121–125.

——. A proof of the lower bound conjecture for convex polytopes. *Pacific J. Math.* **46** (1973), 349–354.

Bellman, R.E. *Dynamic Programming.* Princeton Univ. Press, Princeton, NJ, 1957.

Ben–Or, M. Lower bounds for algebraic computation trees. *In* "Proc. 15th Ann. ACM Sympos. Theory Comput. 1983", 80–86.

Bentley, J.L. and Saxe, J.B. Decomposable searching problems I: static to dynamic tranformations. *J. Algorithms* **1** (1980), 301–358.

Bentley, J.L. and Shamos, M.I. Divide and conquer for linear expected time. *Inform. Process. Lett.* **7** (1978), 87–91.

Bieri, H. and Nef, W. A recursive plane–sweep algorithm, determining all cells of a finite division of R^d. *Computing* **28** (1982), 189–198.

Blum, M., Floyd, R.W., Pratt, V.R., Rivest, R.L. and Tarjan, R.E. Time bounds for selection. *J. Comput. System Sci.* **7** (1972), 448–461.

Bondy, J.A. and Murty, U.S.R. *Graph Theory with Applications.* Maxmillan Press, Hong Kong, 1976.

Borsuk, K. Drei Sätze über die n–dimensionale euklidische Sphäre. *Fund. Math.* **20** (1933), 177–190.

Bowyer, A. Computing Dirichlet tessellations. *Comput. J.* **24** (1981), 162–166.

Bronsted, A. *An Introduction to Convex Polytopes.* Grad. Texts in Math., Springer–Verlag, New York, 1983.

Brostow, W., Dussault, J.–P. and Fox, B.L. Construction of Voronoi polyhedra. *J.Comput. Phys.* **29** (1978), 81–92.

Brown, K.Q. Voronoi diagrams from convex hulls. *Inform. Process. Lett.* **9** (1979), 223–228.

——. Geometric transforms for fast geometric algorithms. Ph.D.Thesis, Rep. CMU–CS–80–101, Dept. Comput. Sci., Carnegie–Mellon Univ., Pittsburgh, Penn., 1980.

Buck, R.C. Partition of space. *Amer. Math. Monthly* **50** (1943), 541–544.

Buck, R.C. and Buck, E.F. Equipartion of convex sets. *Math. Mag.* **22** (1948/49), 195–198.

Burr, S.A., Grünbaum, B. and Sloane, N.J.A. The orchard problem. *Geom. Dedicata* **2** (1974), 397–424.

Canham, R.J. A theorem on arrangements of lines in the plane. *Israel J. Math.* **7** (1969), 393–397.

——. Arrangements of hyperplanes in projective and Euclidean space. Ph.D.Thesis, Dept. Math., Univ. East Anglia, Wash., 1971.

Carathéodory, C. Über den Variabilitätsbereich der Koeffizienten von Potenzreihen, die gegebene Werte nicht annehmen. *Math. Ann.* **64** (1907), 95–115.

——. Über den Variabilitätsbereich der Fourierschen Konstanten von positiven harmonischen Funktionen. *Rend. Circ. Mat. Palermo* **32** (1911), 193–217.

Chand, D.R. and Kapur, S.S. An algorithm for convex polytopes. *J. Assoc. Comput. Mach.*

17 (1970), 78–86.

Chazelle, B. On the convex layers of a convex set. *IEEE Trans. Inform. Theory* **IT–31** (1985a), 509–517.

——. How to search in history. *Inform. Control* **64** (1985b), 77–99.

——. Fast searching in real algebraic manifold with applications to geometric complexity. *In* "Proc. Coll. on Trees in Algebra and Progr. 1985c", 145–156, Lecture Notes in Comput. Sci. 185, Springer–Verlag, Berlin.

——. Filtering search: a new approach to query–answering. *SIAM J. Comput.* **15** (1986a), 703–724.

——. Reporting and counting segment intersections. *J. Comput. System Sci.* **32** (1986b), 156–182.

Chazelle, B. and Dobkin, D.P. Detection is easier than computation. *In* "Proc. 12th Ann. ACM Sympos. Theory Comput. 1980", 146–153.

Chazelle, B. and Edelsbrunner, H. Optimal solutions for a class of point retrieval problems. *J. Symbolic Comput.* **1** (1985a), 47–56.

——. An improved algorithm for constructing k^{th}–order Voronoi diagrams. *In* "Proc. 1st ACM Sympos. Comput. Geom. 1985b", 228–234.

Chazelle, B. and Guibas, L.J. Fractional cascading: I. a data structuring technique. *Algorithmica* **1** (1986a), 133–162.

——. Fractional cascading: II. applications. *Algorithmica* **1** (1986b), 163–191.

Chazelle, B., Guibas, L.J. and Lee, D.T. The power of geometric duality. *BIT* **25** (1985), 76–90.

Chazelle, B. and Preparata, F.P. Halfspace range search: an algorithmic application of k–sets. *Discrete Comput. Geom.* **1** (1986), 83–93.

Cheriton, D. and Tarjan, R.E. Finding minimum spanning trees. *SIAM J. Comput.* **5** (1976), 724–742.

Chvátal, V. *Linear Programming.* W.H.Freeman and Company, New York, 1983.

Chvátal, V. and Klincsek, G. Finding largest convex subsets. *In* "Proc. 11th Southeastern Conf. on Combin., Graph Theory and Comput. 1980", 453–460.

Clarkson, K.L. Linear programming in $O(n \cdot 3^{d^2})$ time. *Inform. Process. Lett.* **22** (1986), 21–24.

——. New applications of random sampling in computational geometry. *Dicrete Comput. Geom.* **2** (1987), 195–222.

Cole, R. Partitioning point sets in 4 dimensions. *In* "Proc. 12th Internat. Colloq. on Autom., Lang. and Progr. 1985", 111–119, Lecture Notes in Comput. Sci. 194, Springer–Verlag, Berlin.

——. Searching and storing similar lists. *J. Algorithms* **7** (1986a), 202–220.

——. Slowing down sorting networks to obtain faster sorting algorithms. *J. Assoc. Comput. Mach.* (1986b), to appear.

——. Partitioning point sets in arbitrary dimensions. *Theoret. Comput. Sci.* (1987), to appear.

Cole, R., Sharir, M. and Yap, C.K. On k–hulls and related problems. *SIAM J. Comput.* **16**

(1987), 61–77.

Coxeter, H.M.S. A problem of collinear points. *Amer. Math. Monthly* **55** (1948), 26–28.

——. A classification of zonohedra by means of projective diagrams. *J. Math. Pures Appl.* **41** (1962), 137–156.

——. *Regular Polytopes.* 3rd edition, Dover, New York, 1973.

Dantzig, G.B. *Linear Programming and Extensions.* Princeton Univ. Press, Princeton, NJ, 1963.

Davenport, H. and Schinzel, A. A combinatorial problem connected with differential equations. *Amer. J. Math.* **87** (1965), 684–694.

Day, W.H.E. and Edelsbrunner, H. Efficient algorithms for agglomerative hierarchical clustering methods. *J. Classification* **1** (1984), 7–24.

Dehn, M. Die Eulersche Formel in Zusammenhang mit dem Inhalt in der nicht–Euklidischen Geometrie. *Math. Ann.* **61** (1905), 561–586.

Delaunay, B. Sur la sphère vide. *Izv. Akad. Nauk SSSR, Otdelenie Matematicheskii i Estestvennyka Nauk* **7** (1934), 793–800.

Dijkstra, E.W. A note on two problems in connexion with graphs. *Numer. Math.* **1** (1959), 269–271.

Dirac, G.A. Collinearity properties of sets of points. *Quart. J. Math. Oxford* **2** (1951), 221–227.

Dirichlet, P.G.L. Über die Reduktion der positiven quadratischen Formen mit drei unbestimmten ganzen Zahlen. *J. Reine Angew. Math.* **40** (1850), 209–227.

Dobkin, D.P. and Edelsbrunner, H. Space searching for intersecting objects. *In* "Proc. 25th Ann. IEEE Sympos. Found. Comput. Sci. 1984", 387–392.

Dobkin, D.P., Edelsbrunner, H. and Overmars, M.H. Searching for empty convex polygons. Manuscript, Dept. Comput. Sci., Univ. Illinois, Urbana, Ill., 1987.

Dobkin, D.P. and Kirkpatrick, D.G. Fast detection of polyhedral intersection. *Theoret. Comput. Sci.* **27** (1983), 241–253.

——. A linear algorithm for determining the separation of convex polyhedra. *J. Algorithms* **6** (1985), 381–392.

Dobkin, D.P. and Lipton, R.J. Multidimensional searching problems. *SIAM J. Comput.* **5** (1976), 181–186.

Dobkin, D.P. and Munro, J.I. Efficient uses of the past. *J. Algorithms* **6** (1985), 455–465.

Dobkin, D.P. and Reiss, S.P. The complexity of linear programming. *Theoret. Comput. Sci.* **11** (1980), 1–18.

Dyer, M.E. Linear algorithms for two– and three–variable linear programs. *SIAM J. Comput.* **13** (1984), 31–45.

——. On a multidimensional search technique and its application to the Euclidean one–center problem. *SIAM J. Comput.* **15** (1986), 725–738.

Edelsbrunner, H. Finding transversals for sets of simple geometric figures. *Theoret. Comput. Sci.* **35** (1985), 55–69.

——. Edge–skeletons in arrangements with applications. *Algorithmica* **1** (1986), 93–109.

Edelsbrunner, H. and Guibas, L.J. Topologically sweeping an arrangement. *In* "Proc. 18th

Ann. ACM Sympos. Theory Comput. 1986", 389–403.

Edelsbrunner, H., Guibas, L.J. and Stolfi, J. Optimal point location in a monotone subdivision. *SIAM J. Comput.* **15** (1986), 317–340.

Edelsbrunner, H. and Haussler, D. The complexity of cells in three-dimensional arrangements. *Discrete Math.* **60** (1986), 139–146.

Edelsbrunner, H. and Huber, F. Dissecting sets of points in two and three dimensions. Rep. F138, Inst. Informationsverarb., Techn. Univ. Graz, Austria, 1984.

Edelsbrunner, H., Kirkpatrick, D.G. and Seidel, R. On the shape of a set of points in the plane. *IEEE Trans. Inform. Theory* **IT–29** (1983), 551–559.

Edelsbrunner, H. and Maurer, H.A. Finding extreme points in three dimensions and solving the post-office problem in the plane. *Inform. Process. Lett.* **21** (1985), 39–47.

Edelsbrunner, H., Maurer, H.A., Preparata, F.P., Rosenberg, A.L., Welzl, E. and Wood, D. Stabbing line segments. *BIT* **22** (1982), 274–281.

Edelsbrunner, H., O'Rourke, J. and Seidel, R. Constructing arrangements of lines and hyperplanes with applications. *SIAM J. Comput.* **15** (1986), 341–363.

Edelsbrunner, H. and Overmars, M.H. Batched dynamic solutions to decomposable searching problems. *J. Algorithms* **6** (1985), 515–542.

Edelsbrunner, H., Overmars, M.H. and Wood, D. Graphics in Flatland: a case study. *In "Advances in Computing Research* **1**: *Computational Geometry*", F.P.Preparata, ed. (1983), 35–59, Jai Press, London.

Edelsbrunner, H., Preparata, F.P. and West, D.B. Tetrahedrizing point sets in three dimensions. Rep. UIUCDCS–R–86–1310, Dept. Comput. Sci., Univ. Illinois, Urbana, Ill., 1986.

Edelsbrunner, H. and Seidel, R. Voronoi diagrams and arrangements. *Discrete Comput. Geom.* **1** (1986), 25–44.

Edelsbrunner, H. and Skiena, S.S. On the number of furthest neighbour pairs in a point set. Rep. UIUCDCS–R–86–1312, Dept. Comput. Sci., Univ. Illinois, Urbana, Ill., 1986.

Edelsbrunner, H. and Stöckl, G. The number of extreme pairs of finite point-sets in Euclidean spaces. *J. Combin. Theory Ser.A* **43** (1986), 344–349.

Edelsbrunner, H. and Waupotitsch, R. Computing a ham-sandwich cut in two dimensions. *J. Symbolic Comput.* **2** (1986), 171–178.

Edelsbrunner, H. and Welzl, E. On the number of line separations of a finite set in the plane. *J. Combin. Theory Ser.A* **38** (1985), 15–29.

——. Constructing belts in two-dimensional arrangements with applications. *SIAM J. Comput.* **15** (1986a), 271–284.

——. On the maximal number of edges of many faces in an arrangement. *J. Combin. Theory Ser.A* **41** (1986b), 159–166.

——. Halfplanar range search in linear space and $O(n^{0.695})$ query time. *Inform. Process. Lett.* **23** (1986c), 289–293.

Edwards, C.H.Jr. and Penny, D.E. *Calculus and Analytic Geometry.* 2nd edition, Prentice Hall, Englewood Cliffs, NJ, 1986.

Eggleston, H.G. *Convexity.* Cambridge Univ. Press, Cambridge, England, 1958.

Erdös, P. Some unsolved problems. *Publ. Math. Inst. Hungar. Acad. Sci.* **6** (1961), 221–254.

——. Nehany elemi geometriai problemarol. *Közepiskolai Matematikai Lapok* **24** (1962), 193–201.

——. On extremal problems of graphs and generalized graphs. *Israel J. Math.* **2** (1964), 183–190.

——. Combinatorial problems in geometry and number theory. *Proc. Sympos. Pure Math.* **34** (1979), 149–162.

Erdös, P., Lovász, L., Simmons, A. and Strauss, E.G. Dissection graphs of planar point sets. *In "A Survey of Combinatorial Theory"*, J.N.Srivastava et al., eds. (1973), 139–149, North-Holland, Amsterdam.

Erdös, P. and Szekeres, G. A combinatorial problem in geometry. *Compositio Math.* **2** (1935), 463–470.

——. On some extremum problems in elementary geometry. *Ann. Univ. Sci. Budapest* **3** (1960), 53–62.

Euler, L. Elementa doctrinae solidorum. *Novi Comm. Acad. Sci. Imp. Petropol.* **4** (1752/53a), 109–140.

——. Demonstratio nonnullarum insignium proprietatum, quibus solida hedris planis inclusa sunt praedita. *Novi Comm. Acad. Sci. Imp. Petropol.* **4** (1752/53b), 140–160.

Fairfield, J. Contoured shape generation: forms that people see in dot patterns. *In* "Proc. IEEE Conf. on Systems, Man and Cybern. 1979", 60–64.

Fedorov, E.S. Elemente der Gestaltenlehre. *Mineralogicheskoe obshchestvo, Leningrad* **21** (1885), 1–279.

Fejes Tóth, L. *Lagerungen in der Ebene, auf der Kugel und im Raum.* 2nd edition, Springer-Verlag, Berlin, 1972.

Floyd, R.W. and Rivest, R.L. Expected time bounds for selection. *Comm. ACM* **18** (1975), 165–172.

Folkman, J. and Lawrence, J. Oriented matroids. *J. Combin. Theory Ser.B* **25** (1978), 199–263.

Fortune, S. A sweepline algorithm for Voronoi diagrams. *In* "Proc. 2nd Ann. ACM Sympos. Comput. Geom. 1986", 313–322.

Fredman, M.L. On the information theoretic lower bound. *Theoret. Comput. Sci.* **1** (1976), 355–361.

——. The inherent complexity of dynamic data structures which accommodate range queries. *In* "Proc. 21st Ann. IEEE Sympos. Found. Comput. Sci. 1980", 191–199.

Fredman, M.L. and Tarjan, R.E. Fibonacci heaps and their uses in improved network optimization algorithms. *In* "Proc. 25th Ann. IEEE Sympos. Found. Comput. Sci. 1984", 338–346.

Fries, O. Zerlegung einer planaren Unterteilung der Ebene und ihre Anwendungen. M.S.Thesis, Inst. Angew. Math. and Inform., Univ. Saarlandes, Saarbrücken, Germany, 1985.

Füredi, Z. and Palásti, I. Arrangements of lines with a large number of triangles. *Proc. Amer. Math. Soc.* **92** (1984), 561–566.

Gale, D. On convex polyhedra. Abstract 794. *Bull. Amer. Math. Soc.* **61** (1955), 556.

Garey, M.R., Johnson, D.S., Preparata, F.P. and Tarjan, R.E. Triangulating a simple polygon. *Inform. Process. Lett.* **7** (1978), 175–179.

Goodman, J.E. and Pollack, R. On the combinatorial classification of nondegenerate configurations in the plane. *J. Combin. Theory Ser.A* **29** (1980a), 220–235.

----. Proof of Grünbaum's conjecture of the stretchability of certain arrangements of pseudolines. *J. Combin. Theory Ser.A* **29** (1980b), 385–390.

----. A combinatorial perspective on some problems in geometry. *Congress. Numer.* **32** (1981a), 383–394.

----. Three points do not determine a (pseudo-) plane. *J. Combin. Theory Ser.A* **31** (1981b), 215–218.

----. Helly–type theorems for pseudoline arrangements in P^2. *J. Combin. Theory Ser.A* **32** (1982a), 1–19.

----. A theorem of ordered duality. *Geom. Dedicata* **12** (1982b), 63–74.

----. Multidimensional sorting. *SIAM J. Comput.* **12** (1983), 484–507.

----. Semispaces of configurations, cell complexes of arrangements. *J. Combin. Theory Ser.A* **37** (1984), 257–293.

----. Upper bounds for configurations and polytopes in R^d. *Discrete Comput. Geom.* **1** (1986), 219–227.

Gowda, I.G. and Kirkpatrick, D.G. Exploiting linear merging and extra storage in the maintenance of fully dynamic geometric data structures. *In* "Proc. 18th Ann. Allerton Conf. Commun., Control, Comput. 1980", 1–10.

Graham, R.L. An efficient algorithm for determining the convex hull of a finite planar set. *Inform. Process. Lett.* **1** (1972), 132–133.

Grömer, H. Abschätzungen für die Anzahl der konvexen Körper, die einen konvexen Körper berühren. *Monatsh. Math.* **65** (1961), 74–81.

Grünbaum, B. On common transversals. *Arch. Math. (Basel)* **9** (1958), 465–469.

----. Measures of symmetry for convex sets. *Proc. Sympos. Pure Math.* **7***: Convexity* (1961), 233–270.

----. *Convex Polytopes.* John Wiley & Sons, London, 1967.

----. Arrangements of hyperplanes. *In* "Proc. 2nd Louisina Conf. on Combin., Graph Theory and Comput. 1971", 41–106.

----. *Arrangements and Spreads.* Regional Conf. Ser. Math., Amer. Math. Soc., Providence, RI, 1972.

----. New views of some old questions of combinatorial geometry. *Atti Accad. Naz. Lincei* **17***: Theorie Combinatorie* (1976), 451–468.

Guibas, L.J. and Stolfi, J. Primitives for the manipulation of general subdivisions and the computation of Voronoi diagrams. *ACM Trans. Graphics* **4** (1985), 74–123.

Hacker, R. Certification of Algorithm 112: position of point relative to polygon. *Comm. ACM* **5** (1962), 606.

Hadwiger, H. Elementare Begründung ausgewählter stetigkeitsgeometrischer Sätze für Kreis und Kugelfläche. *Elem. Math.* **14** (1959), 49–72.

----. Simultane Vierteilung zweier Körper. *Arch. Math. (Basel)* **17** (1966), 274–278.

Hadwiger, H. and Debrunner, H. *Kombinatorische Geometrie in der Ebene.* Monogr. de

L'Enseignement Math., Geneva, Italy, 1959.

Hadwiger, H., Debrunner, H. and Klee, V. *Combinatorial Geometry in the Plane.* Holt, Rinehart and Winston, New York, 1964.

Halmos, P. *Finite-dimensional Vector Spaces.* 2nd edition, D. Van Nostrand, Princeton, NJ, 1958.

----. *Naive Set Theory.* D. Van Nostrand, Princeton, NJ, 1960.

Hansen, S. Contributions to the Sylvester-Gallai theory. Ph.D.Thesis, Univ. Copenhagen, Denmark, 1981.

Harary, F. *Graph Theory.* Addison-Wesley, Reading, Mass., 1969.

Harborth, H. Konvexe Fünfecke in ebenen Punktmengen. *Elem. Math.* **33** (1978), 116-118.

Hardy, G. and Wright, E. *The Theory of Numbers.* 4th edition, Oxford Univ. Press, London, 1965.

Harel, D. A linear time algorithm for the lowest common ancestor problem. *In* "Proc. 21st Ann. IEEE Sympos. Found. Comput. Sci. 1980", 308-319.

Hart, S. and Sharir, M. Nonlinearity of Davenport-Schinzel sequences and of a generalized path compression scheme. *Combinatorica* **6** (1986), 151-177.

Haussler, D. and Welzl, E. ϵ-nets and simplex range queries. *Discrete Comput. Geom.* **2** (1987), 127-151.

Helly, E. Über Mengen konvexer Körper mit gemeinschaftlichen Punkten. *Jahresber. Deutsch. Math.-Verein.* **32** (1923), 175-176.

Heppes, A. Beweis einer Vermutung von A.Vázsonyi. *Acta Math. Acad. Sci. Hungar.* **7** (1956), 463-466.

Hodder, I. and Orton, C. *Spatial Analysis in Archaeology.* Cambridge, 1976.

Hopcroft, J.E. and Ullman, J.D. *Introduction to Automata Theory, Languages, and Computation.* 2nd edition, Addison-Wesley, Reading, Mass., 1979.

Horton, J.D. Sets with no empty convex 7-gon. *Canad. Math. Bull.* **26** (1983), 482-484.

Imai, H., Iri, M. and Murota, K. Voronoi diagrams in the Laguerre geometry and its applications. *SIAM J. Comput.* **14** (1985), 93-105.

Jarvis, R.A. On the identification of the convex hull of a finite set of points in the plane. *Inform. Process. Lett.* **2** (1973), 18-21.

Jung, H.W.E. Über die kleinste Kugel, die eine räumliche Figur einschliesst. *J. Reine Angew. Math.* **123** (1901), 241-257.

----. Über den kleinsten Kreis, der eine ebene Figur einschließt. *J. Reine Angew. Math.* **137** (1910), 310-313.

Kalbfleisch, J.D., Kalbfleisch, J.G. and Stanton, R.G. A combinatorial problem on convex n-gons. *In* "Proc. 1st Louisiana Conf. on Combin., Graph Theory and Comput. 1970", 180-188.

Karmarkar, N. A new polynomial-time algorithm for linear programming. *Combinatorica* **4** (1984), 373-395.

Katchalski, M., Lewis, T. and Zaks, J. Geometric permutations for convex sets. *Discrete Math.* **54** (1985), 271-284.

Kelly, L.M. and Moser, W.O.J. On the number of ordinary lines determined by n points.

Canad. J. Math. **10** (1958), 210–219.

Khachiyan, L.G. Polynomial algorithm for linear programming. *Dokl. Akad. Nauk SSSR* **244** (1979), 1093–1096 (in Russian).

Kirkpatrick, D.G. Optimal search in planar subdivisions. *SIAM J. Comput.* **12** (1983), 28–35.

Kirkpatrick, D.G. and Seidel, R. The ultimate planar convex hull algorithm? *SIAM J. Comput.* **15** (1986), 287–299.

Klee, R. *Über die einfachen Konfigurationen der euklidischen und der projektiven Ebene.* Focken and Oltmanns, Dresden, 1938.

Klee, V. and Minty, G.L. How good is the simplex algorithm? *In "Inequalities III"*, O.Shisha, ed. (1972), 159–179, Academic Press, New York.

Knuth, D.E. *Fundamental Algorithms: The Art of Computer Programming I.* Addison–Wesley, Reading, Mass., 1968.

----. *Seminumerical Algorithms: The Art of Computer Programming II.* Addison–Wesley, Reading, Mass., 1969.

----. *Sorting and Searching: The Art of Computer Programming III.* Addison–Wesley, Reading, Mass., 1973.

Lawson, C.L. Generation of a triangular grid with applications to contour plotting. Memo. 299, Calif. Inst. Jet Propul. Lab., 1972.

Lee, D.T. Proximity and reachability in the plane. Ph.D.Thesis, Rep. R–831, Coördinated Sci. Lab., Univ. Illinois, Urbana, Ill., 1978.

----. On k–nearest neighbor Voronoi diagrams in the plane. *IEEE Trans. Comput.* **C–31** (1982), 478–487.

Lee, D.T. and Drysdale, R.L.III Generalization of Voronoi diagrams in the plane. *SIAM J. Comput.* **10** (1981), 73–87.

Lee, D.T. and Lin, A.K. Generalized Delaunay triangulation for planar graphs. *Discrete Comput. Geom.* **1** (1986), 201–217.

Lee, D.T. and Preparata, F.P. Location of a point in a planar subdivision with applications. *SIAM J. Comput.* **6** (1977), 594–606.

----. The all nearest–neighbor problem for convex polygons. *Inform. Process. Lett.* **7** (1978), 189–192.

----. Euclidean shortest paths in the presence of rectilinear barriers. *Network* **14** (1984a), 393–410.

----. Computational geometry – a survey. *IEEE Trans. Comput.* **C–33** (1984b), 1072–1101.

Lewis, T. Two counterexamples concerning transversals for convex subsets in the plane. *Geom. Dedicata* **9** (1980), 461–465.

Li, H., Lavin, M.A. and LeMaster, R.J. Fast Hough transform: A hierarchical approach. *Comput. Vision, Graphics, Image Process.* **36** (1986), 139–161.

Lipton, R.J. and Tarjan, R.E. Applications of a planar separator theorem. *In* "Proc. 18th Ann. IEEE Sympos. Found. Comput. Sci. 1977", 162–170.

Liu, C.L. *Introduction to Combinatorial Mathematics.* McGraw–Hill, New York, 1968.

----. *Elements of Discrete Mathematics.* 2nd edition, McGraw-Hill, New York, 1985.

Lozano–Perez, T. and Wesley, M. An algorithm for planning collision free paths among polyhedral obstacles. *Comm. ACM* **22** (1979), 560–570.

Lovász, L. On the number of halving lines. *Ann. Univ. Sci. Budapest, Eötvös, Sect. Math.* **14** (1971), 107–108.

Lyusternik, L.A. *Convex Figures and Polyhedra.* D.C. Heath and Company, Boston, 1966.

Martin, G.E. *Transformation Geometry: An Introduction to Symmetry.* Springer–Verlag, New York, 1982.

Matula, D.W. and Sokal, R.R. Properties of Gabriel graphs relevant to geographic variation research and the clustering of points in the plane. *Geogr. Analysis* **12** (1980), 205–222.

Maurer, H.A. and Ottmann, Th. Dynamic solutions of decomposable searching problems. *In* "*Discrete Structures and Algorithms*", U.Pape, ed. (1979), 17–24, Carl Hanser Verlag.

McKenna, M. and Seidel, R. Finding the optimal shadow of a convex polytope. *In* "Proc. 1st ACM Sympos. Comput. Geom. 1985", 24–28.

McMullen, P. On zonotopes. *Trans. Amer. Math. Soc.* **159** (1971a), 91–109.

——. The maximum number of faces of a convex polytope. *Mathematika* **17** (1971b), 179–184.

McMullen, P. and Shepard, G.C. *Convex Polytopes and the Upper Bound Conjecture.* London Math. Soc. Lecture Notes Ser.3, Cambridge Univ. Press, Cambridge, 1971.

Megiddo, N. Applying parallel computation algorithms in the design of serial algorithms. *J. Assoc. Comput. Mach.* **30** (1983a), 852–865.

——. Linear–time algorithms for linear programming in R^3 and related problems. *SIAM J. Comput.* **12** (1983b), 759–776.

——. Linear programming in linear time when the dimension is fixed. *J. Assoc. Comput. Mach.* **31** (1984), 114–127.

——. Partitioning with two lines in the plane. *J. Algorithms* **3** (1985), 430–433.

Mehlhorn, K. *Data Structures and Algorithms 1: Sorting and Searching.* Springer–Verlag, Berlin, 1984a.

——. *Data Structures and Algorithms 2: Graph Algorithms and NP–Completeness.* Springer–Verlag, Berlin, 1984b.

——. *Data Structures and Algorithms 3: Multidimensional Searching and Computational Geometry.* Springer–Verlag, Berlin, 1984c.

Mendelsohn, B. *Introduction to Topology.* Allyn & Bacon, Boston, 1962.

Milnor, J. On the Betti numbers of real varieties. *Proc. Amer. Math. Soc.* **15** (1964), 275–280.

Moser, W.O.J. and Pach, J. Research problems in discrete geometry. Manuscript, Dept. Math., McGill Univ., Montreal, Quebec, 1986.

Moss, W.W. Some new analytic and graphic approaches to numerical taxonomy, with an example from the dermanyssidae (acari). *System. Zool.* **16** (1967), 177–207.

Motzkin, T. The lines and planes connecting the points of a finite set. *Trans. Amer. Math. Soc.* **70** (1951), 451–464.

Mücke, E.P. Ein Modell zur Elimination degenerierter Fälle in geometrischen Algorithmen. M.S.Thesis, Inst. Informationsverarb., Techn. Univ. Graz, Austria, 1985.

Nef, W. Zur Einführung der Eulerschen Charakteristik. *Monatsh. Math.* **92** (1981), 41–46.

——. Ein einfacher Beweis des Satzes von Euler–Schläfli. *Elem. Math.* **39** (1984), 1–6.

Nering, E.D. *Linear Algebra and Matrix Theory.* John Wiley & Sons, New York, 1963.

O'Rourke, J. An on-line algorithm for fitting straight lines between data ranges. *Comm. ACM* **24** (1981), 574–578.

——. The signature of a plane curve. *SIAM J. Comput.* **15** (1986), 34–51.

Overmars, M.H. Dynamization of order decomposable set problems. *J. Algorithms* **2** (1981), 245–260.

——. *The Design of Dynamic Data Structures.* Lecture Notes in Comput. Sci. 156, Springer–Verlag, Berlin, 1983.

Overmars, M.H. and van Leeuwen, J. Maintenance of configurations in the plane. *J. Comput. System Sci.* **23** (1981), 166–204.

Pach, J. and Sharir, M. The upper envelope of piecewise linear functions and the boundary of a region enclosed by convex plates: combinatorial analysis. Manuscript, Courant Inst. of Math. Sci., New York Univ., 1987.

Papadimitriou, C.H. and Steiglitz, K. *Combinatorial Optimization: Algorithms and Complexity.* Prentice Hall, Englewood Cliffs, NJ, 1982.

Paschinger, I. Konvexe Polytope und Dirichletsche Zellenkomplexe. Ph.D.Thesis, Inst. Math., Univ. Salzburg, Austria, 1982.

Paterson, M.S. and Yao, F.F. Point retrieval for polygons. *J. Algorithms* **7** (1986), 441–447.

Peck, G.W. On 'k-sets' in the plane. *Discrete Math.* **56** (1985), 73–74.

Perrin, R. Sur le problème des aspects. *Bull. Soc. Math. France* **10** (1881/82), 103–127.

Pollack, R., Sharir, M. and Sifrony, S. Separating two simple polygons by a sequence of translations. *Dicrete Comput. Geom.* (1987), to appear.

Preparata, F.P. A new approach to planar point location. *SIAM J. Comput.* **10** (1981), 473–482.

Preparata, F.P. and Hong, S.J. Convex hulls of finite sets of points in two and three dimensions. *Comm. ACM* **20** (1977), 87–93.

Preparata, F.P. and Shamos, M.I. *Computational Geometry – an Introduction.* Springer–Verlag, New York, 1985.

Prim, R.C. Shortest connection networks and some generalizations. *Bell System Tech. J.* **36** (1957), 1389–1401.

Purdy, G.B. Triangles in arrangements of lines. *Discrete Math.* **25** (1979), 157–163.

——. On the number of regions determined by n lines in the projective plane. *Geom. Dedicata* **9** (1980), 107–109.

Radon, J. Mengen konvexer Körper, die einen gemeinsamen Punkt enthalten. *Math. Ann.* **83** (1921), 113–115.

Raynaud, H. Sur l'enveloppe convexe des nuages des points aléatoires dans R_n, I. *J. Appl. Probab.* **7** (1970), 35–48.

Reif, J. and Storer, J.A. Shortest paths in Euclidean spaces with polyhedral obstacles. Rep. CS–85–121, Dept. Comput. Sci., Brandeis Univ., Waltham, Mass., 1985.

Reingold, E.M. and Hansen, W.J. *Data Structures in Pascal.* Little, Brown and Company, Boston, 1986.

Rényi, A. and Sulanke, R. Über die konvexe Hülle von n zufällig gewählten Punkten. *Z. Wahrscheinlichkeitstheorie* **2** (1963), 75–84.

Ringel, G. Teilungen der Ebene durch Geraden und topologische Geraden. *Math. Z.* **64** (1956), 79–102.

Rogers, C.A. *Packing and Covering.* Cambridge Univ. Press, Cambridge, England, 1964.

Roudneff, J.-P. Cells with many facets in arrangements of hyperplanes. Manuscript, E.R. Combinaroire U.E.R., Univ. Pierre et Marie Curie, Paris, 1986.

Salomaa, A. *Formal Languages.* Academic Press, New York, 1973.

Sarnak, N. and Tarjan, R.E. Planar point location using persistent search trees. *Comm. ACM* **29** (1986), 669–679.

Schläfli, L. Theorie der vielfachen Kontinuität. *Denkschr. Schweiz. naturf. Ges.* **38** (1901), 1–237.

Schönhage, A.M., Paterson, M.S. and Pippenger, N. Finding the median. *J. Comput. System Sci.* **13** (1976), 184–199.

Schurle, A.W. *Topics in Topology.* North-Holland, New York, 1979.

Schütte, K. Das Problem der dreizehn Kugeln. *Math. Ann.* **125** (1953), 325–334.

Scott, P.R. On the sets of directions determined by n points. *Amer. Math. Monthly* **77** (1970), 502–505.

Seidel, R. A convex hull algorithm optimal for point sets in even dimensions. Rep. 81–14, Dept. Comput. Sci., Univ. British Columbia, Vancouver, BC, 1981.

——. The complexity of Voronoi diagrams in higher dimensions. *In* "Proc. 20th Ann. Allerton Conf. Commun., Control, Comput. 1982", 94–95.

——. A method for proving lower bounds for certain geometric problems. *In* "*Computational Geometry*", G.T.Toussaint, ed. (1985), 319–334, North-Holland, Amsterdam.

——. Constructing higher-dimensional convex hulls at logarithmic cost per face. *In* "Proc. 18th Ann. ACM Sympos. Theory Comput. 1986", 404–413.

Shamos, M.I. Geometric complexity. *In* "Proc. 7th Ann. ACM Sympos. Theory Comput. 1975", 224–233.

——. Geometry and statistics: problems at the interface. *In* "*Algorithms and Complexity*", J.F.Traub, ed. (1976), 251–280, Academic Press, New York.

——. Computational geometry. Ph.D.Thesis, Dept. Comput. Sci., Yale Univ., New Haven, Conn., 1978.

Shamos, M.I. and Hoey, D. Closest-point problems. *In* "Proc. 16th Ann. IEEE Sympos. Found. Comput. Sci. 1975", 151–162.

Shimrat, M. Algorithm 112: position of a point relative to polygon. *Comm. ACM* **5** (1962), 434.

Sibson, R. Locally equiangular triangulations. *Comput. J.* **21** (1978), 243–245.

Sommerville, D.M.Y. The relations connecting the angle-sums and volume of a polytope in space of n dimensions. *Proc. Roy. Soc. London, Ser.A* **115** (1927), 103–119.

Steele, J.M. and Yao, A.C. Lower bounds for algebraic decision trees. *J. Algorithms* **3** (1982), 1–8.

Steinberg, R. Solution to Problem 4065. *Amer. Math. Monthly.* **51** (1944), 169–171.

Steiner, J. Einige Gesetze über die Theilung der Ebene und des Raumes. *J. Reine Angew. Math.* **1** (1826), 349–364.

Steinitz, E. Polyeder und Raumeinteilungen. *Enz. Math. Wiss.* **3**, 1. Teil, 2. Hälfte (1939), 1–139, Teubner, Leipzig, Germany.

Steinitz, E. and Rademacher, H. *Vorlesungen über die Theorie der Polyeder.* Julius Springer, Berlin, 1934.

Stöckl, G. Gesammelte und neue Ergebnisse über extreme k–Mengen für ebene Konfigurationen. M.S.Thesis, Inst. Informationsverarb., Techn. Univ. Graz, Austria, 1984.

Strommer, T.O. Triangles in arrangements of lines. *J. Combin. Theory Ser.A* **23** (1977), 314–320.

Supowit, K.J. The relative neighborhood graph, with an application to minimum spanning trees. *J. Assoc. Comput. Mach.* **30** (1983), 428–448.

Sutherland, J.W. Solution to Problem 167. *Jahresber. Deutsch. Math.-Verein.* **45** (1935), 33–34.

Swart, G. Finding the convex hull facet by facet. *J. Algorithms* **6** (1985), 17–48.

Sylvester, J.J. Mathematical question 11851. *Educational Times* **59** (1893), 98.

Szemerédi, E. On a problem by Davenport and Schinzel. *Acta Arith.* **25** (1974), 213–224.

Szemerédi, E. and Trotter, W.T.Jr. Extremal problems in discrete geometry. *Combinatorica* **3** (1983), 381–392.

Tarjan, R.E. Depth–first search and linear graph algorithms. *SIAM J. Comput.* **2** (1972), 146–160.

----. *Data Structures and Network Algorithms.* CBMS–NSF Regional Conf. Ser. Applied Math., SIAM, Philadelphia, Penn., 1983.

Tarjan, R.E. and van Wyk, Ch.J. An $O(n \log\log n)$–time algorithm for triangulating simple polygons. Rep. CS–TR–052–86, Dept. Comput. Sci., Princeton Univ., NJ, 1986.

Toussaint, G.T. The relative neighbourhood graph of a finite planar set. *Pattern Recognition* **12** (1980), 261–268.

Ungar, P. $2N$ noncollinear points determine at least $2N$ directions. *J. Combin. Theory Ser.A* **33** (1982), 343–347.

Vaidya, P.M. An optimal algorithm for the all–nearest–neighbors problem. *In* "Proc. 27th Ann. IEEE Sympos. Found. Comput. Sci. 1986", 117–122.

van Leeuwen, J. Problem P20. *Bulletin of the EATCS* **19** (1983), 150.

van Leeuwen, J. and Wood, D. Dynamization of decomposable searching problems. *Inform. Process. Lett.* **10** (1980), 51–56.

Voronoi, G. Nouvelles applications des paramètres continus à la théorie des formes quadratiques. Premier Mémoire: Sur quelques propriétés des formes quadratiques positives parfaites. *J. Reine Angew. Math.* **133** (1907), 97–178.

----. Nouvelles applications des paramètres continus à la théorie des formes quadratiques. Deuxième Mémoire: Recherches sur les parallélloèdres primitifs. *J. Reine Angew. Math.* **134** (1908), 198–287.

Welzl, E. Constructing the visibility graph for n line segments in $O(n^2)$ time. *Inform. Process. Lett.* **20** (1985), 167–171.

——. More on k-sets of finite sets in the plane. *Discrete Comput. Geom.* **1** (1986), 95–100.

White, H.S. The convex cells formed by seven planes. *Proc. Nat. Acad. Sci. U.S.A.* **25** (1939), 147–153.

Wiernik, A. Planar realizations of nonlinear Davenport–Schinzel sequences by segments. *In* "Proc. 27th Ann. IEEE Sympos. Found. Comput. Sci. 1986", 97–106.

Willard, D.E. Polygon retrieval. *SIAM J. Comput.* **11** (1982), 149–165.

——. New data structures for orthogonal range queries. *SIAM J. Comput.* (1985), 232–253.

Wood, D. *Paradigms and Programming with PASCAL.* Computer Science Press, Rockville, Md., 1984.

Yaglom, I.M. and Boltyanskii, V.G. *Convex Figures.* English translation, Holt, Rinehart and Winston, New York, 1961.

Yao, A.C. An $O(|E|\log\log|V|)$ algorithm for finding minimum spanning trees. *Inform. Process. Lett.* **4** (1975), 21–23.

——. A lower bound for finding convex hulls. *J. Assoc. Comput. Mach.* **28** (1981), 780–787.

Yao, A.C. and Yao, F.F. A general approach to geometric queries. *In* "Proc. 17th Ann. ACM Sympos. Theory Comput. 1985", 163–168.

Yao, F.F. A 3-space partition and its applications. *In* "Proc. 15th Ann. ACM Sympos. Theory Comput. 1983", 258–263.

Zaslavsky, Th. *Facing up to Arrangements: Face-Count Formulas for Partitions of Space by Hyperplanes.* Memoirs Amer. Math. Soc. 154, Providence, RI, 1975.

APPENDIX A

DEFINITIONS

The main purpose of the appendix is to relieve the other parts of this book of fundamental but common definitions. We decided to collectively present them in this appendix, since it might be necessary to occasionally look up some of these definition. All definitions listed below are more or less standard.

A.1. Arithmetic

All numbers used in this book are either *integers* or *reals*. Let x and y be two real or two integer numbers. If $x>0$ then x is *positive*, and x is *negative* if $x<0$. We define the unary operator $\operatorname{sign} x$ equal to -1 if x is negative, equal to 0 if $x=0$, and equal to $+1$ if x is positive. The *absolute value of* x is defined to be equal to x if x is non–negative, and equal to $-x$ if x is negative; it is denoted as $|x|$. As is common in the mathematical literature, we write

$$
\begin{gathered}
x+y \text{ for the } \textit{sum of } x \text{ and } y, \\
xy \text{ or } x \cdot y \text{ for the } \textit{product of } x \text{ and } y, \\
x-y=x+(-1)y \text{ for the } \textit{difference of } x \text{ and } y, \\
x^y \text{ for } x \textit{ raised to the power } y, \text{ and} \\
\frac{x}{y}=x/y=x(y^{-1}) \text{ for } x \textit{ divided by } y.
\end{gathered}
$$

The power x^y is not defined if $x=0$ and $y\leq 0$ or if $x<0$ and y is not integral, and therefore the ratio $\frac{x}{y}$ is not defined if $y=0$. In all cases, $x+y$, xy, $x-y$, x^y, and $\frac{x}{y}$ are real numbers if x and y are real, $x+y$, xy, and $x-y$ are integer numbers if x and y are integral, and $x+y$, xy, x^y, and $\frac{x}{y}$ are positive if x and y are positive. For any real number x, we let the *floor* $\lfloor x \rfloor$ *of* x be the largest integer not greater than x, and we let the *ceiling* $\lceil x \rceil$ *of* x be the smallest integer not less than x. Two positive integer numbers x and y are *relatively prime* if there is no positive integer $i\geq 2$ such that both $\frac{x}{i}$ and $\frac{y}{i}$ are integral, and x is *prime* if $\frac{x}{i}$ is non–integral for all $2\leq i<x$. For two positive integer numbers x and y, y *modulo* x is equal to $y-x\lfloor y/x \rfloor$. If y modulo x is zero, then y is called a *multiple of* x.

In addition to the above operators, we write

$$
\begin{gathered}
\log_y x \text{ for the } \textit{logarithm of } x \textit{ to the base } y, \\
\sin x \text{ for the } \textit{sine of } x, \text{ and}
\end{gathered}
$$

$\cos x$ for the *cosine of* x.

We let $\log_y x$ be undefined if $x \leq 0$ or if $y \leq 0$; in all defined cases, $\log_y x = z$ if $y^z = x$. It is useful to notice that

$$\log_y x = \frac{\log_a x}{\log_a y},$$

for any positive real number a, and that

$$\log_y x = \frac{1}{\log_x y} \text{ and } a^{\log_y x} = x^{\log_y a}.$$

To denote the ith power of $\log_y x$, we write $(\log_y x)^i$ or, alternately, $\log_y^i x$. The sine and the cosine functions are defined for all real numbers x and obey the rules

$$\sin x = \sin(x+\pi) \text{ and}$$
$$\cos x = \cos(x+\pi).$$

It is therefore sufficient to know the values of these functions for all arguments in the interval $[0,\pi)$. For $x \geq y$ two non-negative integer numbers, we write

$$x! = 1 \cdot 2 \cdot \ldots \cdot (x-1) \cdot x \text{ for } x \textit{ factorial}, \text{ and}$$
$$\binom{x}{y} = \frac{x!}{y!(x-y)!} \text{ for } x \textit{ choose } y.$$

For convenience, we define $0! = 1$ which implies that $\binom{x}{y} = 1$ if $x = y$ or if $y = 0$.

The unary operators defined above are functions that map x to some real value. Another class of important functions on real parameters are the so-called polynomials defined as follows. A *polynomial* $p(x)$ *of degree* k is given by a $(k+1)$-tuple of real numbers $(a_0, a_1, \ldots, a_k)$ and it maps a real number x to

$$p(x) = a_0 + a_1 x + a_2 x^2 + \ldots + a_k x^k$$

which is again a real number. A real number r is a *root of* a polynomial $p(x)$ if $p(r) = 0$. One of the most fundamental and important results in algebra is that a polynomial of degree k has at most k roots.

Sums and products are easily extended from two numbers to arbitrarily many numbers by the use of the associative law. For notational reasons, the numbers are then indexed, and we write

$$\sum_{i \in I} a_i \text{ for the } \textit{sum} \text{ and}$$

$$\prod_{i \in I} a_i \text{ for the } \textit{product}$$

of all numbers a_i, with i in the index-set I. For S a non-empty set of numbers, $\min S$ is the smallest element in S and $\max S$ is the largest element in S, if they exist.

A.2. Sets and Words

A *set* is an unordered collection of pairwise different elements. The notation specifies a particular set either by enumerating its elements between braces, or by stating conditions which are necessary and sufficient for an element to be in the set. If a set S contains an element x

then we say that x *belongs to* S, and we write $x \in S$. For two sets S and T,

$$S \cup T = \{x \mid x \in S \text{ or } x \in T\} \text{ is the } \textit{union},$$
$$S \cap T = \{x \mid x \in S \text{ and } x \in T\} \text{ is the } \textit{intersection}, \text{ and}$$
$$S - T = \{x \mid x \in S \text{ and } x \notin T\} \text{ is the } \textit{difference}$$

of S and T. If $S \cap T = \emptyset$, with $\emptyset$ the *empty set*, then S and T are said to be *disjoint*. A *partition of* a set S is a collection of disjoint sets such that S is the union of all sets in this collection. T is a *subset of* S if every element in T also belongs to S; in this case, S is said to *contain* T, and we write $T \subseteq S$. If S contains T and at least one element of S does not belong to T, then T is a *proper subset of* S which is synonymous to saying that S *properly contains* T. If S contains T, then the set $\mathrm{compl}_S T = S - T$ is called the *complement of* T *with respect to* S. We omit the index if the reference set is understood. The number of elements in a set S is called the *cardinality of* S, denoted as $\mathrm{card} S$. Like a set, a *multiset* is a collection of elements; however, they are not necessarily pairwise different. Thus, a multiset may have many instances of one element. A multiset can be transformed to a set by indexing different instances on one element from 1 to the number of instances there are. All definitions for sets can now be extended to multisets by means of this transformation and its inverse.

A set is *finite* if its cardinality is an integer number. A finite set Σ is sometimes called an *alphabet*. Each element of an alphabet is called a *letter*, and a *word* w *over* Σ is an ordered sequence of letters from Σ (each letter may appear an arbitrary number of times). Sometimes we refer to w as a *string* or a *sequence*. The *length of* w is the number of letters used to form w. We write ϵ for the word of length 0 which is called the *empty word*. If the length of a word is an integer number, then the word is said to be *finite*. For two finite words

$$v = a_1 a_2 \ldots a_n \text{ and } w = b_1 b_2 \ldots b_m,$$
$$vw = a_1 a_2 \ldots a_n b_1 b_2 \ldots b_m,$$

where vw is called the *concatenation of* v and w. Let S and T be two sets of words over some alphabet Σ; then $ST = \{vw \mid v \in S \text{ and } w \in T\}$. Furthermore, we define $S^0 = \{\epsilon\}$, where ϵ is the empty word, and $S^i = S^{i-1} S$, for $i \geq 1$. S^+ is the notation for the union of all sets S^i, for $i \geq 1$, and $S^* = S^+ \cup S^0$.

A.3. Relations and Functions

The *Cartesian product* $A \times B$ *of* two sets A and B is the set of ordered pairs (a,b), with $a \in A$ and $b \in B$. Here, an ordered pair is the same as a sequence or word of length two. For example, the Euclidean plane can be thought of as the Cartesian product $\mathbf{R} \times \mathbf{R}$, where $\mathbf{R}$ denotes the set of real numbers. More generally, the d-dimensional Euclidean space $\mathbf{E}^d$ is the Cartesian product $\mathbf{R} \times \mathbf{R} \times \ldots \times \mathbf{R}$ of length d which is the set of all ordered sequences $(x_1, x_2, \ldots, x_d)$, with $x_i \in \mathbf{R}$ for $1 \leq i \leq d$.

A *relation* $\mathcal{R}$ *from* A *to* B is a subset of $A \times B$, and $\mathcal{R}$ is a *relation on* A if $A = B$. The more interesting relations are those which are defined for one set. Thus, assume for the following discussion that $\mathcal{R}$ is a relation on A. Then $\mathcal{R}$ is said to be *reflexive* if $(a,a) \in \mathcal{R}$, for every a in A, $\mathcal{R}$ is *symmetric* if $(a,b) \in \mathcal{R}$ whenever $(b,a) \in \mathcal{R}$, and $\mathcal{R}$ is *transitive* if $(a,b) \in \mathcal{R}$ and $(b,c) \in \mathcal{R}$ implies $(a,c) \in \mathcal{R}$. If a relation is reflexive, symmetric, and transitive, then it is called an *equivalence relation*. Equivalence relations on A are interesting since they define partitions of A in a natural way. The partition

$$A = A_1 \cup A_2 \cup \ldots$$

defined by an equivalence relation $\mathcal{R}$ is such that two elements a and b are in the same set A_i if and only if $(a,b) \in \mathcal{R}$. In fact, the correspondence works in both directions since every partition of A is defined by a unique equivalence relation. The sets A_i, $i \geq 1$, are called the *equivalence classes of* A defined by $\mathcal{R}$.

We obtain another interesting type of relation if we replace symmetry by its opposite: antisymmetry. A relation $\mathcal{R}$ is said to be *antisymmetric* if (a,b) and (b,a) are both in $\mathcal{R}$ only if $a = b$. We call $\mathcal{R}$ a *partial order* if it is reflexive, antisymmetric, and transitive. It is common to write $a \leq b$ in order to indicate that (a,b) is in $\mathcal{R}$. The name "partial order" is not accidental. Indeed, a partial order induces some kind of order on the elements of A. This order is only partial since there can be pairs of elements such that neither $a \leq b$ nor $b \leq a$; in this case, a and b are said to be *incomparable*. A set A with a partial order $\mathcal{R}$ on A is called a *partially ordered set* or a *poset*, for short. A subset C of a poset is called a *chain* if no two elements in C are incomparable, and C is an *antichain* if the elements in C are mutually incomparable. By a theorem of Dilworth, there is a trade-off between the size of a minimum chain and the size of a minimum antichain. In particular, if A is a set of at least $m \cdot n+1$ elements, then there is a chain of size $m+1$ or an antichain of size $n+1$.

If a poset itself is a chain, then it is called a *totally ordered set* and its partial order is called a *total order*. For every two elements a and b of a totally ordered set we either have $a \leq b$ or $b \leq a$. Thus, we can represent the total order as a linear list of the elements with the understanding that $a \leq b$ if a and b coincide or the location of a in the list is to the left of the location of b. It is true that every partial order can be extended to a total order that does not contradict the partial order. It follows that the elements of every poset can be sorted into a linear sequence such that $a \leq b$ only if $a = b$ or the position of a in the sequence is to the left of the position of b. This operation is referred to as a *topological sort of* the poset.

An element a in a poset is called a *minimum* if there is no b such that $b \leq a$, and a is said to be a *maximum* if there is no b such that $a \leq b$. If $a \leq c$ and $a \leq d$, then a is a *lower bound of* c and d, and a is a *greatest lower bound of* c and d if there is no lower bound b of c and d with $a \neq b$ and $a \leq b$. Similarly, a is an *upper bound of* c and d if $c \leq a$ and $d \leq a$, and a is a *least upper bound of* c and d if there is no upper bound b of c and d with $a \neq b$ and $b \leq a$. Notice that a lower or upper bound of c and d is not necessarily different from c or from d. A poset is said to be a *lattice* if every two elements have a unique greatest lower bound and a unique least upper bound.

Finally, we discuss so-called *functions* which are relations from a set A to a set B such that for each $a \in A$ there is at most one $b \in B$, with a and b in relation. A function $\mathcal{F}$ can be interpreted as a mapping from A to B, and we write $\mathcal{F}(a) = b$ to indicate that $(a,b) \in \mathcal{F}$. The set A is called the *domain of* $\mathcal{F}$, B is the *range of* $\mathcal{F}$, and b is said to be the *image of* a *(under* $\mathcal{F}$*)*. We call $\mathcal{F}$ a *surjection* or an *onto function* if every element b in B is the image of at least one element a in A. Function $\mathcal{F}$ is called an *injection* or a *one-to-one function* if the images of no two different elements in A are the same. Finally, $\mathcal{F}$ is said to be a *bijection* or a *one-to-one onto function* if it is a surjection and an injection. A bijection from a set A to a set B exists if and only if both sets are equally large, that is, if $\text{card}A = \text{card}B$. The *inverse* $\mathcal{F}^{-1}$ *of* a bijection $\mathcal{F}$ from A to B is again a bijection, namely, the one that maps b to a if $\mathcal{F}(a) = b$. A bijection from A to B and its inverse establish a *one-to-one correspondence between* A and B.

A.4. Topology, Geometry, and Linear Systems

The *d-dimensional Euclidean space* is denoted by E^d, for $d \geq 0$. A *point* in E^d is specified by the vector of its d coordinates in a Cartesian system, that is, every point p corresponds to the vector anchored at the *origin* o that has p as its endpoint. A point and its corresponding vector are often used interchangeably. For two points

$$p = (\pi_1, \pi_2, \ldots, \pi_d) \text{ and } q = (\psi_1, \psi_2, \ldots, \psi_d),$$
$$p+q = (\pi_1+\psi_1, \pi_2+\psi_2, \ldots, \pi_d+\psi_d),$$

where $p+q$ is the *sum of* p and q. For every real number λ, we define $\lambda p = (\lambda\pi_1, \lambda\pi_2, \ldots, \lambda\pi_d)$, and $p-q = p+(-1)q$ is called the *difference of* p and q. For sets of points P and Q, these definitions extend to

$$P+Q = \{p+q \mid p \in P \text{ and } q \in Q\}, \text{ and}$$
$$\lambda P = \{\lambda p \mid p \in P\}.$$

If $Q = \{q\}$, the simpler notation $P+q = P+\{q\}$ is used. Q is called a *translate of* P if $Q = P+p$, for some point p, and Q is a *homothet of* P if $Q = \lambda P+p$, for some real number $\lambda \neq 0$ and some point p. P is *centrally symmetric* if $P = -P+p$, for some point p.

Let $P = \{p_0, p_1, \ldots, p_k\}$ be a finite set of points in E^d. A point x is a *linear combination of* P if

$$x = \sum_{i=0}^{k} \lambda_i p_i,$$

for suitable real numbers λ_i. If $\sum_{i=0}^{k} \lambda_i = 1$ then x is also called an *affine combination of* P, and furthermore, if $0 \leq \lambda_i \leq 1$ for $0 \leq i \leq k$, then x is a *convex combination of* P. P is said to be *linearly* or *affinely dependent* if there is a point p_i in P which is a linear or affine combination of $P-\{p_i\}$, respectively. Otherwise, P is *linearly* or *affinely independent*. The set of all linear, affine, and convex combinations of P is called the *linear hull* $\text{lin}P$, *affine hull* $\text{aff}P$, and *convex hull* $\text{conv}P$ *of* P, respectively. A point p in P is said to be *extreme* if $\text{conv}P \neq \text{conv}(P-\{p\})$. The set of all extreme points of P is written as $\text{ext}P$.

One method for testing whether or not a set P of points $p_i = (\pi_{i,1}, \pi_{i,2}, \ldots, \pi_{i,d})$, for $1 \leq i \leq d$, in d dimensions is linearly independent computes the determinant $\det M$ of the matrix

$$M = \begin{pmatrix} \pi_{1,1} & \pi_{1,2} & \cdots & \pi_{1,d} \\ \pi_{2,1} & \pi_{2,2} & \cdots & \pi_{2,d} \\ \cdots & \cdots & \cdots & \cdots \\ \pi_{d,1} & \pi_{d,2} & \cdots & \pi_{d,d} \end{pmatrix},$$

that is, if $\det M = 0$ then P is linearly dependent and it is linearly independent, otherwise. A collection of $d+1$ points $p_1, p_2, \ldots, p_{d+1}$ is affinely independent if and only if the set of d points $\{p_1-p_{d+1}, p_2-p_{d+1}, \ldots, p_d-p_{d+1}\}$ is linearly independent.

To define the *determinant* $\det M$ *of* M, we can use the *Laplace expansion by minors* that works as follows. For any real number $\pi_{i,j}$ in M define $\Delta_{i,j}$ as the determinant of the matrix obtained from M by deleting the ith row and the jth column; $\Delta_{i,j}$ is called the *minor of* $\pi_{i,j}$ and $(-1)^{i+j}\Delta_{i,j}$ is the *cofactor of* $\pi_{i,j}$. Then

$$\Delta = \det M = \sum_{i=1}^{d} (-1)^{i+j} \Delta_{i,j} \pi_{i,j},$$

for any choice of j, $1 \leq j \leq d$, and also

$$\Delta = \sum_{j=1}^{d} (-1)^{i+j} \Delta_{i,j} \pi_{i,j},$$

for any choice of i, $1 \leq i \leq d$. Thus, Δ is a sum of products each one consisting of exactly d entries of M. The coefficient of an entry $\pi_{i,j}$ in this sum is its cofactor, that is, $(-1)^{i+j}\Delta_{i,j}$. Now let $I=(i_1,i_2,\ldots,i_k)$ and $J=(j_1,j_2,\ldots,j_k)$ be two sequences of pairwise different indices between 1 and d, limits included, such that $i_1 < i_2 < \ldots < i_k$. Let i be the sum of all indices in I, and let j be the sum of all indices in J. Furthermore, let $\Delta_{I,J}$ be the determinant of the matrix that we obtain from M when we delete each row whose index is in I and each column whose index is in J. Then the coefficient of the product $\pi_{i_1,j_1} \cdot \pi_{i_2,j_2} \cdot \ldots \cdot \pi_{i_k,j_k}$ in Δ is $(-1)^{i+j+\tau}\Delta_{I,J}$, where τ is the number of transpositions needed to sort J into increasing order.

Sometimes it is useful to know what the effect of simple matrix operations on the determinant of a square matrix is. If we exchange two rows or two columns of a matrix, then the determinant of the matrix gets multiplied by -1. If we multiply each element of either a row or of a column with some real number $\rho \neq 0$, then the determinant of the new matrix is ρ times the determinant of the old matrix. Both results can be easily obtained from the definition of a determinant given above.

Let $P=\{p_0,p_1,\ldots,p_k\}$ be an affinely independent set of points in E^d (note that $k \leq d$). We call aff$P$ a *k-flat* and k the *dimension* dim aff$P$ *of* aff$P$. More generally, we define dim$S$ = dim aff$S$, for S any set in E^d. A $(d-1)$-flat in E^d is also called a *hyperplane*. For two vectors

$$u=(v_1,v_2,\ldots,v_d) \text{ and } w=(\omega_1,\omega_2,\ldots,\omega_d),$$

$$\langle u,w \rangle = \sum_{i=1}^{d} v_i \omega_i$$

is the *scalar* or *inner product of* u and w. If $\langle u,w \rangle = 0$ then u and w are *orthogonal* to each other. The scalar product allows us to define a hyperplane h alternately as a set of points x with $\langle x,v \rangle = \alpha$, for $v \neq o$ a vector and α a real number. In this case, v is a *normal vector of* h. A k-flat f and an m-flat g are *orthogonal* if f and g intersect in a point p such that $\langle p_1-p, p_2-p \rangle = 0$, for all points p_1 of f and p_2 of g. Thus, there is a unique $(k+m)$-flat which contains f and g, which in turn implies $k+m \leq d$. Let S be some set in E^d. The *orthogonal projection of* S *onto* a k-flat f is the intersection of f and $S+g$, where g is the unique $(d-k)$-flat which contains the origin and is orthogonal to f. The reader is asked to note that orthogonal, *normal*, and *perpendicular* are used interchangeably. A k-flat f and an m-flat g, with $k \leq m$, are *parallel* if g contains a translate of f. If a set P of points is contained in a 1-flat then P is said to be *collinear*, and if P is contained in a 2-flat then P is *coplanar*. If all hyperplanes in a set H have a common point then H is *concurrent*.

Let now $u_i=(v_{i,1},v_{i,2},\ldots,v_{i,d})$, $1 \leq i \leq d$, be d non-zero vectors defining d hyperplanes h_i: $\langle x,u_i \rangle = \alpha_i$. If the vectors $u_1,u_2,\ldots,u_d$ are linearly independent, then the hyperplanes $h_1,h_2,\ldots,h_d$ intersect in a common point p whose coordinates $(\pi_1,\pi_2,\ldots,\pi_d)$ can be computed using *Cramer's rule*. That is,

$$\pi_i = \frac{\Delta_i}{\Delta},$$

where

$$\Delta = \det \begin{vmatrix} v_{1,1} & v_{1,2} & \cdots & v_{1,d} \\ v_{2,1} & v_{2,2} & \cdots & v_{2,d} \\ \cdots & \cdots & \cdots & \cdots \\ v_{d,1} & v_{d,2} & \cdots & v_{d,d} \end{vmatrix}$$

and Δ_i is the determinant of the matrix that is obtained by exchanging the ith row from the left by the vector

$$\begin{pmatrix} \alpha_1 \\ \alpha_2 \\ \cdots \\ \alpha_d \end{pmatrix}.$$

For two points

$$p = (\pi_1, \pi_2, \ldots, \pi_d) \text{ and } q = (\psi_1, \psi_2, \ldots, \psi_d),$$
$$d(p,q) = (\sum_{i=1}^{d} (\pi_i - \psi_i)^2)^{1/2}$$

is the *Euclidean distance between* p and q, and for two sets of points, $d(P,Q)$ is the largest lower bound on $d(p,q)$, $p \in P$ and $q \in Q$. For every positive real number ρ, the set $\{x | d(x,p) < \rho\}$ is the *open ball* with *center* p and *radius* ρ. A set S in E^d is said to be *bounded* if there is an open ball that contains S, and S is *unbounded*, otherwise. S is *open* if, for every point p in S, there is an open ball with center p that is contained in S. S is *closed* if its complement with respect to E^d is open. The *interior* $\text{int}S$ *of* S is the largest open subset of S, and the *closure* $\text{cl}S$ *of* S is the smallest closed set that contains S. The set $\text{bd}S = \text{cl}S - \text{int}S$ is called the *boundary of* S. Note that a set can be unbounded even if its boundary is not empty. The intersection of $\text{aff}S$ with $\text{int}(S+g)$, where g is the orthogonal $(d - \dim \text{aff}S)$-flat of $\text{aff}S$ which contains the origin, is called the *relative interior* $\text{ri}S$ *of* S. Thus, the relative interior of a closed line segment is the line segment without endpoints and the relative interior of a point is the point itself. If $S = \text{ri}S$ then S is said to be *relatively open*. A hyperplane h, given by $\langle x,v \rangle = \alpha$, *supports* a set S (or synonymously, is *tangent to* S) if h intersects $\text{cl}S$, and $\langle p,v \rangle \leq \alpha$ (or $\langle p,v \rangle \geq \alpha$) for all points p of S.

The set of all points on one side of a hyperplane h is an *open half-space*, and if we add the points of h, then it is a *closed half-space*. Note that there is a non-zero vector v and a real number α such that a half-space can be written as the set of points x with

$$\langle x,v \rangle < \alpha \quad \text{or} \quad \langle x,v \rangle \leq \alpha.$$

The intersection of a finite number of half-spaces is a *(convex) polyhedron*, and if it is bounded, then it is also called a *(convex) polytope*. Alternatively, a polytope can be defined as the convex hull of a finite set of points.

A collection C of relatively open subsets of E^d is called a *packing* if no two sets in C intersect, and C is a *covering* if E^d is the union of all sets in C. Furthermore, C is a *cell complex* if C is a packing and a covering, and if the closure of any set in C is the union of all sets in some subset of C. A cell complex in E^2 is also called a *subdivision of* E^2. Two points p and q of some set S in E^d are *connected in* S if, for every positive real number ϵ, there is a sequence of points $p = r_0, r_1, \ldots, r_n = q$ in S such that $d(r_i, r_{i+1}) < \epsilon$, for $0 \leq i \leq n-1$. The set S itself is *connected* if every two points of S are connected in S. The *connected components of* S are the maximal connected subsets of S. If S is connected and non-empty, then S itself is its only connected component. The set S is *simply connected* if each connected component of the complement $E^d - S$ is unbounded.

A.5. Graph Theory

A *graph* $\mathcal{G} = (N, A)$ is a pair of sets N and A, where the elements of N are called *nodes*, and A is a set of pairs of nodes called *arcs*. If $N = \emptyset$ then $\mathcal{G}$ is called the *empty graph*, and if A contains all pairs of nodes in N then $\mathcal{G}$ is called the *complete graph with* card$N$ *nodes*. If A is a multiset then $\mathcal{G}$ is called a *multigraph*. If every arc $a = \{v, w\}$ in A is a set of nodes, that is, a is an unordered pair of nodes, then $\mathcal{G}$ is said to be *undirected*. $\mathcal{G}$ is *directed*, or is a *digraph*, if every arc $a = (v, w)$ is an ordered pair of nodes. In both the undirected and the directed case,

arc a is *incident upon* nodes v and w,
nodes v and w are *endpoints of* arc a,
a *connects* v and w, and
v and w are *adjacent* to each other.

In the directed case, v is called the *origin* of a and w is the *destination* of a. Arc a is thus an *outgoing* arc *of* v and an *incoming* arc *of* w, and a is said to *lead from* v *to* w. The *degree* $deg(v)$ *of* node v is the number of arcs that are incident upon v. If $\mathcal{G}$ is directed, then the *out-degree* $deg^+(v)$ *of* v is the number of outgoing arcs of node v, while the *in-degree* $deg^-(v)$ *of* v is the number of incoming arcs of v. Consequently, we have $deg^+(v) + deg^-(v) = deg(v)$. If $deg^+(v) = 0$, then v is a *sink of* $\mathcal{G}$, and if $deg^-(v) = 0$, then v is a *source of* $\mathcal{G}$.

A sequence $P = (v_0, v_1, \ldots, v_n)$ of nodes is a *path* in $\mathcal{G}$ if $\{v_i, v_{i+1}\}$ (respectively, (v_i, v_{i+1}), if $\mathcal{G}$ is a digraph), for $0 \leq i \leq n-1$, is an arc of $\mathcal{G}$. If $\mathcal{G}$ is directed, then there is no ambiguity if the path P is written as a sequence

$$((v_0, v_1), (v_1, v_2), \ldots, (v_{n-1}, v_n))$$

of arcs. Path P *connects* nodes v_0 and v_n, and we say that P *leads from* v_0 *to* v_n if $\mathcal{G}$ is directed. In the undirected case, $\mathcal{G}$ is *connected* if every two nodes of $\mathcal{G}$ can be connected by a path. If $\mathcal{G}$ is directed then it is *strongly connected* if there is a path from each node to every other node; $\mathcal{G}$ is *weakly connected* if the graph that is obtained by ignoring the directions of the arcs is connected. Path P is said to be a *tour* if $v_0 = v_n$, and P is a *eulerian tour* if every arc of A is contained in P exactly once. If P is a tour and contains no node twice then P is a *cycle*. If $\mathcal{G}$ is a connected undirected graph and if no path in $\mathcal{G}$ is a cycle, then $\mathcal{G}$ is a *free tree*. The counterpart in the directed case is a *directed acyclic graph*, for short *DAG*, which is a weakly connected digraph without (directed) cycles. For two graphs $\mathcal{G}_1 = (N_1, A_1)$ and $\mathcal{G}_2 = (N_2, A_2)$, $\mathcal{G}_2$ is called a *subgraph of* $\mathcal{G}_1$ if $N_2 \subseteq N_1$ and $A_2 \subseteq A_1$. Furthermore, if $N_1 = N_2$, then $\mathcal{G}_2$ *spans* $\mathcal{G}_1$. If $\mathcal{G}_2$ is a tree and if it spans $\mathcal{G}_1$ then it is called a *spanning tree of* $\mathcal{G}_1$.

A special kind of tree plays an important role in data structuring. It is best described as a directed graph. A directed graph $\mathcal{T} = (N, A)$ is called a *rooted tree* if $\mathcal{T}$ is weakly connected and each node is the destination of exactly one arc, except for one node, called the *root of* $\mathcal{T}$, which has no incoming arc. This implies that card$A$ =card$N$−1. A node of $\mathcal{T}$ is a *leaf* if it has no outgoing arc, and it is an *internal node*, otherwise. The *rank of* a node is the number of outgoing arcs of that node. If every node in $\mathcal{T}$ has rank at most two then $\mathcal{T}$ is a *binary tree*; it is *complete* if every internal node has rank exactly two. If (u,v) is an arc in A, then u is the *parent* or *father of* v, and v is a *child* or *son of* u. If, furthermore, (v,w) is in A, then u is the *grandparent* or *grandfather of* w, and w is a *grandchild* or *grandson of* u. In general, we say that a node y is a *descendent of* a node x if $x=y$, or if y is a descendent of a child of x. If $x=y$, then y is an *improper descendent of* x. If y is a descendent of x, then x is called an *ancestor of* y. Two nodes are *siblings* or *brothers* if they have a common parent.

A *subtree (with root* u*) of* $\mathcal{T}$ is a subgraph of $\mathcal{T}$ whose nodes are all descendents of node u of $\mathcal{T}$ and whose arcs are all arcs of $\mathcal{T}$ incident upon two such nodes. Notice that the root of $\mathcal{T}$ is an ancestor of every node in $\mathcal{T}$ and that there is a unique path from the root of $\mathcal{T}$ to every node u of $\mathcal{T}$. If this path consists of i arcs then node u has depth i. The ith *level of* $\mathcal{T}$ is the set of nodes with depth $i-1$. The *height of* $\mathcal{T}$ is defined as the largest depth of any node in $\mathcal{T}$, and the *height of* a node u is the height of the subtree of $\mathcal{T}$ with root u.

A binary tree is *perfect* if it is complete and the depths of any two leaves are the same. It follows that a perfect binary tree of height k has exactly 2^k-1 internal nodes and exactly 2^k leaves. A binary tree $\mathcal{T}$ is *ordered* if each child is either a *left* or a *right child*, and if a node has two children then one is left and one is right. For ordered binary trees, there are several special orderings of its nodes defined. A sequence of its nodes is in *preorder* if a node u precedes all proper descendents, and its descendents precede the right sibling of u, if it exists. It is in *postorder* if u succeeds all proper descendents and it precedes all descendents of its right sibling, if it exists. The sequence is in *inorder* if u succeeds all descendents of its left child and it precedes all descendents of its right child. A binary tree is *sorted* if each node is labeled with an element of some totally ordered set such that the labels of the nodes taken in inorder are sorted.

A graph $\mathcal{G}_1=(N_1,A_1)$ is a *weighted graph* if each arc a in A_1 is associated with a real number $w(a)$ called the *weight* or *length* or *cost of* a. The *weight*, *length*, or *cost* of a subgraph $\mathcal{G}_2=(N_2,A_2)$ of $\mathcal{G}_1$ is defined as the sum of weights of all arcs in A_2. If $\mathcal{G}_2$ is a spanning tree of $\mathcal{G}_1$ with minimum weight, then it is called a *minimum spanning tree*, or *MST*, *of* $\mathcal{G}_1$. If all weights are positive real numbers then there is a *shortest path* between every pair of connected nodes in $\mathcal{G}_1$, which is a path with smallest possible weight that connects the two nodes.

A.6. Data Structures and Data Types

Loosely speaking, data structures and data types are mainly distinguished by the way in which they are specified: a *data type* is defined by its functions, while a *data structure* is defined by its structural properties. A data type is thus a rather abstract concept and serves conveniently as a black box. A data structure is said to *implement* a data type if it functions as required. Data types as well as data structures can be modeled hierarchically, that is, they can be used as components of more complex data types or data structures. In fact, the components of complex data structures are often specified as data types which allows us to concentrate on the essential parts of the structure.

For the description of data types and of data structures, we use PASCAL–like terminology and notation. The basic data types used are

integer, which represents integer numbers,
real, which uses floating-point to represent real numbers,
boolean, which represents **true** and **false**,
character, which represents letters, and
pointer, which represents addresses in internal storage.

For all data types, the obvious operations are also supported. Unfortunately, every implementation of the data type integer is dependent on the limitations of the host machine. On the other hand, for most applications, it is enough to use a mere computer-word of storage (which is typically in the neighborhood of 32 bits long) to represent an integer. It seems fair in theses cases to assume that simple operations, like addition or multiplication of two integers, can be performed in a constant amount of time. Throughout this book, we make this assumption; in fact, if the input data consists of integers, each of which can be represented by one computer-word, then any number used by any algorithm in this book fits into some constant number of computer-words. We are less concerned about the necessarily imperfect implementation of real numbers, since we believe that the consistent use of integers is a better practice. Pointers are used to explicitly establish relations between records, where a *record* is either a computer-word or the agglomeration of a constant number of computer-words. If a pointer p holds the address of a record which represents some node, then we say that p *points at* or *to* the node, and we use p as a name of the node and of the record. A pointer p that does not point at any node is called the *empty pointer*, and we write $p =$**nil**.

More advanced data types are implemented either by implicit or by explicit data structures. An *implicit* data structure is a *linear array* of records which is a sequence of records in consecutive storage. The relationship between two records of a linear array is implicitly expressed by the values of their addresses, and most often, a relationship exists if the two records lie adjacent to each other. There are notable exceptions to this rule, however. *Explicit* data structures use pointers to explicitly establish relationships between records. The simplest explicit data structure is probably the *linear list* which is a sequence of records where two consecutive records are interconnected by one or two pointers. A linear list is *singly linked* if each record stores only one pointer which connects it to the next record in the list; it is *doubly linked* if each record contains a pointer to its successor and a pointer to its predecessor. If a record does not have a successor or a predecessor then the pointer is *nil* which means that it points nowhere. It is simple enough to *embed* a linear list in a linear array which saves on the storage for the pointers. There are also disadvantages to such an embedding: it is no longer easy to insert or delete records in the middle of the list, and it is a waste of storage if the list shrinks and thus vacates many records of the host array. Notice that integer numbers are used to represent addresses of records in linear arrays. Sometimes, we will use the term pointer to designate an integer which is mainly used to describe addresses of records in a linear array.

In the remainder of this section, we describe in turn each of the more advanced data types which are used but not developed in this book. We neglect the minor functions supported by such data types which are typically the initialization and the test as to whether or not the represented collection of data is empty. We also indicate efficient implementations and their complexities. Each such data type can be used to represent any finite collection of elements and the properties of the elements are of little significance. We assume that basic data types take care of the elements, that a constant amount of storage is needed for each element, and that constant time suffices to perform a comparison between two such elements.

A *bit-vector* B stores a collection of boolean variables or bits. The operations supported by B are

1. to determine the value of the ith bit, and
2. to change the value of the ith bit to true or to false.

Each operation can be performed in constant time if a linear array of boolean variables is used.

A *stack* S stores elements of any domain and supports the operations

1. PUSH, which adds another element to S, and
2. POP, which removes the element last added to the stack.

The easiest implementation of a stack is a singly linked linear list with an extra pointer to the last record in the list which stores the element added the latest. This implementation takes constant time for each of the above operations.

A *queue* Q differs from a stack by the way an element is removed. More precisely, Q supports operations

1. ENQUEUE, which adds another element to Q, and
2. DEQUEUE, which removes the oldest element in Q.

A singly linked linear list with two additional pointers, one to the "head" and one to the "tail" of Q, guarantees the execution of both operations in constant time. For some applications, we need to be able to add and to remove on both ends of a queue. A data structure that supports these operations is called a *deque* and it can be implemented using a doubly linked list such that each operation costs only constant time.

A strategy for removing elements which is more sophisticated than the one followed by a stack and a queue is supported by a *priority queue* Q which stores elements of a totally ordered domain. The operations to be performed are

1. INSERT, which adds another element to Q, and
2. MIN_DELETE, which removes a smallest element from Q.

There is a great variety of implementations available which support each operation in time $O(\log n)$. Some of the implementing data structures also support the deletion of a given element in time $O(\log n)$, and some are more efficient for one of the two standard operations. Since a priority queue can be used for sorting, at least one of the two standard operations must take time $\Omega(\log n)$.

A *dictionary* D stores elements of a totally ordered domain and supports the following operations:

1. INSERT, which adds another element to D,
2. DELETE, which removes a given element from D, and
3. SEARCH, which determines whether or not a given element is in D.

Using standard implementations, each operation can be executed in time $O(\log n)$ if n elements are currently stored. Many of the standard implementations use a binary tree as their underlying structure. The logarithmic performance is achieved by keeping the height of the tree proportional to the smallest possible height. Such trees are also called *balanced trees*.

There are some more complicated data structures implementing data types which support geometric functions described in this book. Such a data structure is *static* if it represents a fixed set of objects and does not allow for changes in the set. It is *dynamic* if it supports updates which are usually insertions of new elements and deletions of old elements. An operation that extracts some information about the stored set but does not change the set is called a *query*. A problem that requires the implementation of a data type specified by some sort of query is called a *search problem*. A typical example of a search problem is the problem solved by a dictionary; in this case, a query is a search for a given element.

A.7. Description and Analysis of Algorithms

To describe algorithms we use a terminology similar to that of standard PASCAL notation. We suppress the declaration of data structures, however, and often indicate sequences of statements by English text. We assume that recursion is available; if the reader works in an environment that does not provide recursion, then he or she can easily simulate recursion using a stack. Below, we offer a list of control structures used in this book. Letters C and S, possibly with indices, stand for generic clauses and groups of executable statements, respectively. A *clause* is an expression that is either true or false.

The simplest control structure provides the possibility of a decision based on the value of some clauses. The syntax used in this book is as follows:

if C_1 **then** S_1
  elseif C_2 **then** S_2
  elseif C_3 **then** S_3
  . . .
  elseif C_k **then** S_k
  else S_{k+1}
endif.

If clause C_1 is true then the group of statements S_1 is executed. If C_1 is false and C_2 is true then S_2 is executed. In general, S_{i+1} is executed if C_1 through C_i are false and C_{i+1} is true. If no clause is true then S_{k+1} is executed. If all optional parts of the statement are removed, we are left with

if C **then** S **endif.**

If convenient, this is written as

unless $\overline{C}$ **do** S **endunless,**

where $\overline{C}$ is the negation of C. If clauses C_1 through C_k are such that there is always exactly one of them true, then we may also list the clauses and the statements as **Case 1**, **Case 2**, etc. If the statements for each case are similar or symmetric or analogous, in some sense, we use the statement

without loss of generality, assume that ...

and describe only the assumed case.

There are three kinds of loops used in this book: two iterate with the iteration controlled by a clause, the third loop iterates once for each element in some set, or for each integer in some given range. The syntax of the first loop is

while C **do** S **endwhile.**

S is executed until clause C becomes false. If C is false at the beginning, then S is never executed. The syntax of the second loop is

repeat S **until** C.

This has the effect that S is executed until clause C becomes true. Even if C is true at the beginning, S is executed at least once. The third loop is written as

for D **do** S **endfor**,

where D specifies an integer and the range of values that it assumes ("$i:=n_1$ **to** n_2" or "$i:=n_2$ **downto** n_1"), or it specifies a set and S is executed for each element in the set. If the range of integers is zero, or the set of elements is empty, then S is never executed.

The *time-complexity of* an algorithm $\mathcal{A}$ is the amount of time needed by $\mathcal{A}$, and the *space-complexity of* $\mathcal{A}$ is the amount of storage it requires. In both cases, the complexity is expressed as a function of the size of the input, n. The size, n, is typically the cardinality of the input set. In all cases discussed in this book, the complexity is defined as the maximum amount of time or storage taken over all inputs of size n.

Since there are many different implementations of one algorithm and since basic operations do not take the same amount of time if they are executed on different machines, there is no such thing as the exact amount of time or storage required by an algorithm. However, it is possible in principle to decide whether or not an algorithm takes time proportional to some function $f(n)$. The following concepts are used to allow for rigorous statements of this kind:

$$\mathrm{O}(f(n))=\{g(n) \mid \text{there are constants } c \text{ and } n_0 \text{ such that } g(n)\leq cf(n), \text{ for all } n\geq n_0\},$$
$$\Omega(f(n))=\{g(n) \mid f(n) \text{ is in } \mathrm{O}(g(n))\},$$
$$\Theta(f(n))=\mathrm{O}(f(n))\cap\Omega(f(n)),$$
$$\mathrm{o}(f(n))=\{g(n) \mid \text{for every constant } c, \text{ there is an } n_0 \text{ such that } cg(n)<f(n) \text{ if } n\geq n_0\}, \text{ and}$$
$$\omega(f(n))=\{g(n) \mid f(n) \text{ is in } \mathrm{o}(g(n))\}.$$

If $g(n)$ is contained in $\mathrm{O}(f(n))$ etc., then we write $g(n)=\mathrm{O}(f(n))$ etc. If $g(n)=\Theta(f(n))$ then $g(n)$ and $f(n)$ are said to be *of the same order* or they are *asymptotically the same*. If $g(n)=\mathrm{o}(f(n))$ then $f(n)$ is *of higher order than* $g(n)$ and we also say that $f(n)$ *grows asymptotically faster than* $g(n)$. The following terms are used if $g(n)$ is contained in special classes of functions:

$g(n)$ is *constant* if $g(n)=\mathrm{O}(1)$,
$g(n)$ is *logarithmic* if $g(n)=\mathrm{O}(\log n)$,
$g(n)$ is *linear* if $g(n)=\mathrm{O}(n)$,
$g(n)$ is *sublinear* if $g(n)=\mathrm{o}(n)$,
$g(n)$ is *superlinear* if $g(n)=\omega(n)$,
$g(n)$ is *quadratic* if $g(n)=\mathrm{O}(n^2)$, and
$g(n)$ is *cubic* if $g(n)=\mathrm{O}(n^3)$.

Notice that the cases are not mutually exclusive since, for example, $\mathrm{O}(n)$ is a subset of $\mathrm{O}(n^2)$.

A.8. Bibliographic Notes

The basic definitions of calculus and of set theory presented in Sections A.1 and A.2 are common knowledge in modern mathematics and computer science. Textbooks that cover this material and much more are Edwards, Penny (1986), Halmos (1960), and Liu (1985). For the terminology used in connection with words and sets of words, we refer to Hopcroft, Ullman

(1979) and Salomaa (1973) which are introductory texts on formal languages and the theory of computation. Further background on relations and functions can be found for example in Liu (1968). The terminology used to describe geometric and topological objects is borrowed from that which is used in Grünbaum (1967) and in Bronsted (1983). Detailed information about determinants of square matrices and related topics can be found in Halmos (1958), Nering (1963), and other textbook in the area. For further details about graphs in general and about algorithms for graphs, we refer to Harary (1969) and to Bondy, Murty (1976). Introductions to the basic concepts of computer science and further details can be found in Knuth (1968, 1969, 1973), Aho, Hopcroft, Ullman (1974), and Mehlhorn (1984a, 1984b, 1984c). These textbooks also include implementations of the data types defined in Section A.6 within the time bounds mentioned there. Introductions to algorithms using PASCAL as the language of description are given by Reingold, Hansen (1986) and Wood (1984).

APPENDIX B

NOTATIONAL CONVENTIONS

This appendix offers a collection of all names used to denote objects that appear throughout the book. We also indicate the general principles that guide us in choosing a name for an object, which can be a function, a set of points, or a vector. As the reader might expect, no rule is without exceptions, although we have tried our best to follow the rules described in this appendix. The names used for the various objects discussed in this book can be categorized as follows:

upper case script,
upper case roman,
upper case greek,
lower case roman,
lower case greek,

and combinations thereof. We treat each category in turn. For each category, we briefly describe the properties of the objects that qualify for names in the category. This description is followed by a list of names used and the objects which are named this way. Combinations of various types of letters via concatenation, subscribing, superscribing, etc. are used extensively to designate combinatorial functions. Since the names chosen for functions are more delicate because they convey more subtle information, we devote the last section of this appendix to combinatorial functions.

B.1. Script Used for Structures

We use upper case scripts for all kinds of objects which have a structure that is more complicated than that of a set or of a sequence. Typical objects of this kind are graphs and data structures. There are two exceptions to this rule: $\mathcal{D}$ and $\mathcal{E}$ are used for geometric transforms.

$\mathcal{A}$	**a**rrangement or **a**lgorithm.
$\mathcal{B}$	**b**it-vector.
$\mathcal{C}$	**c**ell **c**omplex, **c**ircular sequence or **c**ontour diagram.
$\mathcal{D}$	**d**ictionary, **d**ual transform or auxiliary for cell complex and for circular sequence.
$\mathcal{E}$	**e**mbedding transform (it maps objects or sets from E^d to E^{d+1}) or graph of **e**dges.

$\mathcal{F}$	function.
$\mathcal{G}$	graph, in general, or Gabriel graph.
$\mathcal{H}$	halfperiod of a circular sequence or hierarchy of sequences.
$\mathcal{I}$	incidence graph.
$\mathcal{L}$	list, level in an arrangement or linear program.
$\mathcal{P}$	polytope, polyhedron or power diagram.
$\mathcal{Q}$	queue or priority queue.
$\mathcal{R}$	relation, relative neighborhood graph, graph of regions or stabbing region.
$\mathcal{S}$	stack, skeleton of edges in an arrangement or stabbing region.
$\mathcal{T}$	triangulation or tree.
$\mathcal{V}$	Voronoi diagram or visibility graph.
$\mathcal{Z}$	zonotope, a special type of polytope.

We violate the one–letter convention by the use of $\mathcal{DT}$ for Delaunay triangulation.

B.2. Upper Case Roman Used for Sets

We use upper case roman (not to be confused with the fonts which are italic) to designate sets that carry little or no additional structure. Other objects which are frequently named using upper case roman letters include sequences and several combinatorial functions. We account for the combinatorial functions in Section B.6.

A	set of arcs in a graph, set of assignments or set of half–spaces.
B	equivalence class of borders, set of balls, union of discs or unit–ball; in the latter case, a superscript is used to specify its dimensionality.
C	clause, subset of sites, curve or circle of points.
D	data or auxiliary for clause.
E	set of edges.
$\mathbf{E}$	Euclidean space; a superscript is used to specify the number of dimensions.
F	set of faces, set of flats or collection of functions.
G	auxiliary for set of faces, flats, and hyperplanes.
H	set of hyperplanes.
I	set of indices, independent set or interval.
K	set of integer numbers.
L	set of lines or label.
M	measure, moment curve, matching, matrix or constant upper bound.
N	set of nodes in a graph or auxiliary for matching.
P	set of points, set of polygons or path in a graph.
Q	auxiliary for set of points.
R	rectangle of points.
$\mathbf{R}$	set of real numbers.

S	generic set, sequence, set of pseudo-lines, set of segments, set of half-spaces, set of spheres or unit-sphere; in the last case, a superscript is used to specify the dimensionality of its surface.
T	tour in a graph or auxiliary for generic set and for set of segments.
U	union or paraboloid; "U" reminds us of the shape of a paraboloid in two dimensions.
V	set of vertices or set of vectors.
X	auxiliary for substring.
Y	substring.
Z	zone or auxiliary for substring.

B.3. Upper Case Greek

Upper case greek letters are used only in exceptional cases if we disregard the frequent use of Σ for writing sums and Π for writing products. We list these exceptional cases.

Δ	determinant or set defined by differences of points.
Γ	matrix.
Π	permutation.
Σ	alphabet or range of solutions.

B.4. Lower Case Roman

Lower case roman letters are mainly used for simple objects, and some are reserved for important numbers. For each lower case roman letter, we list only the typical objects designated by the letter; they are used so frequently that there is no point in being exhaustive. We do not mention any combinatorial functions designated by a lower case roman letter; they can be found in Section B.6.

a	arc, assignment, answer, anchor point or anchor line, number of hyperplanes above a point or auxiliary for real number.
b	ball, disc, bisector, positive valued function, auxiliary for real number or number of hyperplanes below a point.
c	cell or center.
d	number of dimensions.
e	edge or auxiliary for face.
f	face, facet, flat or auxiliary for hyperplane.
g	auxiliary for face, for hyperplane and for flat.
h	hyperplane or height.
k	auxiliary for number of dimensions or for cardinality.

ℓ	integer.
l	line.
m	move, auxiliary for cardinality or constant lower bounds.
n	cardinality of the set of interest which is often a measure of the size of the problem at hand.
o	origin or number of hyperplanes a point lies on.
p	point, position, or polynomial.
q	auxiliary for point.
r	region or ray.
s	site, segment, half-space, pseudo-line, or auxiliary for ray.
t	triangle, time, transversal, pivot, test hyperplane, test in general, parameter, or auxiliary for site, for segment, for skeleton, and for ray.
u	auxiliary for vertex or for vector.
v	vertex, node, vector or auxiliary for word.
w	word, weight, double wedge or auxiliary for vertex or vector.
x	generic element, generic number, coordinate or auxiliary for point.
y	auxiliary for point.

To denote chordales, which are special hyperplanes defined for pairs of balls, we use the two letter combination ch.

B.5. Lower Case Greek

Almost all uses of lower case greek letters are for real variables and for coordinates of points or parameters that specify other objects. If they are used to designate coordinates or parameters then they are usually indexed and they match the roman letter used for the point or other object. For example, the coordinates of a point p in Euclidean d dimensions are written as $\pi_1, \pi_2, \ldots, \pi_d$.

α	real number or angle.
β	real number.
γ	real number, one-to-one correspondence, or generator.
δ	distance function or real number; in the second case it is usually small and positive.
ϵ	empty word, real number (in this case, it is usually small and positive), or embedding of a graph in the plane.
ι	interval.
κ	integer.
λ	rank of a point, rank of a sequence of points or real number; in the last case it is often used as a factor.
μ	mass distribution.
ν	node in a graph or a tree.

π	180 degrees or projection.
ϕ	the Euler function; $\phi(n)$ denotes the number of relatively prime positive integer numbers that do not exceed n.
ρ	radius of a circle, root of a tree or real number.
σ	infeasibility.
τ	number of transpositions or parameter.
χ	binary function on the dimensions of faces.

B.6. Combinatorial Functions

This section provides a collection of combinatorial functions defined and investigated in this book. The subtlety of denoting combinatorial functions appropriately is the reason for placing them into a separate section of this appendix. Very often, an argument, a subscript, or a superscript is used to distinguish related (but nevertheless different) functions.

$A(n)$	Ackermann's function defined as $A_n(n)$.
$A_k(n)$	defined recursively as follows: $A_k(1)=2$, for $k \geq 1$, $A_1(n)=A_1(n-1)+2$, for $n \geq 2$, and $A_k(n)=A_{k-1}(A_k(n-1))$, for $k \geq 2$.
$\alpha(n)$	$\min\{m \mid A(m) \geq n\}$; the inverse of Ackermann's function.
$a_{C,k}(H)$	sum of $deg_k(c)$, for all cells c is a collection C of cells in the arrangement defined by a set H of hyperplanes.
$a_{m,k}(H)$	$\max\{a_{D,k}(H) \mid D$ a collection of m cells in $\mathcal{A}(H)\}$; if $m=1$ then the first subscript is omitted.
$a_{m,k}^{(d)}(n)$	$\max\{a_{m,k}(G) \mid G$ a set of n hyperplanes in $E^d\}$; if $m=1$ then we omit the first subscript.
$C^{(d)}(n)$	number of combinatorially different arrangements of n hyperplanes in E^d.
$C_s^{(d)}(n)$	number of combinatorially different simple arrangements of n hyperplanes in E^d.
$c_k^{(d)}(n)$	maximum number of cells c in a simple arrangement of n hyperplanes in E^d such that $deg_{d-1}(c)=k$.
$DI(n)$	smallest integer k such that every set of n lines in E^2, not all concurrent or parallel, contains a line which intersects the other lines in at least k different points.
$E_k(n)$	maximum number of collinear k-tuples in any set of n points in E^2 which contains no collinear $(k+1)$-tuple.
$ES(m)$	minimum number n such that every set of n points in E^2, with no three points collinear, contains the vertices of a convex m-gon.
$e_k(P)$	number of k-sets of point set P.
$e_k^{(d)}(n)$	$\max\{e_k(Q) \mid Q$ a set of n points in $E^d\}$.
$f_k(H)$	number of k-faces in the arrangement defined by the set H of hyperplanes.
$f_k^{(d)}(n)$	$\max\{f_k(G) \mid G$ a set of n hyperplanes in $E^d\}$.

$f_i(H,k)$	number of i-faces f in $\mathcal{A}(H)$ with $a(f)=k$; H is omitted if it is clear from the context.
$G_k(C)$	sum of $g_i(C)$, for i ranging from 0 to k.
$G_k(n)$	$\max\{G_k(D) \mid D$ a circular sequence of n numbers$\}$.
$g_k(C)$	number of k-sets of the circular sequence C.
$g_k(n)$	$\max\{g_k(D) \mid D$ a circular sequence of n numbers$\}$.
$H(m)$	smallest integer n such that every set of n points in E^2 contains the vertices of a convex m-gon that has no point of the set in its interior.
$h_k^{(d)}$	infimum over all real numbers α such that for any set of n hyperplanes in E^d there are k hyperplanes with the property that no (open) cell defined by the k hyperplanes intersects more than αn hyperplanes of the set.
$i_k(H)$	number of incident pairs (f,g), where f is a k-face and g is a $(k+1)$-face of the arrangement $\mathcal{A}(H)$.
$i_k^{(d)}(n)$	smallest number m such that $i_k(H) \leq m$, for every set H of n hyperplanes in d dimensions.
$i(n)$	largest integer j such that every planar graph with n nodes has an independent set of j nodes.
$i_k(n)$	largest integer j such that every planar graph with n nodes has an independent set of j nodes each of which is incident upon at most k arcs.
$k(d)$	smallest integer j such that $h_j^{(d)}<1$.
$m(d)$	largest number n such that there is an arrangement of n hyperplanes in E^d and each cell in this arrangement is incident upon n facets.
$SC(n)$	smallest integer k such that every set of n non-collinear points in E^2 define at least k directions.
$s(C)$	maximum number of cells in cell complex C that intersect (or are stabbed by) a common hyperplane.
$s(P)$	$\min\{s(D) \mid D$ an erasing cell complex of point set $P\}$.
$s^{(d)}(n)$	$\max\{s(Q) \mid Q$ a set of n points in $E^d\}$.
$\bar{s}(M)$	maximum number of pairs of points in the matching M that can be separated by a single line.
$\bar{s}(P)$	$\min\{s(N) \mid N$ a perfect matching of point set $P\}$.
$\bar{s}(n)$	$\max\{\bar{s}(Q) \mid Q$ a set of $2n$ points in $E^2\}$.
$s_k(H,h)$	sum of $deg_k(c)$, for all cells c in $\mathcal{A}(H)$ which lie above hyperplane $h \in H$ and are supported by h.
$s_k^{(d)}(n)$	$\max\{s_k(G,g) \mid G$ a set of $n+1$ hyperplanes in E^d and g a hyperplane in $G\}$.
$t(n)$	maximum number of non-equivalent simultaneous bisectors of two sets each of n points in E^3, where n is odd and the two sets are separable by a plane.
$u(H,k)$	number of unbounded cells c in $\mathcal{A}(H)$ with $a(c)=k$; H is omitted if it is clear from the context.
$v_V(H)$	sum of $deg(v)$, for all vertices v in the collection V of vertices taken from the arrangement defined by the set H of hyperplanes.
$v_m(H)$	$\max\{v_V(H) \mid V$ a collection of m vertices of $\mathcal{A}(H)\}$.

$v_m(n)$	$\max\{v_m^{(d)}(G) \mid G$ a set of n hyperplanes in $E^d\}$.
$\bar{x}(n)$	maximum number of edges of an arrangement of n non-vertical lines in the plane such that the union of the closures of these edges intersects every vertical line in exactly one point.
$\hat{x}(n)$	maximum number of turns of the union of closures of edges in an arrangement of n lines in the plane such that every vertical line intersects this union in exactly one point.
$z(S)$	number of edges of the stabbing region of the set S of line segments in the plane.
$z(n)$	$\max\{z(T) \mid T$ a set of n segments in $E^2\}$.

INDEX